T0319670

ELECTROCHEMICAL PROCESSES IN BIOLOGICAL SYSTEMS

WILEY SERIES ON ELECTROCATALYSIS AND ELECTROCHEMISTRY

Andrzej Wieckowski, Series Editor

Fuel Cell Catalysis: A Surface Science Approach, Edited by Marc T. M. Koper

Electrochemistry of Functional Supramolecular Systems, Margherita Venturi, Paola Ceroni, and Alberto Credi

Catalysis in Electrochemistry: From Fundamentals to Strategies for Fuel Cell Development, Elizabeth Santos and Wolfgang Schmickler

Fuel Cell Science: Theory, Fundamentals, and Biocatalysis, Andrzej Wieckowski and Jens Norskov

Vibrational Spectroscopy at Electrified Interfaces, Edited by Andrzej Wieckowski, Carol Korzeniewski and Bjorn Braunschweig

ELECTROCHEMICAL PROCESSES IN BIOLOGICAL SYSTEMS

Edited by

ANDRZEJ LEWENSTAM
LO GORTON

Wiley Series on Electrocatalysis and Electrochemistry

Published by John Wiley & Sons, Inc., Hoboken, New Jersey
Published simultaneously in Canada

Library of Congress Cataloging-in-Publication Data

Electrochemical processes in biological systems / edited by Andrzej Lewenstam, Lo Gorton.
pages cm. – (Wiley series on electrocatalysis and electrochemistry)
Includes bibliographical references and index.
ISBN 978-0-470-57845-2 (cloth : alk. paper)
1. Bioenergetics. 2. Ion exchange. I. Lewenstam, Andrzej. II. Gorton, L. (Lo)
QP517.B54E44 2015
612′.01421–dc23

2014049433

Set in 10/12pt Times by SPi Global, Pondicherry, India

Printed in the United States of America

10 9 8 7 6 5 4 3 2 1

CONTENTS

CONTRIBUTORS

Julea N. Butt, School of Chemistry and School of Biological Sciences, University of East Anglia, Norwich, UK

Krzysztof Dołowy, Laboratory of Biophysics, Warsaw University of Life Sciences (SGGW), Warsaw, Poland

Andrew J. Gates, School of Biological Sciences, University of East Anglia, Norwich, UK

Harry B. Gray, Beckman Institute, California Institute of Technology, Pasadena, CA, USA

Wiesław I. Gruszecki, Department of Biophysics, Institute of Physics, Maria Curie-Skłodowska University, Lublin, Poland

Jan Halámek, Department of Chemistry and Biomolecular Science, and NanoBio Laboratory (NABLAB), Clarkson University, Potsdam NY, USA

David E. Heppner, Department of Chemistry, Stanford University, Stanford, CA, USA

Michael G. Hill, Department of Chemistry, Occidental College, Los Angeles, CA, USA

Evgeny Katz, Department of Chemistry, University at Albany, SUNY, Albany, NY, USA

Christian H. Kjaergaard, Department of Chemistry, Stanford University, Stanford, CA, USA

Ramya Kolli, Department of Chemical and Biomolecular Engineering, New York University Polytechnic School of Engineering, Six Metrotech Center, Brooklyn, USA

Yaara Lefler, Department of Neurobiology, The Institute of Life Sciences and Edmond and Lily Safra Center for Brain Sciences (ELSC), The Hebrew University, Jerusalem, Israel

Kalle Levon, Department of Chemical and Biomolecular Engineering, New York University Polytechnic School of Engineering, Six Metrotech Center, Brooklyn, USA

Krzysztof Maksymiuk, Faculty of Chemistry, University of Warsaw, Warsaw, Poland

Sophie J. Marritt, School of Chemistry, University of East Anglia, Norwich, UK

Aabhas Martur, Department of Chemical and Biomolecular Engineering, New York University Polytechnic School of Engineering, Six Metrotech Center, Brooklyn, USA

Agata Michalska, Faculty of Chemistry, University of Warsaw, Warsaw, Poland

David J. Richardson, School of Biological Sciences, University of East Anglia, Norwich, UK

Tomasz Sokalski, Laboratory of Analytical Chemistry, Faculty of Science and Engineering, Åbo Akademi University, Turku, Finland

Edward I. Solomon, Department of Chemistry, Stanford University, Stanford, CA, USA

Andrew K. Udit, Department of Chemistry, Occidental College, Los Angeles, CA, USA

Marylka Yoe Uusisaari, Department of Neurobiology, The Institute of Life Sciences and Edmond and Lily Safra Center for Brain Sciences (ELSC), The Hebrew University, Jerusalem, Israel

Yanyan Wang, Department of Chemical and Biomolecular Engineering, New York University Polytechnic School of Engineering, Six Metrotech Center, Brooklyn, USA

Qi Zhang, Department of Chemical and Biomolecular Engineering, New York University Polytechnic School of Engineering, Six Metrotech Center, Brooklyn, USA

PREFACE

This series covers recent advancements in electrocatalysis and electrochemistry and depicts prospects for their contribution into the present and future of the industrial world. It aims to illustrate the transition of electrochemical sciences from its beginnings as a solid chapter of physical chemistry (covering mainly electron transfer reactions, concepts of electrode potentials, and structure of electrical double layer), to the field in which electrochemical reactivity is shown as a unique chapter of heterogeneous catalysis; is supported by high-level theory; connects to other areas of science; and includes focus on electrode surface structure, reaction environment, and interfacial spectroscopy.

The scope of this series ranges from electrocatalysis (practice, theory, relevance to fuel cell science, and technology), to electrochemical charge transfer reactions, biocatalysis, and photoelectrochemistry. While individual volumes may appear quite diverse, the series promises updated and overall synergistic reports providing insights to help further our understanding of the properties of electrified solid/liquid systems. Readers of the series will also find strong reference to theoretical approaches for predicting electrocatalytic reactivity by such high-level theories as density functional theory. Beyond the theoretical perspective, further vehicles for growth are such significant topics such as energy storage, syntheses of catalytic materials via rational design, nanometer-scale technologies, prospects in electrosynthesis, new instrumentation, and surface modifications. In this context, the reader will notice that new methods being developed for one field may be readily adapted for application in another.

Electrochemistry and electrocatalysis have both benefited from numerous monographs and review articles due to their depth, complexity, and relevance to the practical world. The Wiley Series on Electrocatalysis and Electrochemistry is dedicated to present the current activity by focusing each volume on a specific topic that is timely and promising in terms of its potential toward useful science and technology. The chapters in these volumes will also demonstrate the connection of electrochemistry to other disciplines beyond chemistry and chemical engineering, such as physics, quantum mechanics, surface science, and biology. The integral goal is to offer a broad-based analysis of the total development of the fields. The progress of the series will provide a global definition of what electrocatalysis and electrochemistry are now, and will contain projections about how these fields will further evolve in time. The purpose is twofold—to provide a modern reference for graduate instruction and for active researchers in the two disciplines, as well as to document that electrocatalysis and electrochemistry are dynamic fields that are expanding rapidly, and are likewise rapidly changing in their scientific profiles and potential.

Creation of each volume required the editor's involvement, vision, enthusiasm, and time. The Series Editor thanks each Volume Editor who graciously accepted his invitation. Special thanks go to Ms. Anita Lekhwani, the Series Acquisitions Editor, who extended the invitation to edit this series to me and has been a wonderful help in its assembling process.

ANDRZEJ WIECKOWSKI
Series Editor

1

MODELING OF RELATIONS BETWEEN IONIC FLUXES AND MEMBRANE POTENTIAL IN ARTIFICIAL MEMBRANES

Agata Michalska and Krzysztof Maksymiuk
Faculty of Chemistry, University of Warsaw, Warsaw, Poland

1.1 INTRODUCTORY CONSIDERATIONS

A membrane can be regarded as a phase, finite in space, which separates two other phases and exhibits individual resistances to the permeation of different species (Schlögl's definition cited in [1]). The membranes can be of different thickness, from thin used typically for biological and artificial bilayers (in the range of a few nanometers) to relatively thick (hundreds of micrometers) used typically in ion-selective electrodes. A particular case is a membrane separating two electrolyte solutions, where ions are transferable species. In such a case, different modes of ion transport are possible: (i) Brownian motion; (ii) diffusion, resulting from concentration gradient; and (iii) migration as transport under the influence of an electrical field.

A general prerequisite related to the presence of charged species is electroneutrality condition of the membrane. However, even if electroneutrality is held on a macroscopic scale, charge separation effects appear, mainly at membrane/solution interfaces, resulting in the formation of potential difference. Taking into account possible chemical and electrical forces present in the system, assuming for simplicity one-dimensional

Electrochemical Processes in Biological Systems, First Edition. Edited by Andrzej Lewenstam and Lo Gorton.

transfer along the x-axis only, the flux of ion "i," J_i, across the membrane can be described as

$$J_i = -kU_i c_i \frac{\partial \bar{\mu}_i}{\partial x} \tag{1.1}$$

where k is a constant; U_i, c_i, and $\bar{\mu}_i$ are electrical mobility, concentration, and electrochemical potential of ion "i," respectively; and x is the distance from the membrane/solution interface. Using a well-known definition of electrochemical potential and assuming that the activity of ion "i" is equal to the concentration, this equation can be transformed to

$$J_i = -kU_i c_i \left(\frac{RT \partial \ln c_i}{\partial x} + z_i F \frac{\partial \varphi}{\partial x} \right) \tag{1.2}$$

with φ as the Galvani potential of the phase.

Since mobility, U_i, is a ratio of the transfer rate, v, and potential gradient ($\partial\varphi/\partial x$), while the flux under influence of the electrical force is $J = vc$, it follows from Equation (1.2) that $k = 1/|z_i|F$. Taking then into account the Einstein relation, concerning diffusion coefficient, $D_i = U_i RT/|z_i|F$, Equation (1.2) can be rewritten as

$$J_i = -D_i \left(\frac{\partial c_i(x,t)}{\partial x} + \frac{z_i F}{RT} c_i(x,t) \frac{\partial \varphi(x,t)}{\partial x} \right) \tag{1.3}$$

This is the Nernst–Planck equation, relating the flux of ionic species, "i," to gradients of potential and concentration, being generally functions of distance, x, and time, t.

The Nernst–Planck equation is a general expression describing transport phenomena in membranes. Unfortunately, as differential equations deal with functions dependent on distance and time, solving of this equation is neither easy nor straightforward. However, under some conditions, simplifications of this equation are possible.

(i) For the equilibrium case, summary fluxes of all ionic species are zero, $J_i = 0$. In such a case,

$$\frac{\partial \ln c_i}{\partial x} = -\frac{z_i F}{RT} \frac{\partial \varphi}{\partial x} \tag{1.4}$$

After rearrangement and integration across the whole membrane (of thickness d), the well-known form is obtained:

$$\varphi_R - \varphi_L = \Delta\varphi_{mem} = \frac{RT}{z_i F} \ln \frac{c_R}{c_L} \tag{1.5}$$

where R and L refer arbitrarily to "right" and "left" hand side (membrane/solution interface) and c_R and c_L are solution concentrations at "right" and "left" interfaces.

This equation describing a membrane potential, $\Delta\varphi_{mem}$, is equivalent of the typical Nernst equation.

(ii) For the case of a neutral substance, $z_i = 0$, or in the absence of electrical driving force $(\partial\varphi/\partial x) = 0$, the Nernst–Planck equation reduces to Fick's equation, describing diffusional transport:

$$J_i = -D_i\left(\frac{\partial c_i}{\partial x}\right) \tag{1.6}$$

Solutions of the Nernst–Planck equation can be more easily obtained for the steady state, when the ionic fluxes $J_i = \text{const}$ and a time-independent version of the equation can be used. In this case, the Nernst–Planck equation can be applied to calculate potential difference in the membrane for given values of concentrations and mobilities. However, this procedure also requires integration, which can be difficult in some cases. Therefore, additional approximations are often used [2, 3]. The most known and used solutions are the Goldman and Henderson approximations.

1.1.1 Goldman Approximation and Goldman–Hodgkin–Katz Equation

This approximation assumes linearity of potential gradient across the membrane (i.e., constant electrical field in the membrane). This approximation is usually applicable to thin biological membranes, where charge prevails only in the surface areas of the membrane. In such a case, the derivative $(\partial\varphi/\partial x)$ can be approximated by the term $(\varphi_R - \varphi_L)/d$ leading to the simplified Nernst–Planck equation:

$$J_i = -\left(D_i\frac{\partial c_i}{\partial x} + \frac{z_iF}{RT}c_i\frac{\varphi_R - \varphi_L}{d}\right) \tag{1.7}$$

Under constant field condition, a steady state is practically obtained ($J_i = \text{const}$) and the Goldman flux equation can be derived:

$$J_i = \frac{z_iFD_i(\varphi_R - \varphi_L)}{dRT} \cdot \frac{c_i(\mathrm{R}) - c_i(\mathrm{L})\exp(-(z_iF(\varphi_R - \varphi_L))/RT)}{\exp(-(z_iF(\varphi_R - \varphi_L))/RT) - 1} \tag{1.8}$$

In the absence of transmembrane potential, $\varphi_R - \varphi_L \sim 0$, this equation simplifies to the well-known diffusion equation in a steady state.

Taking into account that the sum of individual ionic contributions to electrical current is zero (denoting the absence of applied external current),

$$\sum_{i=1}^{n} z_iJ_i = 0 \tag{1.9}$$

further rearrangements are possible. For a simplified case of solution of ions of ±1 charge (e.g., Na^+, K^+, Cl^-), the equation describing the potential difference across the membrane (under steady-state conditions) can be obtained:

$$\varphi_R - \varphi_L = \frac{RT}{F}\ln\frac{D_{Na^+}c_{Na^+}(L) + D_{K^+}c_{K^+}(L) + D_{Cl^-}c_{Cl^-}(R)}{D_{Na^+}c_{Na^+}(R) + D_{K^+}c_{K^+}(R) + D_{Cl^-}c_{Cl^-}(L)} \quad (1.10)$$

It can be also assumed that ions take part in ion-exchange equilibrium, between the membrane and bathing electrolyte solution (sol) from the right or left hand side, (Eq. 1.11a) and (Eq. 1.11b), respectively:

$$c_i(R) = k_i c_i(\text{sol}, R) \quad (1.11a)$$

$$c_i(L) = k_i c_i(\text{sol}, L) \quad (1.11b)$$

with partition coefficients k_i of the species "i" between the solution and membrane phases. Then, introducing the permeability coefficient, P_i, $P_i = U_i k_i / |z_i| Fd$, Equations (1.10) and (1.11) can be transformed to the Goldman–Hodgkin–Katz equation, expressing the membrane potential as a function of ion concentrations in bathing solutions on both sides of the membrane and partition (permeability) coefficients:

$$\varphi_R - \varphi_L = \frac{RT}{F}\ln\frac{P_{Na^+}c_{Na^+}(L) + P_{K^+}c_{K^+}(L) + P_{Cl^-}c_{Cl^-}(R)}{P_{Na^+}c_{Na^+}(R) + P_{K^+}c_{K^+}(R) + P_{Cl^-}c_{Cl^-}(L)} \quad (1.12)$$

This equation is applicable, for example, to describe resting potentials of biological membranes.

1.1.2 Henderson Approximation

This approximation assumes linear concentration gradient across the membrane, while the electrical field need not be constant [4]. This approximation is usually applied to describe diffusion (liquid junction) potentials, particularly for the case of ion-selective electrodes. This potential can be approximated by the equation

$$\Delta\varphi_{LJ} = -\frac{RT}{F}\cdot\frac{\sum z_i u_i (c_i(R) - c_i(L))}{\sum z_i^2 u_i (c_i(R) - c_i(L))}\ln\frac{\sum z_i^2 u_i c_i(R)}{\sum z_i^2 u_i c_i(L)} \quad (1.13)$$

where u_i is $U_i/|z_i|F$.

Membrane processes related to charge separation and transport of charged species concern both biological membranes in cell biology (or artificial membranes having significant importance in separation processes) and membranes used in electroanalytical chemistry, for example, in ion-selective electrodes. However, in contrast to similarity of physicochemical phenomena occurring in all membranes containing mobile charged species, the description related to biological or separation membranes is different from that applicable to membranes of ion-selective electrodes. Therefore, the considerations in the following were divided into two sections: (i) related to more general description typical for separation and biological membranes where typically the Nernst–Planck equation is applicable and (ii) related to membranes used in ion-selective electrodes. In case (ii), practical and historical conditions result in dominance of simple empirical equations for the membrane potentials; however, in the last decade, the role of a more general theory using the Nernst–Planck equation is increasing.

1.2 GENERAL CONSIDERATIONS CONCERNING MEMBRANE POTENTIALS AND TRANSFER OF IONIC SPECIES

1.2.1 Boundary and Diffusion Potentials

Separation membranes are important both in biology and various technological areas: fuel cells, dialysis, reverse osmosis, separation of mixtures components, etc. These membranes can be generally described as neutral or charged membranes. For the former class of membranes, size exclusion and specific chemical interactions are the main factors responsible for selective permeability, while for charged membranes with incorporated ionic sites, electrostatic interactions are of substantial significance.

For the charged membranes, the membrane potential, understood as a potential difference between two electrolyte solutions (of different concentration or/and composition) separated by the membrane, is an important parameter characterizing their properties [5]. Measurements of membrane potential offer also a straightforward method for studying transport processes of charged species. Due to difficulties in solving the Nernst–Planck equation in a general case, simplifications are used, as they were shortly described in the previous section.

In the discussion given in the following, only cases with no external current flow are considered. In the description of membrane potentials, mostly a simplified formalism is used, expressed in terms of classical model proposed by Sollner [6], Teorell, Meyer, and Sievers [7, 8]. This model postulates splitting the potential prevailing in the system into three components: boundary potentials on the membrane (left and right side)/bathing solution interfaces and membrane bulk diffusion potential resulting from different mobility of ionic species as well as ion-exchange reactions leading to inhomogeneities in the interior of the membrane. It should be noted that the Teorell–Meyer–Sievers theory is applicable to fixed-site membranes with one kind of monovalent cations and anions as transferable species. It can be not valid for liquid membranes, where ionic sites are mobile.

Boundary potentials are regarded as equilibrium potentials resulting from ion exchange on the interface between the charged membrane and solution. In the case of equilibrium, assuming that activity coefficients are equal to 1, the change of chemical potential of transferable ion, "i," is equal to zero:

$$\Delta\mu_{\mathrm{i}} = 0 \tag{1.14}$$

The chemical standard potentials, $\mu_{i,m}^0$ and $\mu_{i,s}^0$, in the membrane and in the solution, respectively, can be different, determining the value of the distribution coefficient of species, "i" and k_i:

$$k_i = \exp\left[\frac{-\left(\mu_{i,m}^0 - \mu_{i,s}^0\right)}{RT}\right] \tag{1.15}$$

Then, assuming the existence of ion-exchanging sites in the membrane (unable to be released from the membrane), the boundary potential at a chosen membrane/solution interface can be represented by the Donnan potential, $\Delta\varphi_D = \varphi^m - \varphi^s$ [9]:

$$RT \ln\frac{k_i c_i^m}{c_i^s} + z_i F \Delta\varphi_D = 0 \tag{1.16}$$

where superscripts "m" and "s" relate to the membrane and solution phases, respectively.

The dependence of the Donnan potential on the electrolyte concentration in solution can be derived from Equation (1.16) taking into account electroneutrality condition in the membrane phase. In the case of a membrane with fixed ionic site concentration, X, and for 1:1 electrolyte (with cation M^+ and anion A^-) of concentration c, assuming the absence of specific interactions of ions with the membrane components ($k_i = 1$), the Donnan potential is expressed by equation [10]

$$\Delta\varphi_D = \varphi^m - \varphi^s = \pm\frac{RT}{F}\ln\left[\frac{X}{2c} + \left(1 + \left(\frac{X}{2c}\right)^2\right)^{\frac{1}{2}}\right] \tag{1.17}$$

with sign "+" or "−" for anion- or cation-exchanging membrane, respectively.

From this equation, two limiting cases follow (Fig. 1.1). For dilute solutions, when $c \ll X$, the concentration of counterions in the membrane is practically equal to X (with negligible concentrations of coions, this is the so-called Donnan exclusion case), and the Donnan potential becomes a linear function of $\ln c$ with positive or negative Nernstian slope, for cation and anion exchanger, respectively. On the other hand, for high electrolyte concentrations, when $c \gg X$, concentrations of counter- and coions in the membrane are almost equal; the Donnan potential is independent of electrolyte concentration and close to zero (Donnan exclusion failure).

In contrast to the equilibrium state observed at the membrane/solution interface, in the membrane interior, similar equilibria are typically not observed. In this case, different rates of ion transfer in the membrane result in diffusion potential formation, necessary to maintain a zero current steady state [1].

In order to calculate exact values of the diffusion potentials, knowledge about concentration profiles of all species in the membrane is needed. However, for practical purposes, some approximations can be used; the most popular is the Henderson

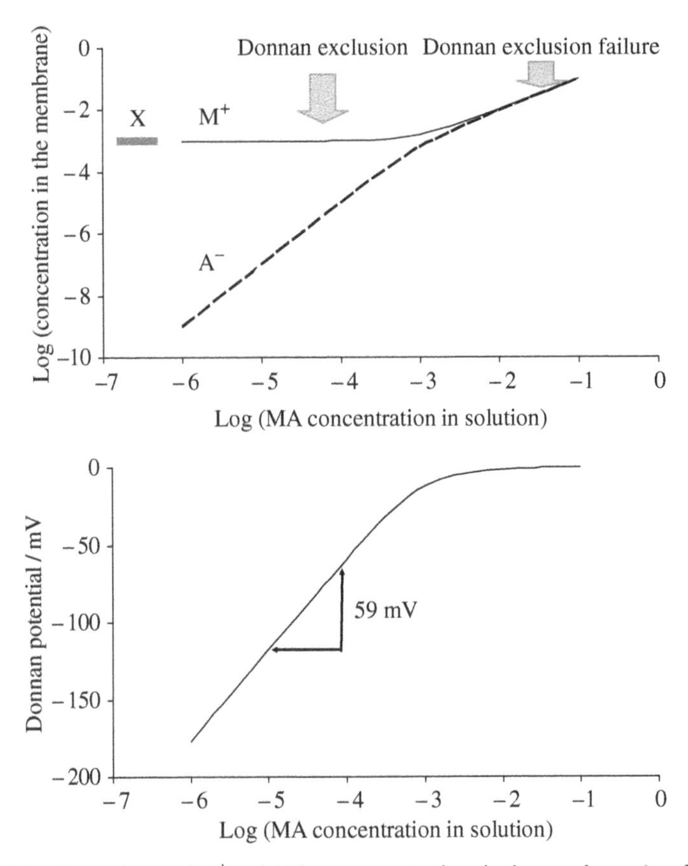

FIGURE 1.1 Dependence of M^+ and A^- ions concentrations in the membrane (top figure) and Donnan potential (bottom figure) on concentration of salt MA in solution, for cation-exchanging membrane with ion-exchange site concentration $X = 10^{-3}$ M.

approximation (see previous section). More advanced treatments are also available, for example, the Planck and Schlögl description is based on a given model of diffusion in the layer [1, 2], in contrast to the model of uniform mixing typical for the Henderson approach. However, the classical Planck approach is less convenient for practical applications than the Henderson approximation; thus, some simplifications facilitating calculations in terms of the Planck model have been proposed, and thus, iterative calculations method can be used [11]. A significant simplification of the Planck description is expected for the case of two equimolar solutions of 1:1 electrolytes, where the Planck equation simplifies to the Goldman equation. On the other hand, the Planck approach and the Goldman equation can be considered as special cases of a more general Teorell equation [7], applicable for any two classes of diffusing ions and for additional fixed sites in the membrane [1]. Assuming that all transferring ions have the same charge, the following equation for the diffusion potential can be obtained:

$$E_D = \frac{RT}{z_i F} \ln \frac{\sum u_i a_i(\mathrm{L})}{\sum u_i a_i(\mathrm{R})} \tag{1.18}$$

1.2.2 Application of the Nernst–Planck Equation to Describe Ion Transport in Membranes

Knowledge about transfer of ionic species in membranes is crucial from the point of view of some general issues concerning membrane potentials (without arbitrary and sometimes controversial splitting into boundary and diffusion potential) and practical applications of membranes in separation processes, dialysis, and selective transfer of charged species. This requirement is connected with the solution of a system of the Nernst–Planck equations for all ionic species, which can be transported in the membrane. Due to mathematical difficulties, calculation of ion fluxes or membrane potential accompanying ionic fluxes by solving the Nernst–Planck equation is usually connected with application of numerical procedures. Additionally, simplifying assumptions, as the aforementioned Henderson or Goldman approximations, are used. The proposed solution is related rather to the steady state (time-independent phenomena).

In 1954, Schlögl proposed a general solution of the Nernst–Planck equation under steady-state conditions, applicable also for thick membranes, for any number of transferring ions, and for fixed ionic sites [1, 12]. He proposed to divide diffusing ions into subgroups—so-called valency classes. For ion-exchanging membrane containing only one class of ions, the equation proposed here simplifies to the Goldman flux equation.

The first numerical solution of the Nernst–Planck equations coupled with the Poisson equation

$$\frac{\partial E(x,t)}{\partial x} = \frac{\rho(x,t)}{\varepsilon}; \quad \rho(x,t) = F \cdot \sum_i z_i \cdot c_i(x,t) \tag{1.19}$$

that is, taking into account the existence of noncompensated charge, was proposed by Cohen and Cooley in 1965 [13] ($\rho(x,t)$ is the charge density, E is the electrical field, while ε is the dielectric permittivity). In this work, a solution of the Nernst–Planck–Poisson equations has been proposed, taking into account also time influence; however, this procedure was not explicit (nonpredictive). Authors have introduced a system of reduced units, often used in electrochemical simulations.

Sandifer and Buck [14] have extended this method, and then a significant progress has been achieved by Brumleve and Buck [15] who have discussed numerical solution both for steady-state and transient phenomena, for species with arbitrary valence, mobility, and interfacial charge transfer rate constants. These authors were using finite difference method and applied iterative Newton–Raphson method coupled with the Gaussian elimination to solve the nonlinear equations. This time also, some numerical procedures for solving the Nernst–Planck–Poisson equation systems have been proposed by other researchers outside the area of membrane physical chemistry and electrochemistry [16–18].

In 1984, Buck discussed critically limitations of the Nernst–Planck equation [19]. The first argument against this theory results from the macroscopic and smooth nature of the medium considered in the model. The second argument arises from the omission of "cross terms"—this means that the flux of any species is not only linearly related to its activity gradient but also to the corresponding gradients of other mobile species. Some alternatives and microscopic models have been discussed in terms of atomic properties. For instance, percolation theory [20] has been presented as complementary one to the Nernst–Planck model, as it provides values of transport parameters in terms of structure and composition. However, in contrast to their significant limitations, the Nernst–Planck equations are very useful, applicable to describe transport in solids, liquids, and gels [19].

On the other hand, the use of transport equations of linear nonequilibrium thermodynamics for the description of membrane transport is difficult owing to a large number of coefficients present in the equations; these coefficients are also functions of the composition of the solution being in contact with the membrane. Due to these difficulties, fundamental classical works of Staverman [21] or Kedem and Katchalsky [22, 23] are not used very often. Moreover, cross-coefficients should be also considered, but there are approaches neglecting some of these coefficients (e.g., [24]).

MacGillivray [25] has also proposed quasianalytical solutions of the Nernst–Planck equations for time-dependent phenomena in the form of asymptotic solutions based on assumptions consistent with reasonable experimental cases. This was analyzed on the example of studying instantaneous potential jump regarded as small compared to potential difference existing before the experiment (e.g., method of "voltage clamp"). MacGillivray [26] and later Seshadri [27] have analyzed also the problem of electroneutrality condition using perturbation theory resulting in the electroneutrality condition as a certain limiting case. It was found that the electroneutrality condition is a consequence of the Poisson equation when a certain dimensionless parameter, interpreted as a ratio of the Debye length and the membrane thickness, is small. It was also shown that under such assumption the Donnan equilibrium can be derived from the Nernst–Planck equation. Later, Mafé et al. [28] in their consideration have taken into account electrical and diffusional relaxation times as alternative to the approach with relation of Debye length to the membrane thickness.

Similar issues have been critically discussed by Kato [29] by relating also applicability of simplifying assumptions with ratio of Debye length to the membrane thickness. Castilla et al. [30] have used a network approach to obtain numerical solutions based on the Nernst–Planck and Poisson equations under non-steady-state conditions, in a liquid membrane. In this approach, the fluxes were regarded as currents in electrical circuits and diffusion or electrical field influences were regarded as circuit elements (resistors, capacitances), and electrical circuit simulation program has been used to simulate concentration profiles, ionic fluxes, and membrane potentials. Robertson et al. [31] have also used a model with hybrid network description, using a network simulation program (SPICE). This procedure was applied to describe ion transport in bilayer lipid membranes tethered to a gold electrode.

The most significant assumption concerning the charged membrane is homogeneous fixed-charge distribution. The role of charge distribution inhomogeneity has been taken into account by Mafé et al. [32] as well as by Tanioka and coworkers [33]. Tanioka et al. [33, 34] have shown that distribution of effective charge can be nonhomogeneous even in the membrane with homogeneous fixed-charge distribution, under conditions of ion-pair formation between counterions and fixed-charge groups.

Besides inhomogeneity in charge distribution, also, the role of surface and bulk parts of the membrane has been taken into account (e.g., [35, 36]). In the surface region, the conductivity results from mobile counterions compensating charge of fixed groups. In the central part, the conductive properties resemble those of the bathing solution. For this system, the results of modeling have been presented showing a clear difference between mechanisms typical for bulk and surface conductivity [37].

Takagi and Nakagaki [38] have analyzed two kinds of membrane asymmetry: with respect to the partition coefficient and with respect to the charge density. For the first case, when the ion concentration gradient within the membrane is larger than within a symmetric membrane, facilitated transport will take place. However, when the direction of the ion concentration gradient within the membrane is opposite to that expected from the concentration difference between two solutions separated by the membrane, the reverse transport will take place. In the case of charge density inhomogeneity, the difference in cation concentration between the two membrane surfaces is not equal to the difference in anion concentration. Here, facilitated or reverse ion transfer can take place. This model, with the Henderson assumption concerning the distribution of ionic components, has been applied to the aforementioned nonhomogeneous systems. Ion transfer against concentration gradient as a phenomenon related to the multi-ionic transport in a membrane has been discussed in other papers (e.g., [39, 40]). Ramirez et al. [41] have solved numerically the Nernst–Planck equations without invoking the Goldman assumption for the case of pH difference as the driving force for ions.

Chou and Tanioka [42, 43] have discussed the role of organic solvent presence resulting in ion-pair formation in the membrane between counterions and fixed-charge groups. The effective membrane charge densities and the cation-to-anion mobility ratios in the membrane were determined by a nonlinear regression method. This issue can be analyzed in terms of Donnan equilibrium taking into account effective charge concentration that is lower than the analytical one, due to interactions with membrane charged groups [44, 45].

The role of acid/base (amphoteric) properties of polymeric membranes composed of a weak electrolyte as well as acid/base properties of transferring species has been also considered, mainly in papers of Ramirez and Tanioka with coworkers [46–50]. Some papers proposed solutions without invoking simplifying approximation as Goldman constant field assumption [46–48]. The aforementioned, so-called continuous models capable of describing ionic transport in membranes with wide pores have been found useful for nanopore membranes [51] (being of special importance for biomembranes).

The Teorell, Meyer, and Sievers model (including Donnan potentials and diffusion potential in the membrane) was used to describe processes of nanofiltration [52], including also the extended Nernst–Planck equation and Gouy–Chapman theory [53]. Nanofiltration membranes have a cavity structure with pore sizes approximately 1–2 nm, where separation results mainly from size exclusion and electrostatic interactions. Hagmeyer and Gimbel [54] utilized the zeta potential measurements to calculate surface charges of nanofiltration membranes; this charge was then incorporated in a model based on the Teorell, Meyer, and Sievers theory, and the salt rejection effect by the membrane was described. Various models have been proposed to describe rejection mechanism for salts and charged organic species in nanofiltration membranes. These models are based mainly on the extended Nernst–Planck equation, and one of the most popular models is the Donnan steric pore model [55–58]. According to this approach, the membrane is considered as a charged porous layer, characterized by average pore radius, volumetric charge density, and effective membrane thickness. It is also assumed that partitioning effects are described taking into account steric hindrances and the Donnan equilibrium. The role of membrane porosity has been also discussed in papers of Revil et al. (e.g., [59, 60]). The proposed model was based on a volume-averaging approach applied to the Stokes and Nernst–Planck equations and uses Donnan equilibria. It can explain the influence of pore water behavior and water saturation on the diffusion coefficient of a salt in the membrane. As described in [61], selectivity of ion transport in nanopore systems was based on charge repulsion [62], size exclusion [63, 64], and polarity [65].

Nanopore membrane mimicking biological ion channels have been also adapted and developed for sensing purposes in chemical analysis ([66, 67] and review article [68]).

The description of transport phenomena and membrane potentials becomes more complicated in the case of bi-ionic systems, where an ion-exchange membrane separates two electrolyte solutions having the same coion but different counterions. In this case, diffusion processes result in appearance of multi-ionic system because the counterions will be present in both solutions. Many papers concentrate on potential measurements (e.g., [44]), and due to system complexity, analysis of transmembrane ionic fluxes is not often presented [69, 70].

Other more complicated systems are bipolar membranes. They consist of a cation-exchange layer in series with an anion-exchange layer. The models describing potential of such membranes are typically extension of the Teorell–Meyer–Sievers model of monopolar membranes [71–73], with application of the Nernst–Planck equation [74, 75].

1.3 POTENTIALS AND ION TRANSPORT IN ION-SELECTIVE ELECTRODES MEMBRANES

One of the most powerful and successful applications of artificial membranes permeable for ionic species are sensors or biosensors [76] and particularly ion-selective electrodes, where the membrane composition is responsible for sensor's sensitivity and selectivity. The recorded response is open-circuit potential, representing, under typical conditions, the membrane potential as a function of sample composition and concentration. This system and particularly the relation between membrane potential, ion concentrations, and fluxes, as in the case of other membranes, are described by the Nernst–Planck equation for a number of ionic species participating in transfer processes. A simplification/condition required for this class of sensors is the summary flux of ionic species (electrical current) equal to zero.

1.3.1 Equilibrium Potential Models

Taking into account all ionic mobile species, one obtains a system of nonlinear differential equations that cannot be solved analytically. Therefore, for practical purposes related to ISE applications, significant simplification of the rigorous Nernst–Planck protocol or even empirical or semiempirical equations are typically used.

These simplifications are based on the aforementioned model of Teorell, Meyer, and Sievers. Assuming then linear ion concentration profiles in the membrane, integration of the electrical field [5] results in the diffusion potential. Then, the Nernst–Donnan equations describing the ion-exchange equilibrium at membrane/solution interfaces are added to obtain the membrane potential. In more general treatments, differential diffusion potential is integrated and the interfacial components can be added.

For practical purposes, the description of membrane potentials is usually based on a simple phase-boundary potential model. Within this approach, the membrane potential is equal to the equilibrium potential prevailing on the membrane/solution interface. Migration effects in the membrane were ignored, and thus, diffusion potential was assumed zero (or constant) (Fig. 1.2). Eisenman has extended this

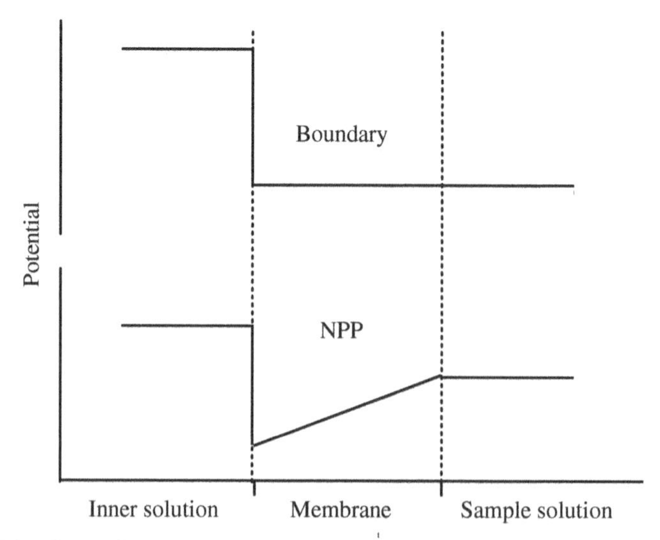

FIGURE 1.2 Comparison of (a) boundary potential profile and (b) approached steady-state Nernst–Planck–Poisson potential profile [77].

model also for nonzero diffusion potential (so-called total membrane potential approach).

Basing on works of Nicolsky and Eisenman, the IUPAC has recommended the semiempirical Nicolsky–Eisenman equation to describe the potential, E, of ion-selective electrode [78, 79]:

$$E = \text{const} + \frac{RT}{z_i F} \ln \left(c_i + K_{ij} c_j^{\frac{z_i}{z_j}} \right) \tag{1.20}$$

derived from separate Nicolsky and Eisenman models. In this equation, K_{ij} is the potentiometric selectivity coefficient for the interferent ion, j. This treatment was based on assumption that the membrane potential response is determined by the sample/membrane interface potential and electrochemical equilibrium prevails at this interface. Moreover, migration effects in the membrane are ignored by assumption that mobilities of all ionic species are equal, and thus, diffusion potential can be neglected.

This relatively simple phase-boundary (equilibrium potential) model is useful and sufficient under typical conditions, denoting ion concentration above 10^{-6} M, membranes saturated with primary ions, and fast ion-exchange processes. Significant advantages of this models have been summarized in a review of Bakker et al. [80], showing a series of its recent applications as a basis for understanding of membrane selectivity, applications of the so-called sandwich membrane method, theoretical advances in optimizing the lower detection limit of ISEs, understanding polyion sensors, potential drifts in the case of ISEs with solid contact (replacing the inner solution), as well as development of galvanostatic ISEs. The problem of selectivity in terms of phase-boundary model has been also discussed in other papers of Bakker et al. [81, 82].

1.3.2 Local Equilibrium Potential Model

However, under some conditions, this simplified description is not sufficient, especially when time-dependent or nonequilibrium phenomena appear. This can concern (i) membranes with low diffusion coefficient values (e.g., polyacrylate-based membranes), where uniform distribution of ions is not attained within reasonable experiment's time; (ii) diluted sample solutions, close to detection limits; (iii) very fast measurement protocols; and (iv) presence of thin membrane, thin transducer phases placed between the membrane and electrode support, or presence of thin aqueous layers between the membrane and support, characterized by significant alteration of their composition in course of measurements. In such cases, the role of transmembrane ion fluxes, under zero current conditions, can be no longer ignored. Moreover, some applications of ISEs directly explore ion flow, for example, determination of polyions as heparin or protamine [83], when the analyte ions are spontaneously accumulated in the membrane phase. Due to the influence of mass transport, the slope of potential versus logarithm of activity is higher than expected from the Nernst relation (concerning equilibrium conditions).

The evolution of models related to responses of ion-selective electrodes has been presented in [84]. Compared to the simple phase-boundary potential model, a more advanced description was offered by a local equilibrium model (or diffusion layer models (DLM)), where a local equilibrium at interfaces was assumed but concentrations of ions in the membrane and the contacting phases are dependent on the distance but are independent of time. Any source of ion fluxes is concentration gradient described by Fick's law, that is, migration is ignored. Steady-state ion fluxes resulting from linear concentration gradients between the interface area and bulk are also assumed. Some cases with any number of ionophores and differently charged ions have been discussed [85].

Steady-state concentration profiles accompanying ion fluxes can be also used in the area of nonequilibrium potentiometry, where reproducible accumulation and depletion processes at ion-selective membranes may be used to gain analytical information about the sample [86].

The diffusion model has been developed to take into account the role of time, in contrast to the equilibrium potential model. The time was introduced to describe attaining of equilibrium via diffusion-controlled ion transport. The role of time can be, for example, reflected in changes of selectivity coefficients [87–90]. The selectivity coefficient (within the frame of this model) can change from the value typical for short times, when ion transport limitations are of crucial importance and the selectivity coefficient represents the ratio of diffusion coefficients of interferent and analyte ion. On the other hand, for a long time, when equilibrium is attained, the selectivity coefficient is the same as predicted by equilibrium potential model.

In 1999, Sokalski et al. [91] used this model to interpret lowering of the detection limit, taking into consideration ion fluxes in a system consisting of a plastic membrane bathed by two solutions. Assumptions of steady-state and linear concentration changes were used. This issue is described in detail in chapter 2 of this book.

Morf et al. [92–94]) have taken into account two monovalent ions (preferred and interfering one) assuming equal diffusion coefficients of these ions. A finite difference method was used in modeling ion fluxes. This approach was also extended assuming that diffusion coefficients can have different values and with other boundary conditions.

1.3.3 Model Based on the Nernst–Planck Equation

A more advanced description, compared to local equilibrium model, accommodates influence of electrical field, using the Nernst–Planck equation, and these approaches do not require equilibrium or steady state as well as electroneutrality condition.

A more rigorous description of phenomena occurring in charged membranes of ISEs can be based on pioneering works proposing first attempts to find numerical solutions of the Nernst–Planck equations, as papers published by Cohen and Cooley [13], Hafemann [95], and MacGillivray [25, 26]. The first important contribution relating numerical solving of the Nernst–Planck equation for the case of ion-selective electrodes, considering space-charge effects using the Poisson equation, has been published by Brumleve and Buck [15], where finite difference simulation method has been used. This method has been developed also by other authors, for example, Manzanares et al. [32, 96], Rudolph [97], Samson and Marchand [98], and Moya et al. (network simulation [99]).

Sokalski and Lewenstam [77, 100] have developed the method proposed by Brumleve and Buck [15] and have proposed a more advanced description useful to numerical solving the Nernst–Planck equations in the case of ion-selective electrodes. In this approach, diffusion and migration of ions were described by the Nernst–Planck equation, while electrical interactions of ionic species were expressed by the Poisson equation (Eq. 1.19).

These equations and the continuity equation (law of mass conservation) relating fluxes, f_{i}, and concentrations, c_{i},

$$\frac{\partial c_{\mathrm{i}}(x,t)}{\partial t} = -\frac{\partial f_{\mathrm{i}}(x,t)}{\partial x} \tag{1.21}$$

formed a system of partial nonlinear differential equations that were solved numerically.

In all calculations, Chang–Jaffe boundary conditions expressing ion transfer kinetics were used:

$$\begin{aligned} f_{\mathrm{i0}}(t) &= \vec{k}_{\mathrm{i}} \cdot c_{\mathrm{i,bL}} - \overleftarrow{k}_{\mathrm{i}} \cdot c_{\mathrm{i0}}(t) \\ f_{\mathrm{id}}(t) &= -\vec{k}_{\mathrm{i}} \cdot c_{\mathrm{i,bR}} + \overleftarrow{k}_{\mathrm{i}} \cdot c_{\mathrm{id}}(t) \end{aligned} \tag{1.22}$$

where $f_{\mathrm{i0}}, f_{\mathrm{id}}, c_{\mathrm{i0}}, c_{\mathrm{id}}$ are the fluxes and concentrations of the ith component at $x = 0$ and $x = d$ (d: membrane thickness), $\vec{k}_{\mathrm{i}}$ and $\overleftarrow{k}_{\mathrm{i}}$ are the forward and backward rate constants, and $c_{i,\mathrm{bL}}$ and $c_{i,\mathrm{bR}}$ are the concentrations in the bathing solutions at the left (L) and right (R) side of the membrane, respectively.

In the calculations, implicit finite difference method was used, and the resulting system of difference equations was solved using the Newton–Raphson method, expressed in the following form:

$$\mathbf{J}\left(\mathbf{x}^{(\mathrm{i})}\right)\Delta\mathbf{x}^{(\mathrm{i}+1)} = -\mathbf{F}\left(\mathbf{x}^{(\mathrm{i})}\right) \tag{1.23}$$

where $\mathbf{J} \equiv \nabla\mathbf{F}$, the Jacobian of $\mathbf{F}$.

The authors managed to avoid splitting the membrane potential into boundary and diffusion potential; it could be shown that both membrane/solution interfaces and the membrane bulk contribute to overall membrane potential. Therefore, analyses of membrane potential distribution and changes covered the whole membrane. The Planck and Henderson equations for liquid junction and the Nicolsky–Eisenman equations for membranes of ISEs were shown as specific cases of the Nernst–Planck–Poisson approach. The former equations give the same results as the general model but for infinite time.

This model can predict significant influence of ion transport (diffusion), distribution and rate parameters, ion charges, dielectric constants, and thickness of the membrane. This model provides also the profile of electrical potential in space and time, which according to DLM (a steady-state model) cannot be modeled. Thus, analyses of transient membrane potentials and its spatial distribution were possible.

The Nernst–Planck–Poisson model offers new solutions related to practical problems of work with ISEs. It can successfully describe concentration and potential profiles in the processes of electrode equilibration, for example, change of the shape of potential profiles in the membrane can be followed, showing approaching a linear dependence of potential on the distance from membrane surface (Fig. 1.2). Moreover, under steady-state conditions (or equilibrium), it can show nonlinearity for ion-exchanger sites, R^-, and migrational effects, which could not be predicted using earlier models (phase-boundary model or DLM). It can be shown that contribution of "diffusion" potential before equilibrium is significant and changes with time; moreover, it is dependent on the magnitude and relation of diffusion coefficients of primary, interfering ions and ion-exchanger sites (Fig. 1.3), and thus, predictions of the present and earlier models can be quite different. These effects are also reflected in the shape of potentiometric calibration plots. Linearity of such plots depends on the distance from the steady state or equilibrium state of the system. However, under steady-state conditions, the shape of the plots can be to some extent affected by diffusion potential.

This model has been later used to analyze time-dependent phenomena in the potentiometric response of ion-selective electrodes related to the role of transmembrane fluxes for the detection limit [101, 102] and selectivity of ISEs [103]. Particularly, it was possible to analyze concentration profiles with time for membranes conditioned

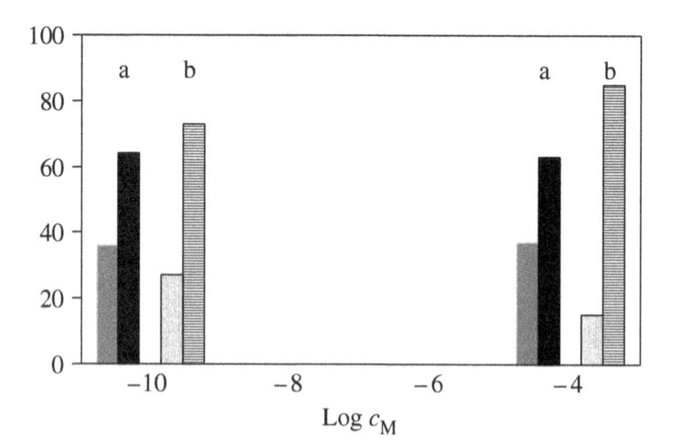

FIGURE 1.3 Contribution (in %) to boundary potential (black and horizontal lines) and inner membrane potential (grey and vertical lines) calculated according to the Nernst–Planck–Poisson model, for $[M^+] = 10^{-4}$ and $[M^+] = 10^{-10}$ M, in the presence of interfering ion ($[N^+] = 10^{-3}$ M) and in the presence of R^- anions in the membrane [77], for diffusion coefficients: (a) $D_M = D_R = 1 \cdot 10^{-7}$, $D_N = 1 \cdot 10^{-6}\ cm^2 \cdot s^{-1}$; (b) $D_M = 1 \cdot 10^{-6}$, $D_R = D_N = 1 \cdot 10^{-7}\ cm^2 \cdot s^{-1}$; $\vec{k}_M = \overleftarrow{k}_M = 10^{-1}$, $\vec{k}_N = 10^{-2}$, $\overleftarrow{k}_N = 10^{-1}$, for X^- anions $\vec{k}_X = X = 0$, $k_N/k_M = 0.1$.

in solutions of highly interfering ions. Moreover, selectivity changes with time could be effectively described. Following changes of concentration and potential in time enabled quantitative explanation of observed selectivity coefficients changes. This model was also used to explain the role of transmembrane ion fluxes and electrical potential drops for the detection limit, its changes, and stabilization in time.

Recently, Morf et al. have analyzed behavior of very thin membranes [104] (below 1 μm), which cannot be regarded as electroneutral phases (as bulk membranes), but they are distinct space-charge systems. The authors applied the theory based on the Nernst–Planck equation and Poisson relationship between space-charge density and electrical field gradient to analyze steady-state potentials and concentration profiles. Solid-contacted thin membranes exhibited sub-Nernstian response, while those contacting aqueous solutions on both sides showed a theoretical Nernstian slope.

Gyurcsanyi with coworkers [105] has applied the Nernst–Planck–Poisson equations to describe potentiometric responses of permselective gold nanopore membranes [61], using the surface charge density of nanopores as a fitting parameter.

Introduction of the Nernst–Planck equation to solve problems related to ion-selective electrodes has brought significant progress in the theoretical description and understanding of response mechanisms of potentiometric sensors. However, there are still problems that could be analyzed in the nearest future as, for example, modeling in 2- or 3-dimensional space, taking into account details of the membrane structure, changes of experimental parameters in dependence on location in the membrane, the role of the membrane inhomogeneity, the role of cross terms, and details of ion-exchange mechanisms.

1.4 SUMMARY

Ionic fluxes and membrane potentials are coupled by the Nernst–Planck equation. This equation is widely used to describe transport phenomena in biological and artificial membranes. In biological membranes, most often, the Goldman approximation is used, assuming constant electrical field in the membrane. Generally, for the description of transport processes in the membranes, also, numerical procedures were proposed, enabling transport description without invoking the Goldman assumption, taking into account, for example, acid/base equilibria, ion-pair formation, and inhomogeneity of charge distribution.

A significant area of membrane applications are potentiometric ion-selective electrodes. Although the nature of transport processes of charged species is similar as for biological membranes, usually, a different method of description prevails in the literature, based on simplified empirical description of boundary potentials (Nicolsky–Eisenman equation), resulting from ion-exchange equilibria on the membrane/solution interface. Additionally, diffusion potential can be taken into account, described usually by considering the Henderson approximation (linear concentration gradient). More advanced descriptions are also proposed as DLM and recently the most advanced approach based on the Nernst–Planck–Poisson equations, without splitting the membrane potential into boundary and diffusion potentials. The last theory is particularly useful to describe the phenomena in the case of, for example, dilute solutions, for short time scale of membrane potential measurements and for thin or highly resistive membranes.

REFERENCES

1. W.E. Morf, *Studies in Analytical Chemistry 2. The Principles of Ion-Selective Electrodes and of Membrane Transport.* Akadémiai Kiadó, Budapest, 1981.
2. J. Koryta, J. Dvorak, L. Kavan, *Principles of Electrochemistry*, Wiley, New York, 1993.
3. H. Ti Tien, A. Ottova-Leitmannova, *Membrane Biophysics. S Viewed from Experimental Bilayer Lipid Membranes (Planar Lipid Bilayers and Spherical Liposomes)*, Elsevier, Amsterdam, 2000.
4. P. Henderson. *Z. Phys. Chem.* 59, 118 (1907)
5. R.P. Buck, E. Lindner. *Acc. Chem. Res.* 31, 257 (1998).
6. K. Sollner. *Z. Elektrochem.* 36, 36 (1930).
7. T. Teorell. *Proc. Soc. Exp. Biol. Med.* 33, 282 (1935).
8. K.H. Meyer, J.F. Sievers. *Helv. Chim. Acta* 19, 649 (1936).
9. F.G. Donnan. *Z. Electrochem.* 17, 572 (1911).
10. K. Vetter. *Electrochemical Kinetics*, Academic Press, New York, 1967.
11. W.E. Morf. *Anal. Chem.* 49, 810 (1977).
12. R. Schlögl. *Z. Phys. Chem.* 1, 305 (1954).
13. H. Cohen, J.W. Cooley. *Biophys. J.* 5, 145 (1965).
14. J.R. Sandifer, R.P. Buck. *J. Phys. Chem.* 79, 384 (1975).
15. T.R. Brumleve, R.P. Buck. *J. Electroanal. Chem.* 90, 1 (1978).
16. A. de Mari. *Solid State Elektron.* 11, 1021 (1968).
17. G.D. Hachtel, R.C. Joy, J.W. Cooley. *Proc. IEEE* 60, 86 (1972).
18. J.W. Slotboom. *IEEE Trans. Electron Devices* ED-20, 669 (1973).
19. R.P. Buck. *J. Membr. Sci.* 17, 1 (1984).
20. S. Kirkpatrick. *Rev. Mod. Phys.* 45, 574 (1973).
21. A.J. Staverman. *Trans. Faraday Soc.* 48, 176 (1952).
22. O. Kedem, A. Katchalsky. *J. Gen. Physiol.* 45, 143 (1961).
23. O. Kedem, A. Katchalsky. *Trans. Faraday Soc.* 59, 1918 (1963).
24. S. Koter, W. Kujawski, I. Koter. *J. Membr. Sci.* 297, 226 (2007).
25. A.D. MacGillivray. *J. Chem. Phys.* 52, 3126 (1970).
26. A.D. MacGillivray. *J. Chem. Phys.* 48, 2903 (1968).
27. M.S. Seshadri. *Ber. Bunsenges. Phys. Chem.* 87, 666 (1983).
28. S. Mafé, J. Pellicer, V.M. Aguilella. *J. Phys. Chem.* 90, 6045 (1986).
29. M. Kato. *J. Theor. Biol.* 177, 299 (1995).
30. J. Castilla, M.T. Garcia-Hernández, A. Hayas, J. Horno. *J. Membr. Sci.* 136, 101 (1997).
31. J.W.F. Robertson, M.G. Friedrich, A. Kibrom, W. Knoll, R.L.C. Naumann, D. Walz. *J. Phys. Chem. B* 112, 10475 (2008).

32. J.A. Manzaranes, S. Mafé, J. Pellicer. *J. Phys. Chem.* 95, 5620 (1991).
33. R. Yamamoto, H. Matsumoto, A. Tanioka. *J. Phys. Chem. B* 107, 10615 (2003).
34. H. Matsumoto, R. Yamamoto, A. Tanioka. *J. Phys. Chem. B* 109, 14130 (2005).
35. K. Kontturi, S. Mafé, J.A. Manzaranes, L. Murtomäki, P. Viinikka. *Electrochim. Acta* 39, 883 (1994).
36. N.P. Berezyna, N. Gnusin, O. Dyomina, S. Timofeyev. *J. Membr. Sci.* 86, 207 (1994).
37. S. Mafé, J.A. Manzaranes, P. Ramirez. *Phys. Chem. Chem. Phys.* 5, 376 (2003).
38. R. Takagi, M. Nakagaki. *J. Membr. Sci.* 27, 285 (1986).
39. P. Schwahn, D. Woermann. *Ber. Bunsenges. Phys. Chem.* 90, 773 (1986).
40. M. Higa, A. Tanioka, K. Miyasaka. *J. Membr. Sci.* 37, 251 (1988).
41. P. Ramirez, A. Alcaraz, S. Mafé. *J. Membr. Sci.* 135, 135 (1997).
42. T.-J. Chou, A. Tanioka. *J. Phys. Chem. B* 102, 7198 (1998).
43. T.-J. Chou, A. Tanioka. *J. Membr. Sci.* 144, 275 (1998).
44. N. Lakshminarayanaiah. *Transport Phenomena in Membranes*, Academic Press, New York, 1969.
45. S. Mafé, P. Ramirez, A. Tanioka, J. Pellicer. *J. Phys. Chem. B* 101, 1851 (1997).
46. P. Ramirez, S. Mafé, A. Tanioka, K. Saito. *Polymer* 38, 4931 (1997).
47. P. Ramirez, A. Alcaraz, S. Mafé. *J. Electroanal. Chem.* 436, 119 (1997).
48. M. Kawaguchi, T. Murata, A. Tanioka. *J. Chem. Soc. Faraday Trans.* 93, 1351 (1997).
49. P. Ramirez, A. Alcaraz, S. Mafé, J. Pellicer. *J. Membr. Sci.* 161, 143 (1999).
50. I. Uematsu, T. Jimbo, A. Tanioka. *J. Colloid Interf. Sci.* 245, 319 (2002).
51. P. Ramirez, S. Mafé, A. Alcaraz, J. Cervera. *J. Phys. Chem. B* 107, 13178 (2003).
52. H. Matsumoto, Y.-Ch. Chen, R. Yamamoto, Y. Konosu, M. Minagawa, A. Tanioka. *J. Mol. Struct.* 739, 99 (2005).
53. J. Garcia-Aleman, J.M. Dickson. *J. Membr. Sci.*, 235, 1 (2004).
54. G. Hagmeyer, R. Gimbel. *Sep. Pur. Technol.* 15, 19 (1999).
55. W.R. Bowen, A.W. Mohammad, N. Hilal. *J. Membr. Sci.* 126, 91 (1997).
56. D. Vezzani, S. Bandini. *Desalination* 149, 477 (2002).
57. A.W. Mohammad, M.S. Takriff. *Desalination* 157, 105 (2003).
58. A. Szymczyk, C. Labbez, P. Fievet, A. Vidonne, A. Foissy, J. Pagetti. *Adv. Colloid Interface Sci.* 103, 77 (2003).
59. P. Leroy, A. Revil, D. Coelho. *J. Colloid Interface Sci.* 296, 248 (2006).
60. A. Revil, D. Jougnot. *J. Colloid Interface Sci.* 319, 226 (2008).
61. G. Jagerszki, A. Takacs, I. Bitter, R.E. Gyurcsanyi. *Angew. Chem. Int. Ed.* 50, 1656 (2011).
62. C.R. Martin, M. Nishizawa, K. Jirage, M. Kang, S.B. Lee. *Adv. Mater.* 13, 1351 (2001).
63. K.B. Jirage, J.C. Hulteen, C.R. Martin. *Science* 278, 655 (1997).
64. K. Motesharei, M.R. Ghadiri. *J. Am. Chem. Soc.* 119, 11306 (1997).
65. E.D. Steinle, D.T. Mitchell, M. Wirtz, S.B. Lee, V.Y. Young, C.R. Martin. *Anal. Chem.* 74, 2416 (2002).
66. M. Sugawara, K. Kojima H. Sazawa, Y. Umezawa. *Anal. Chem.* 59, 2842 (1987).
67. H. Minami, M. Sugawara, K. Odashima, Y. Umezawa, M. Uto, E.K. Michaelis, T. Kuwana. *Anal. Chem.* 63, 2787 (1991).

68. R.E. Gyurcsanyi. *Trends Anal. Chem.* 27, 627 (2008).
69. L. Dammak, C. Larchet, V.V. Nikonenko, V.I. Zabolotsky, B. Auclair. *Eur. Polym. J.* 32, 1199 (1996).
70. S. Mokrani, L. Dammak, C. Larchet, B. Auclair. *New J. Chem.* 23, 375 (1999).
71. N. Lakshminarayanaiah. *Equations of Membrane Biophysics*, Academic Press, New York, 1984.
72. F. Helfferich. *Ion Exchange*, McGraw-Hill, New York, 1962.
73. A. Higuchi, T. Nakagawa. *J. Membr. Sci.* 32, 267 (1987).
74. P. Ramirez, S. Mafé, J.A. Manzaranes, J. Pellicer. *J. Electroanal. Chem.* 404, 187 (1996).
75. T.-J. Chou, A. Tanioka. *J. Colloid Interf. Sci.* 212, 293 (1999).
76. E. Lindner, R. Gyurcsányi, R.P. Buck. Membranes in electroanalytical chemistry: Membrane-based chemical and biosensors. in *Encyclopedia of Surface and Colloid Science*, Marcel Dekker, New York, 2002.
77. T. Sokalski, P. Lingenfelter, A. Lewenstam. *J. Phys. Chem. B* 107, 2443 (2003).
78. K.N. Mikhelson. *Ion-Selective Electrodes*, Springer, Berlin, Heidelberg, 2013.
79. IUPAC. *Pure Appl. Chem.* 66, 2527 (1994).
80. E. Bakker, P. Bühlmann, E. Pretsch. *Talanta* 63, 3 (2004).
81. E. Bakker, E. Pretsch, P. Bühlmann. *Anal. Chem.* 72, 1127 (2000).
82. E. Bakker. *J. Electroanal. Chem.* 639, 1 (2010).
83. B. Fu, E. Bakker, J.H. Yun, V.C. Yang, M.E. Meyerhoff. *Anal. Chem.* 66, 2250 (1994).
84. J. Bobacka, A. Ivaska, A. Lewenstam. *Chem. Rev.* 108, 329 (2008).
85. W.E. Morf, M. Badertscher, T. Zwickl, N.F. de Rooij, E. Pretsch. *J. Phys. Chem. B* 103, 11346 (1999).
86. K. Tompa, K. Birbaum, A. Malon, T. Vigassy, E. Bakker, E. Pretsch. *Anal. Chem.* 77, 7801 (2005).
87. A. Hulanicki, A. Lewenstam. *Anal. Chem.* 53, 1401 (1981).
88. A. Hulanicki, A. Lewenstam. *Talanta* 29, 661 (1982).
89. W.E. Morf. *Anal. Chem.* 55, 1165 (1983).
90. A. Lewenstam, A. Hulanicki, T. Sokalski. *Anal. Chem.* 59, 1539 (1987).
91. T. Sokalski, T. Zwickl, E. Bakker, E. Pretsch. *Anal. Chem.* 71, 1204 (1999).
92. W.E. Morf, E. Pretsch, N.F. de Rooij. *J. Electroanal. Chem.* 602, 43 (2007).
93. W.E. Morf, E. Pretsch, N.F. de Rooij. *J. Electroanal. Chem.* 614, 15 (2008).
94. W.E. Morf, E. Pretsch, N.F. de Rooij. *J. Electroanal. Chem.* 633, 137 (2009).
95. D.R. Hafemann. *J. Phys. Chem.* 69, 4226 (1965).
96. J.A. Manzaranes, W.D. Murphy, S.S. Mafé, H. Reiss. *J. Phys. Chem.* 97, 8524 (1993).
97. M. Rudolph. *J. Electroanal. Chem.* 375, 89 (1994).
98. E. Samson, J. Marchand. *J. Colloid Interface Sci.* 215, 1 (1999).
99. A.A. Moya, J. Horno. *J. Phys. Chem. B* 103, 10791 (1999).
100. T. Sokalski, A. Lewenstam. *Electrochem. Commun.* 3, 107 (2001).
101. T. Sokalski, W. Kucza, M. Danielewski, A. Lewenstam. *Anal. Chem.* 81, 5016 (2009).

102. J.J. Jasielec, T. Sokalski, R. Filipek, A. Lewenstam. *Electrochim. Acta* 55, 6836 (2010).
103. P. Lingenfelter, I. Bedlechowicz-Sliwakowska, T. Sokalski, M. Maj-Żurawska, A. Lewenstam. *Anal. Chem.* 78, 6783 (2006).
104. W.E. Morf, E. Pretsch, N.F. de Rooij. *J. Electroanal. Chem.* 641, 45 (2010).
105. I. Makra, G. Jagerszki, I. Bitter, R.E. Gyurcsanyi. *Electrochim. Acta* 73, 70 (2012).

2

TRANSMEMBRANE ION FLUXES FOR LOWERING DETECTION LIMIT OF ION-SELECTIVE ELECTRODES

TOMASZ SOKALSKI
Laboratory of Analytical Chemistry, Faculty of Science and Engineering, Åbo Akademi University, Turku, Finland

2.1 INTRODUCTION

The two important qualities of any analytical method are **specificity** (the possibility to determine the substance of interest in the presence of other substances) and **detection limit** (DL) (the ability of the method to enable the determination of as small quantities of the substance as possible).

The formation of the membrane potential of ion-selective electrodes (ISEs) depends on the thermodynamic and kinetic properties of the membrane/solution system, and it is strongly time dependent. Selectivity (K_{IJ}) and DL are essential parameters of all ISEs.

This chapter is devoted to the description of the DL of ISEs with special attention to the role of transmembrane fluxes.

During the last decade, the DL of ISEs has been radically improved, that is, lowered. This has led to the prospect of significantly expanding the use of ISEs in such important fields as clinical or environmental analysis.

Electrochemical Processes in Biological Systems, First Edition. Edited by Andrzej Lewenstam and Lo Gorton.

2.2 DEFINITION OF THE DL

A general definition of the DL for an analytical method will be followed by the specific definition commonly used in the ISE field.

2.2.1 Statistical DL

The criterion of detection of any analytical method is defined as the level of the measured signal (point Q in Fig. 2.1) at which the probability of error equals γ. The error is defined as the incorrect analytical conclusion; in this case, the judgment that the determined substance *is* present in the sample when in reality it is not and the high signal level are caused by random errors. The criterion of detection is defined as

$$Q = \Phi_{1-\gamma}\sigma_Q \tag{2.1}$$

where σ_Q is the standard deviation of a single measurement of the difference between the sample and the blank signals:

$$\sigma_Q = \sigma(\mathrm{A}-\mathrm{B}) = \sqrt{\sigma_\mathrm{A}^2 + \sigma_\mathrm{B}^2} \tag{2.2}$$

Assuming that the standard deviations of the signals for the sample and the blank are equal ($\sigma_\mathrm{A} = \sigma_\mathrm{B}$), we obtain

$$Q = \Phi_{1-\gamma}\sqrt{2}\cdot\sigma_\mathrm{B} \tag{2.3}$$

If we assume that the probability of error γ equals 0.05,

$$Q = 1.645\cdot\sqrt{2}\cdot\sigma_\mathrm{B} = 2.33\cdot\sigma_\mathrm{B} \tag{2.4}$$

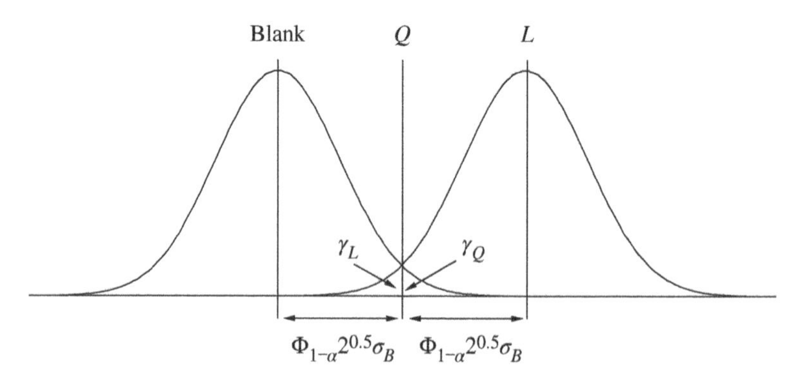

FIGURE 2.1 The statistical detection limit. Reproduced from Ref. [1] by permission of Elsevier.

The DL can be defined in an analogical way (point L in Fig. 2.1). In this case, the wrong analytical conclusion will be that the substance to be determined is *not* present in the sample when in reality it is present, and the small difference between the signals of the sample and the blank is caused by random errors:

$$L = Q + \Phi_{1-\gamma}\sigma_L \tag{2.5}$$

Assuming that $\sigma_A = \sigma_B$, we obtain

$$L = Q + \Phi_{1-\gamma} \cdot \sqrt{2} \cdot \sigma_B = 2 \cdot \sqrt{2} \cdot \Phi_{1-\gamma} \cdot \sigma_B \tag{2.6}$$

Assuming, as before, that the error $\gamma = 0.05$, we eventually obtain

$$L = 4.65 \cdot \sigma_B \tag{2.7}$$

2.2.1.1 Application of the Statistical DL to ISEs The DL of ISEs is defined in a specific way that differs from the definition used for other analytical techniques. The statistical DL, which is valid for any analytical method, is also used for ISEs, although much less frequently [1].

In the case of potentiometry, we may concretize the criterion of detection as the potential difference between a blank and a sample with the concentration C_Q that fulfills the equation

$$E_Q = E_B + z_i \cdot Q \tag{2.8}$$

where E_Q is the potential in the sample, E_B is the potential in the blank, and z_i is the charge of the preferred ion.

By analogy, the DL is given by

$$E_L = E_B + z_i \cdot L \tag{2.9}$$

assuming that the concentration in the sample is C_L.

Equations (2.8) and (2.9) express the criterion of detection and the DL in terms of potential. In terms of concentration, they may be obtained from the intersection of the calibration curve with the straight lines parallel to the concentration axis for the values $E_Q - E_B = \Phi_{1-\gamma} \cdot \sqrt{2} \cdot \sigma_B$ and $E_L - E_B = \Phi_{1-\gamma} \cdot \sqrt{2} \cdot \sigma_B$.

2.2.2 Nonstatistical DL of ISEs

As mentioned earlier, the DL for ISEs is defined in a specific way.

Commonly, the response of ISEs is described by the semiempirical Nikolskii–Eisenman (N–E) equation postulated by the IUPAC [2, 3]. In the IUPAC equation, it is assumed that (i) equilibrium applies, that is, time is not involved, and (ii) concentrations in the membrane are constant and are implicitly included in the term "const." The equation is strictly true only for ions of the same charge.

To acknowledge the role of the DL, a generalized semiempirical equation for ISEs has been proposed [4]:

$$E = \text{const} + S \cdot \log\left(a_\mathrm{I} + K_\mathrm{IJ}^\mathrm{pot} a_\mathrm{J}^{z_\mathrm{I}/z_\mathrm{J}} + L\right) \tag{2.10}$$

where const is an empirical constant, S is the empirical slope, a_I and a_J are the activities of the preferred and discriminated ions in the bulk of the solution, z_I and z_J are their respective charges, $K_\mathrm{IJ}^\mathrm{pot}$ is the selectivity coefficient, and L is the DL.

In the original IUPAC recommendation [5], the DL for ISEs was defined as the activity where the calibration curve deviates by $S \cdot \log(2)$ mV from the extrapolated prolongation of its linear part (point A in Fig. 2.2). In practice, a later IUPAC definition [2, 3] is more often used. According to this definition, the DL is the activity that corresponds to the intersection of the extrapolated linear parts of the calibration curve (point B in Fig. 2.2). If the electrode has a theoretical (Nernstian) slope, points A and B will of course be in the same place. The later IUPAC definition takes into account the nonideal slopes of real electrodes. Because of the specific definition, it is possible to determine the concentrations of ions below the DL [1, 6].

Interestingly, this definition was adequate until the discovery of a method of substantially lowering the DL of ISEs [7]. A redefinition of the DL (Fig. 2.3) has consequently been proposed. According to this, the DL is the bulk sample concentration of the preferred ion at which the measured potential begins to deviate from the Nernstian value by $S \cdot \log(2)$ in *either* direction [8].

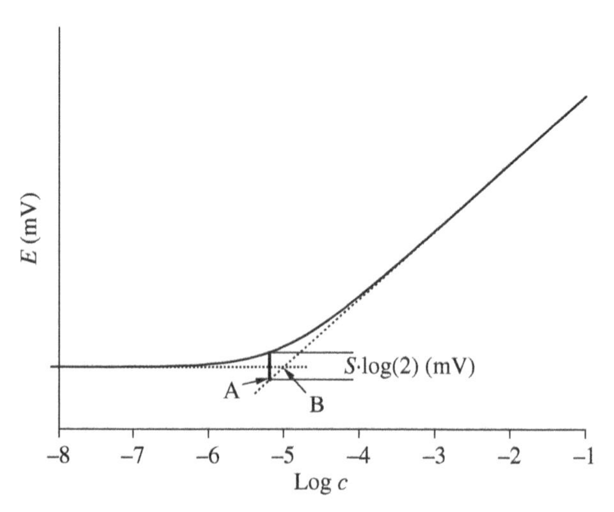

FIGURE 2.2 Definition of the lower detection limit according to the IUPAC. Reproduced from Refs. [2, 3, 5] by permission of Pergamon/De Gruyter.

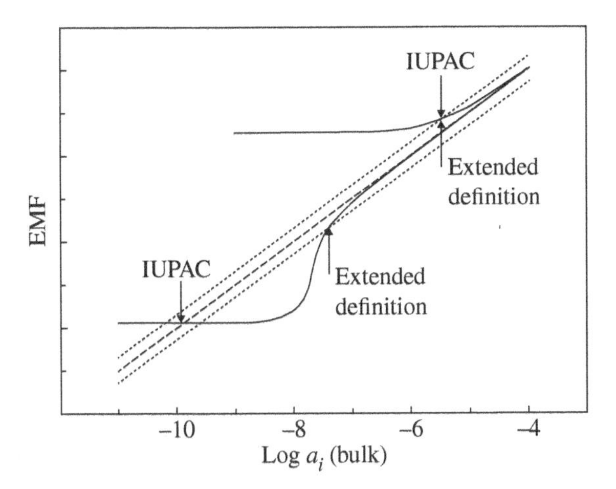

FIGURE 2.3 Definition of the lower detection limit according to the IUPAC and the new extended definition. Reproduced from Ref. [8] by permission of ACS Publications.

2.2.3 Comparison between the Statistical and the Nonstatistical DL of ISEs

Obviously, the same physical factors influence both the statistical and the nonstatistical method, the only difference being the numerical value of the DL. For instance, for solid-state electrodes composed of 1:1 salts, the nonstatistical DL equals

$$C_L = \sqrt{K_{so}} \tag{2.11}$$

(cf. Eq. 2.13)

In this case, the statistical DL is given by the following equation [1]:

$$C_L(\text{stat}) = \sqrt{K_{so}}\, 10^{L/s}\left(1 - 10^{-2L/s}\right) \tag{2.12}$$

The difference between the results obtained by using the nonstatistical approach (2.11) and the statistical approach (2.12) depends on the properties of the electrode. The more reproducible the electrode potentials are (i.e., the smaller L and the standard deviation, σ, are), the smaller the statistical DL will be and the more it will differ from the nonstatistical DL.

2.3 SIGNIFICANT REDUCTION OF THE DL

The lower DL of ion-selective polymeric membrane electrodes are in the micromolar range, except when ion buffers are used to keep the activities of the ions in the sample at low and constant levels through complexation or solubility equilibria [9]. It is a well-known hypothesis that preferred ions leaching from the sensor membrane determine the DL of the ISE by causing a local increase of the concentration of analyte ions in the sample close to the membrane surface. Preferred ions most likely leach into the sample because, together with their counterions, they are extracted from the internal electrolyte and transported across the membrane. Preferred ions are also partially replaced by discriminated ions due to ion exchange [10–12].

The driving force of these fluxes is the huge difference between the activity of the analyte in the diluted samples and in the internal solution (IS), where the concentration of the analyte ions is in the millimolar concentration range or even higher.

This leakage also explains the lack of response toward highly discriminated ions at low concentrations of the preferred ion, which prevents the determination of true selectivity coefficients. By buffering the preferred ion in the sample without influencing the activity of the discriminated ion, a theoretical (Nernstian) response and true selectivity coefficients can be obtained [13]. A Nernstian slope and unbiased selectivity coefficients can be obtained also without buffering, even for highly discriminated ions, as long as contact of the electrode with the preferred ion is avoided [14].

Although such methods are important for understanding selectivity behavior and for the determination of unbiased selectivities, they cannot be applied to real-world samples and so do not improve the response range in analytical applications.

A breakthrough concerning the DL of ISEs occurred in 1996. A novel idea, which made it possible to lower the DL of ISEs by several orders of magnitude, was presented in the work by Sokalski et al. [7]. Since then, works relating to potentiometric measurements in subnanomolar and even lower concentration ranges have become the mainstream of ISE research.

2.3.1 Chemical Lowering of the DL

The first approach aimed at reducing the transmembrane fluxes was invented in 1996 and described in reference [7]. In this seminal work, the authors' basic idea was to prevent preferred ions from leaching into the sample by building up a concentration gradient in the membrane as shown in Figure 2.4.

The concentration gradient is established by choosing an internal electrolyte with a low activity of the preferred ion and a sufficiently high activity of a discriminated ion. Since any activity change in the inner compartment would induce a potential change, the low activity of the preferred ion is kept constant by using an ion buffer. If no suitable buffer exists, a low activity of analyte in the IS may be kept by using, for example, ion-exchange resins [15, 16].

Under these conditions, the analyte ions in the ISE membrane close to the internal surface are largely replaced by the interfering ions, producing a concentration gradient of the analyte ions across the membrane.

The responses of two Pb^{2+} ISEs with the same membrane but different inner reference systems are shown in Figure 2.5. With the conventional inner filling solution, the DL for the Pb^{2+} ISE is $4 \cdot 10^{-6}$ M, with no Nernstian response being observed for the discriminated Na^{+} ion since it is not potential determining under these conditions. On the other hand, with the new IS, the linear Nernstian response range for the Pb^{2+} ISE is extended to about 10^{-11} M and the DL is lowered to 5×10^{-12} M.

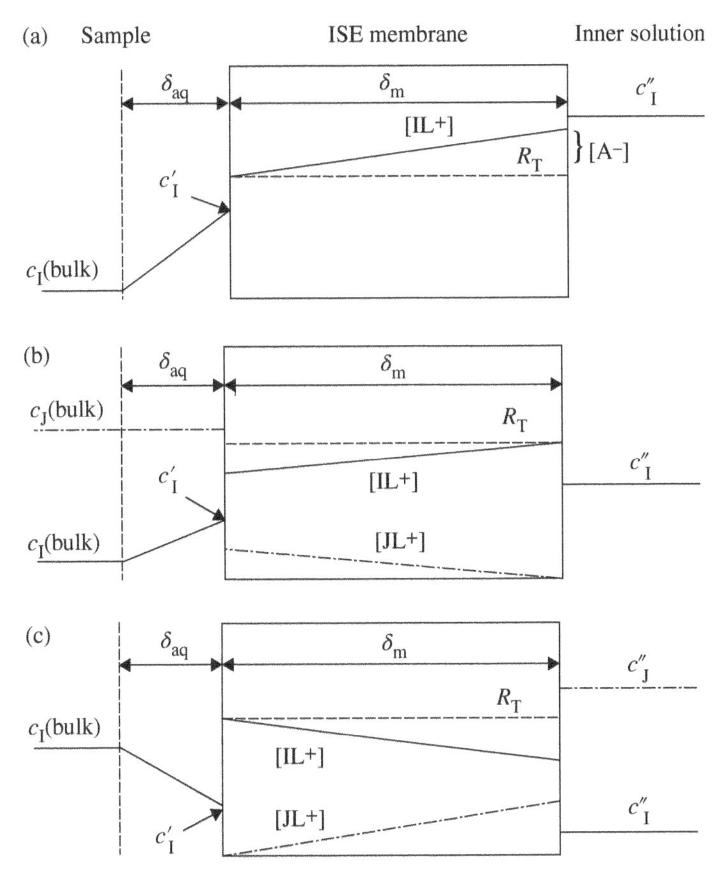

FIGURE 2.4 Schematic representation of the processes influencing the lower detection limit of ISEs based on an ionophore (*L*) forming 1:1 complexes with the monovalent primary (I^+) and interfering ions (J^+). Gradients are generated in the aqueous Nernstian phase boundary (thickness δ_{aq}) because of coextraction of I^+ and A^- from the inner solution (a) or partial exchange of primary ions by interfering ones at the sample or reference side (b or c, respectively). Reproduced from Ref. [8] by permission of ACS Publications.

At the same time, the response to Na^+ is now Nernstian between 10^{-2} and 10^{-3} M, indicating that Na^+ is the potential-determining ion at these concentrations. Thus, true selectivity coefficients can be determined.

The concentration of free Pb^{2+} in the buffered IS, calculated at the measured pH 4.34, is 10^{-12} M and that of Na^+ (counterion of EDTA) is 10^{-1} M. From the experimentally determined value of $\log K^{pot}_{PbNa} = -4.7$, it is obvious that a_{Pb} is much smaller than $K^{pot}_{PbNa} \cdot a^2_{Na}$ in the internal electrolyte, which means that Na^+ is the potential-determining ion at the inner reference side. Hence, Na^+ nearly quantitatively replaces Pb^{2+} at that boundary so that, at steady state, a concentration gradient of Pb^{2+}

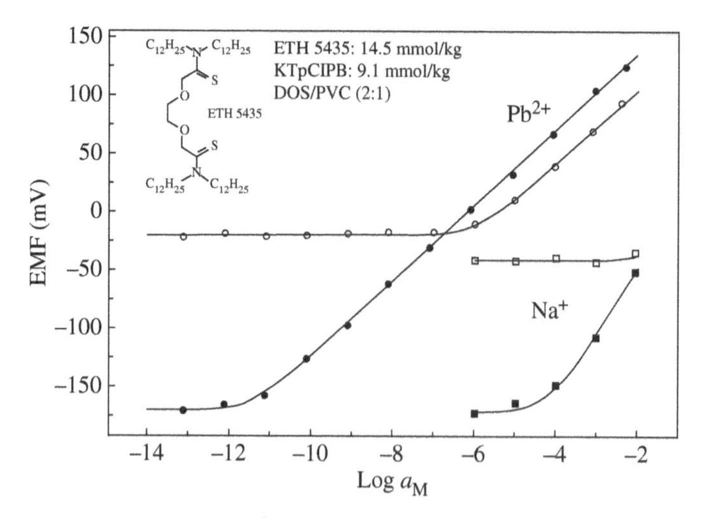

FIGURE 2.5 Response of two Pb^{2+} ISEs with the same membrane but different internal electrolytes (22–23°C). Conventional (empty symbols): 1:1 mixture of 10^{-3} M $PbCl_2$ and 0.1 M $MgCl_2$. New (full symbols): 1 mL of 0.1 M $Pb(NO_3)_2$ in 100 mL of 0.05 M EDTA-Na_2; measured pH 4.34. Calculated activities: 10^{-12} M Pb^{2+}, 10^{-1} M Na^{+}. Reproduced from Ref. [7] by permission of ACS Publications.

(and a corresponding, inverse one of Na^+) is built up in the membrane. The importance of at least a partial displacement of primary ions at the inner membrane surface is also confirmed by the fact that when Pb^{2+} in the internal electrolyte is buffered to 10^{-7} M with nitrilotriacetic acid in the presence of ca. 10^{-3} M Ca^{2+}, no improvement of the lower DL is found [17]. At the concentrations used, Pb^{2+} remains the dominating ion also on the inner side of the membrane, since Ca^{2+} is strongly discriminated $\left(\log K_{\mathrm{PbCa}}^{\mathrm{pot}} = -8.6\right)$.

The ionic fluxes across ion-selective membranes have been registered experimentally in [18]. The internal reference system of a solid contact ISE, for example, based on a conducting polymer, may also be a source of analyte ions, which eventually contaminate sample solutions, although to somewhat lesser extent than in the case of ISEs with internal aqueous solution [19–21].

The disadvantage of this (chemical) approach is that the transmembrane fluxes of ions are eliminated at only one concentration of the analyte in the sample. From a practical viewpoint, this means that the slope in even more diluted solutions is super-Nernstian because there the flux is directed toward the IS and the sample in the vicinity of the membrane is depleted of analyte ions.

2.3.2 Manipulation of Membrane Parameters

More advanced approaches are based on the theory of diffusion and suggest modifications of the membrane geometry or composition should influence the DL [22]. The magnitude of the transmembrane flux can be decreased by lowering the driving force, the gradient of the analyte ion concentration, for example, by using thicker membranes [23, 24]. Also, the flux can be decreased by reducing the diffusion coefficients of the ions in the membrane, for example, by using a higher polymer content [25]. The transportation of ions across membranes may also be minimized by dispersing silica gel microparticles in the membrane [26]. Thus deviations from Nernstian response in diluted samples can be minimized at the cost of increased sensor resistance and some loss of selectivity [26]. Acceleration of ion transport in the sample phase can be achieved using a rotating disk electrode or simply by stirring [27]. Several of these approaches have been critically analyzed and evaluated elsewhere [12].

2.3.3 Electrochemical Lowering of the DL

An alternative way of eliminating transmembrane fluxes is based on galvanostatic polarization of the ISE. This electrochemical approach was first proposed by Lindner et al. [28] and has since then been utilized by several groups of researchers [20, 21, 29–32]. By applying a suitable current, one can eliminate the transmembrane ion flux. Galvanostatic polarization seems to be a more flexible way of improving the DL than modifying the composition of the IS or the composition and/or the geometry of the membrane.

The net flux across the membrane is zero if a particular concentration gradient is counterbalanced by the electric field induced by a particular current. This approach works ideally only in a very narrow concentration range, similarly to the chemical approach, which relies on buffering the IS.

A Nernstian response can be obtained in a broad concentration range by applying currents of different magnitudes suitable for particular concentrations of the analyte in the sample. Another possibility is using a potential recorded at a particular time of the chronopotentiometric experiment as the analytical signal, in which case the measuring time must be adjusted to the particular concentration of the analyte.

These options are challenging and have been only sparsely explored. ISE membrane potentials at nonzero-current conditions have recently been thoroughly analyzed [33–35]. However, it is not completely clear what exactly is the voltage measured when an ISE is galvanostatically polarized. In particular, it is not clear whether only transportation of ions is altered by passing current or also the electrochemical equilibrium at the membrane/solution interface is disturbed because of the polarization. From a practical point of view, using different currents or polarization times for different concentrations of the analyte means having two unknown variables instead of one: if the concentration is not known, the suitable magnitude of current (or suitable time) is not known either. Thus, the analytical relevance of such measurements may seem questionable.

These challenges are addressed experimentally [36] using a calcium-selective ISE based on the neutral ionophore ETH 1001. This particular ISE was chosen primarily because of its excellent performance in zero-current potentiometric measurements, so this electrode is a suitable object for studying a novel measurement technique. The possibility of getting linear Nernstian response within a broad concentration range (from 0.1 to 10^{-10} M) by means of tuned galvanostatic polarization of ISEs was demonstrated for the first time. Moreover, the procedure was developed, which allows for analytical application of the novel technique.

Linear Nernstian response was obtained for a neutral ionophore-based Ca^{2+}-selective electrode down to 10^{-10} M $CaCl_2$ by means of galvanostatic polarization.

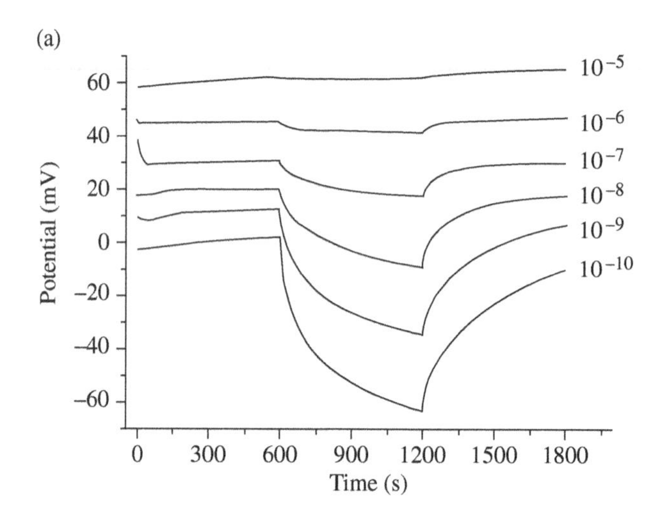

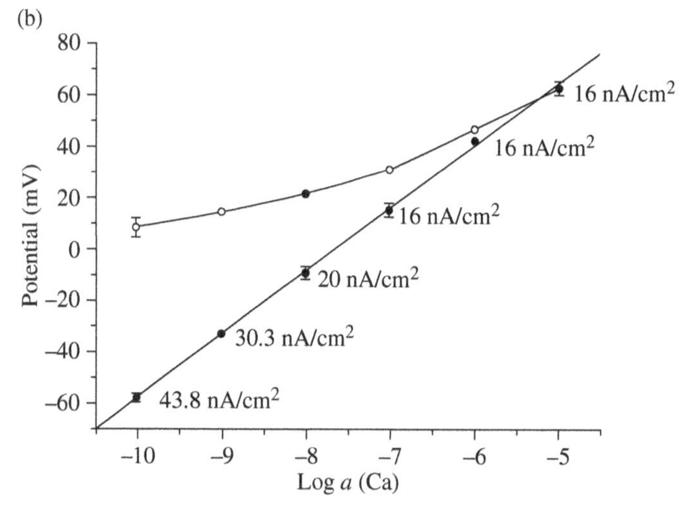

FIGURE 2.6 Chronopotentiometric curves (a) and calibration curves (b) obtained for a Ca ISE polarized by different cathodic currents, in solutions with different concentrations of $CaCl_2$ containing 0.001 M KCl background. (a) Curve 1, 10^{-5} M, 16.0 nA/cm^2; curve 2, 10^{-6} M, 16.0 nA/cm^2; curve 3, 10^{-7} M, 16.0 nA/cm^2; curve 4, 10^{-8} M, 20.0 nA/cm^2; curve 5, 10^{-9} M, 30.3 nA/cm^2; curve 6, 10^{-10} M, 43.8 nA/cm^2. (b) Open circles refer to zero-current potential values (taken immediately before current is on). Filled circles refer to potential values taken immediately after the positive ohmic drop. Reproduced from Ref. [36] by permission of ACS Publications.

The densities of the applied cathodic current were tuned for particular concentrations of Ca^{2+}. The procedure included recording the potential at zero current, followed by measurements when current is passed through the electrode, and then again at zero current. The respective chronopotentiometric curves included negative ohmic drop immediately after turning the current on, the polarization domain, and positive ohmic drop when the current was turned off, followed with the relaxation domain. The potentials immediately after the positive ohmic drops were used as analytical signals. These potentials make a straight line with Nernstian slope when currents are tuned (optimized) for each particular concentration. An iteration procedure was proposed that allows for simultaneous optimization of the current density and accessing analyte concentration in the sample [36]. The results are presented in Figure 2.6. The same technique (tuned galvanostatic polarization) was used for solid-state lead-selective electrode [37].

When analyzing a sample with this technique there are two unknowns: the activity of the target analyte and the current density that is optimal for that particular activity. However, it has been shown previously how to deal with this matter, also mathematically [36]. To overcome this issue, a simple procedure of at least three measurements is required. Firstly, the electrode is polarized using a current density that is optimal to an arbitrarily chosen particular activity of lead, for example, i_{opt} for $pPb^{2+} = 7.2$. If the actual target analyte concentration is lower, the obtained potential is below the Nernstian value for $pPb^{2+} = 7.2$. In this case, another current density must be applied but this time corresponding to much lower lead activity, for example, i_{opt} for $pPb^{2+} = 8.5$. If the potential obtained lies on the other side of the calibration curve, the target analyte concentration is somewhere in between $pPb^{2+} = 7.2$ and $pPb^{2+} = 8.5$. None of the potentials belong to the calibration curve, and when connected with a straight line, they cross it. It has been shown [36] that the intersection point must correspond to the lead activity in the sample solution (1.07×10^{-8} mol/dm^3 Pb^{2+}). To confirm this, an additional third run must be performed with the current density optimal to the lead activity in the sample corresponding to the intersection point. The results in Figure 2.7 [37] show that the potential recorded for the electrode polarized by passing this optimal current was in good agreement with the potential at the intersection point, which confirms the correctness of the analysis.

2.4 THEORETICAL DESCRIPTION OF DL

There is an ongoing and lively debate in the ISE literature concerning models describing ISE behavior in general and K_{IJ} and DL in particular.

The ISE models can be roughly divided into three categories with increasing generality and decreasing idealization level (as illustrated in Fig. 2.8): phase-boundary models, diffusion layer models (DLM), and models including migration. They might also be divided into time-independent (equilibrium or steady state) and time-dependent models [38]. There are two main schools of thought. The first one opts for simplicity, while the second one stresses generality [38].

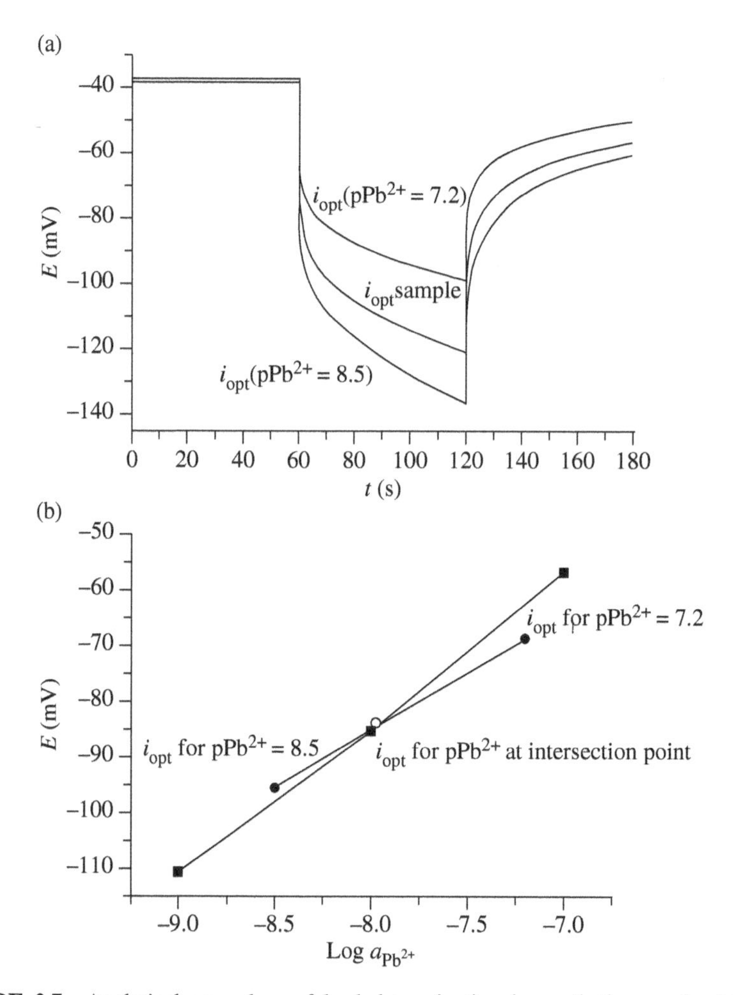

FIGURE 2.7 Analytical procedure of lead determination in synthetic sample. (a) Three chronopotentiometric curves recorded with the two arbitrarily chosen current densities corresponding to pPb = 7.2 and 8.5, and the third one corresponding to the estimated actual lead activity. (b) Calibration curve recorded using tuned polarization in solutions from 10^{-7} to 10^{-9} mol/dm^3 of Pb^{2+} and current-off potentials recorded in a sample of 1.07×10^{-8} mol/dm^3 Pb^{2+} with ISE polarized with the three respective currents (labeled in the figure), all with 10^{-3} mol/dm^3 KNO_3 background. Reproduced from Ref. [37] by permission of Elsevier.

Usually, supporters of the simple models restrict these in the following ways: (i) the diffusion coefficients of the ions in the membrane are assumed to be equal, (ii) the migration of ions is disregarded, (iii) only two (or, at the most, three) ions are considered, (iv) only ions of the same charge are taken into account, and

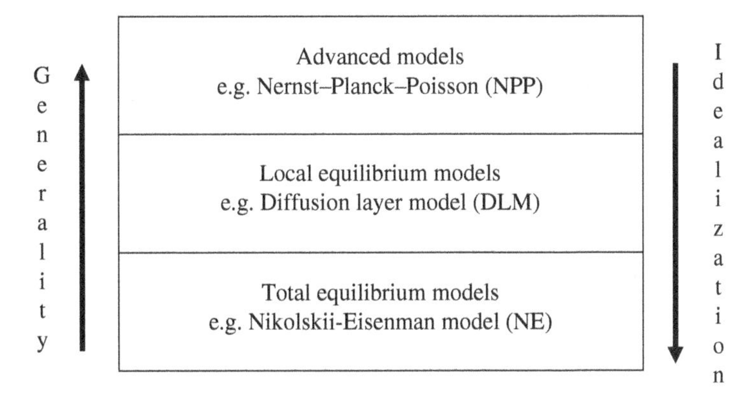

FIGURE 2.8 Methodology used in models of potentiometric ion sensors response.

(v) steady-state conditions are assumed. Some proponents of the simple models claim that the models are general, that is, that they can describe the time-dependent response of ISEs in the presence of ions of different charges. They either assume the presence of only two ions of the same charge and time dependence or ions of different charges and steady state (no time dependence). From these two cases, they draw the incorrect conclusion that the models (simultaneously) apply to both circumstances, that is, ions of different charges *and* time dependence.

The more general models take into account diffusion (each ion having its own diffusion coefficient both in water and in the membrane), migration, an unlimited number of ions of any charge, and time dependence.

All three classes of models are useful in the description of ISE response. Depending on the complexity of the problem and the requirements of the scientist or user, an appropriate model should be chosen.

2.4.1 Phase-Boundary Models

Phase-boundary models are described first. Although they are the oldest and most idealized, their usefulness cannot be denied. In everyday laboratory practice and as a support for commonsense thought experiments, they are most useful.

2.4.1.1 Solid-State ISEs In general, the DL of solid-state electrodes depends on the solubility of the membrane-forming salt [39].

For the solid-state chalcogenide electrode, Buck [39] proposed the following formula for describing electrode response at low concentrations:

$$E = E^0_{\mathrm{AgX}} - \frac{RT}{F}\ln\left(\frac{[\mathrm{X}^-] + \sqrt{[\mathrm{X}^-]^2 + 4K_{\mathrm{so}}}}{2}\right) \tag{2.13}$$

where F is the Faraday constant, R and T denote the gas constant and the absolute temperature, K_{so} is the solubility product of the sparingly soluble salt forming the membrane, and X is the anion to which the electrode is sensitive. Similar equations have also been suggested by other authors [40–43]. According to this approach, the DL is determined exclusively by the solubility product of the membrane material.

In real samples, the DL is influenced also by other factors, such as sample contamination or leakage from the membrane. Morf et al. [44] attributed the leakage (α) of silver ions from the membrane to interstitial ions. Buck [45, 46] did not dispute the formula of Morf et al., but he criticized their explanation of the source of the leakage and stressed the role of (i) impurities during the process of membrane preparation and (ii) chemisorption.

According to Morf [44], when $\alpha \ll K_{SO}$, we obtain a formula analogous to (2.13). When $\alpha \gg K_{SO}$, we obtain the following formula:

$$E = E^0_{Ag} - \frac{RT}{F}\ln([Ag^+] + \alpha) \tag{2.14}$$

2.4.1.2 Ion-Exchanger Electrodes The DL of liquid membrane electrodes mainly depends on the coextraction of the preferred ion. Kamo et al. proposed the following formula for the description of the DL of anionic electrodes based on a liquid ion exchanger [47, 48]:

$$E = E^0 - \frac{RT}{F}\ln\left(\frac{C + \sqrt{C^2 + A_X}}{2}\right) \tag{2.15}$$

where C is the concentration of the preferred ion, $A_X = (4\sigma^2/b_X)$ (if the ion exchanger in the membrane is completely dissociated), σ is the total concentration of the ion exchanger in the membrane, and the parameter b_X depends on the difference between the standard chemical potential of the preferred anion and the ion exchanger in the membrane and water phases.

The DL is described by the value of A_X, which in turn depends on the parameters σ and b_X. According to Equation (2.15), the higher the concentration of the ion exchanger in the membrane is, the higher the DL should be. The more lipophilic (bigger b_X) the anion, the lower the DL. This was confirmed experimentally by the authors.

2.4.2 DLM

The roots of the diffusion models can be traced back to the original formulation of the DLM, which was first used to model changes in the selectivity coefficients of solid-state ISEs [49, 50]. Some descriptions of the DL of solid-state electrodes are also based on the DLM [51, 52]. Later, the model was developed in a number of papers [53–60].

All these implementations of the DLM share three main assumptions: (i) the use of local concentrations in the ISE potential (described by the Nernst (Nikolskii–Eisenman) equation), in the ion-exchange equilibrium, in the mass balance, and in the charge balance (electroneutrality); (ii) that steady state has been reached; and (iii) that only the Fickian diffusion is considered, that is, the only source of ion fluxes is the concentration gradient (migration is ignored). This is a rough approximation, since ions are charged and therefore both create and interact with an electrical field.

Some models based on the DLM include the possibility of a time-dependent response [49, 53–56, 58, 59], although in a slightly ad hoc manner.

The solution of the diffusion problem, along with suitable initial and boundary conditions, belongs to the canon of diffusion and materials engineering and can be found, for instance, in the seminal book by Crank [61] and in other books [62, 63].

2.4.2.1 Solid-State Electrodes According to the DLM, the ISE potential is given by the equation

$$E = E^0 + \frac{RT}{z_I F}\cdot \ln\frac{a_I(0)}{c_I(0)} = E^0 + \frac{RT}{z_I F}\cdot \ln\frac{a_I(0) + K_{IJ} a_J(0)}{c_I(0) + c_J(0)} \tag{2.16}$$

where $a(0)$ denotes the surface activities of the ions in the solution, $c(0)$ is the surface concentrations of the ions in the membrane, $c_I(0) + c_J(0) = \text{constant}$, E^0 is the standard potential, and K_{IJ} is the ion-exchange constant.

In the absence of discriminated ions, $a_J(0)$, the potential of a solid-state ISE depends only on the activity of the preferred ion, $a_I(0)$, in the solution near the sensor's membrane surface. This concentration and, consequently, the DL of solid-membrane ISEs result from various contributions: the analytical concentration in the bulk of the sample (a_I^A), the solubility of the membrane (K_{so}), redox reactions that release the preferred ion from the membrane (β), redox reactions that consume the preferred ion (α), and the adsorption of the preferred ion at the membrane surface (γ). If all of these contributions are taken into account, the following semiempirical equation can be derived [52]:

$$E = E^0 + \frac{RT}{z_I F}\cdot \ln\frac{1}{2}\left[\left(a_I^A + \beta - \alpha + \gamma\right) + \left[\left(a_I^A + \beta - \alpha + \gamma\right)^2 + 4K_{so}\right]^{1/2}\right] \tag{2.17}$$

Figure 2.9 shows the influence of the parameters listed earlier on the DL of the iodide solid-state ISE. Incidentally, the responses observed with solid-state electrodes resemble those observed for plastic membranes.

2.4.2.2 Ion-Exchanger Electrodes The principles of the DLM have been successfully used to describe also the behavior of ion-exchanger electrodes. The Jyo–Ishibashi model [64], applying the principles of the DLM, describes the response of ion-exchanger electrodes subjected to a solution of strongly preferred ion (the Hulanicki effect), but does not interpret the DL. The model was modified by

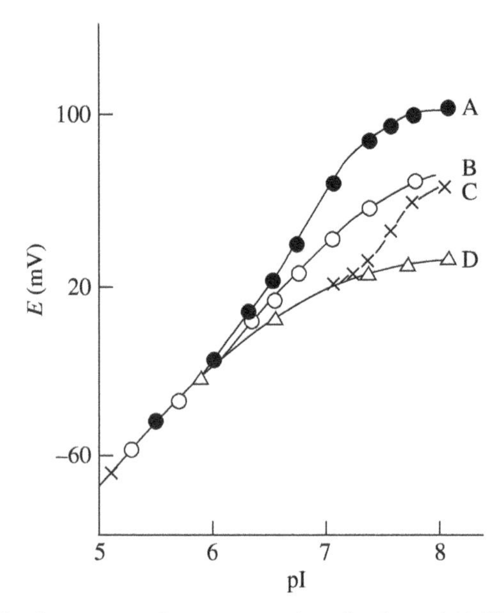

FIGURE 2.9 Calibration curves at low concentrations for electrode 1. The points correspond to experimental values recorded in 0.1 M $NaNO_3$ media. (a) Potential reading after 30 min at 25°C, (b) after 15 min with 4% ascorbic acid added at 20°C, (c) after 30 min with 4% ascorbic acid added at 20°C, and (d) after 24 h with 4% ascorbic acid. The lines shown are calculated for the following parameters: $K_{so} = 9 \cdot 10^{-16}$; (a) $\beta - \alpha = -8 \cdot 10^{-8}$, $\gamma = 0$; (b) $\beta - \alpha = 0$, $\gamma = 0$; (c) $\beta - \alpha = 0$, $\gamma > 0$; (d) $\beta - \alpha = 0$, $\gamma = 1.5 \cdot 10^{-7}$. Reproduced from Ref. [52] by permission of Elsevier.

Maj-Zurawska et al. [57] in order to explain the response of ion-exchanger ISEs at low-concentration levels of both the primary and discriminated ions. This modified model was based on the assumption that if the external solution contains no excess of ions, which may exchange with the plastic membrane, then the concentration of the exchanger at the interface decreases, and this decrease is responsible for the formation of a diffusion layer inside the membrane.

The modified model is derived from the following equations:

Exchange between preferred and discriminated ion:

$$K_{IJ} = \frac{a_I(0)c_J(0)}{a_J(0)c_I(0)} \tag{2.18}$$

Mass balance and electroneutrality:

$$c_I(l) + c_J(l) = c_S(l) \tag{2.19}$$

where $c_S(l) = c_S - \Delta c_S(l)$ and c_S denotes the initial concentration of the ion exchanger, $c(l)$ the space-dependent concentration of ions in the membrane, $c_s(l)$

the space-dependent concentration of the ion exchanger in the membrane, $\Delta c_S(l)$ the space-dependent decrease in the ion-exchanger concentration owing to leakage from the membrane, l the distance inside the membrane ($0 \le l \le \delta_m$), and δ_m the membrane thickness.

Equation (2.19) can be expressed in terms of molar fractions:

$$x_I(l) + x_J(l) + x_\Delta(l) = 1 \tag{2.20}$$

At steady state, the fluxes through the membrane and the diffusion layer can be expressed as

$$D_{aq}\left[\frac{a(0) - a}{\delta_{aq}}\right] = D_m\left[\frac{c(\delta) - c(0)}{\delta_m}\right] \tag{2.21}$$

where D_m and D_{aq} are the diffusion coefficients of the ions in the membrane and in the external solution and δ_m and δ_{aq} are the thicknesses of the membrane and the water diffusion layer.

Combining Equations (2.18), (2.19), (2.20), and (2.21) for a membrane preconditioned in a solution of the ion "I," we obtain

$$a_I + K_{IJ}^{Pot} a_J = a_I(0) + K_{IJ} a_J(0) = \frac{1}{2}(a_I + K_{IJ} a_J - K_{IJ} C) + \frac{1}{2} x_\Delta(0) C(K_{IJ} + 1) + \frac{1}{2}\sqrt{\Delta} \tag{2.22}$$

where

$$\begin{aligned}\Delta = {} & (a_I + K_{IJ} a_J - K_{IJ} C)^2 + 2K_{IJ} C(2 - x_\Delta(0))(a_I + a_J) + \\ & C^2\left[(x_\Delta(0) K_{IJ} - x_\Delta(0))^2 + 2x_\Delta(0) K_{IJ}(1 - K_{IJ})\right] + 2x_\Delta(0) C\left(K_{IJ}^2 a_J + a_I\right)\end{aligned} \tag{2.23}$$

where

$$C = \frac{D_m \delta_{aq}}{D_{aq} \delta_m} c_S$$

The DL can be expressed as

$$L = x_\Delta(0) C = x_\Delta(0) c_S \frac{D_m \delta_{aq}}{D_{aq} \delta_m} \tag{2.24}$$

Figure 2.10 illustrates the experimental data and theoretical curves according to Equation (2.22) and the Jyo–Ishibashi model.

2.4.2.3 Liquid Neutral Carrier Electrodes Currently, the most widely used type of ISE is the neutral carrier electrode. The discovery of the method of significantly lowering the DL (see Section 2.3) was made using this type of electrode, although

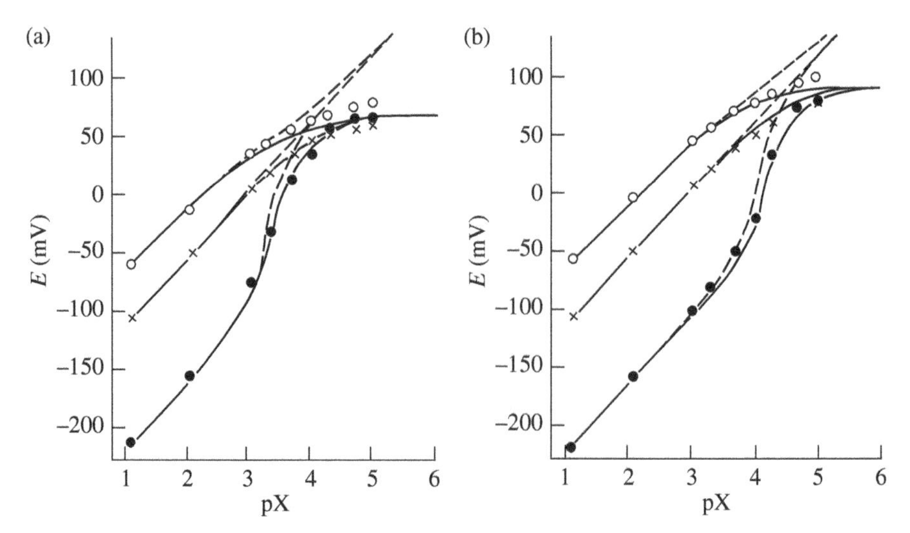

FIGURE 2.10 Response pattern of bromide liquid membrane electrode, Aliquat 336S-Br, and PVC: (a) unstirred solution and (b) maximal stirring rate (v_3). Experimental points, O-Cl^-, x–Br^-, O-SCN^-; solid lines, theoretical curves calculated from Equation (2.22); dashed lines, theoretical curves calculated from the Jyo and Ishibashi equation. Set of parameters: (a) $x_\Delta(0) = 0.14$, $C = 5.3 \times 10^{-4}$; (b) $x_\Delta(0) = 0.32$, $C = 9.0 \times 10^{-5}$. Reproduced from Ref. [57] by permission of Elsevier.

the method may be applied also to other types of ISE. Hence, in the present-day literature, many models describing the behavior of neutral carrier electrodes at low concentrations are presented. A detailed comparison of the predictions of several different models is given in [65]. The models can be divided into two main categories: time-independent and time-dependent models.

Remark: The names of the models presented in the following are not generally adopted. However, the abbreviations proposed in [65] are used here to provide consistency with the descriptions in the original figure captions.

Time-Independent Diffusion Models (SDM) The first model, the steady-state DLM (here called the **SDM1**), was developed in order to obtain the analytical solution of the system of equations presented in the following [8, 22, 66]. It is based on the following assumptions: (i) mass balance, (ii) charge balance, and (iii) the anionic sites are uniformly distributed; (iv) only monovalent preferred and discriminated ions are present; and (v) the stoichiometry of the preferred ion/ionophore complex is 1:1.

Exchange and coextraction:

$$K_{\text{exch}} = \frac{[\text{JL}^+]' c_\text{I}'}{[\text{IL}^+]' c_\text{J}'} = \frac{[\text{JL}^+]'' c_\text{I}''}{[\text{IL}^+]'' c_\text{J}''} \tag{2.25}$$

$$K_{\text{coex}} = \frac{[\text{IL}^+]'[\text{A}^-]'}{c'_{\text{I}}c'_{\text{A}}[\text{L}]'} = \frac{[\text{IL}^+]''[\text{A}^-]''}{c''_{\text{I}}c''_{\text{A}}[\text{L}]''} \tag{2.26}$$

where c_{I}, [L], and $[\text{IL}^+]$ are the sample concentrations of the preferred ion and the membrane concentrations of free ionophore and its preferred ion complex, respectively. The sample boundary layer and the inner side of the ISE membrane are symbolized by prime and double prime, respectively. The same convention is used for the discriminated (J) ion.

Mass balance:

$$[\text{IL}^+]'' - [\text{IL}^+]' + [\text{JL}^+]'' - [\text{JL}^+]' + [\text{L}]'' - [\text{L}]' = 0 \tag{2.27}$$

Charge balance:

$$[\text{IL}^+]'' - [\text{IL}^+]' + [\text{JL}^+]'' - [\text{JL}^+]' + [\text{A}^-]'' - [\text{A}^-]' = 0 \tag{2.28}$$

The concentration changes in the diffusion layer as described by Fick's II law:

$$\begin{aligned} (c'_{\text{I}} - c_{\text{I}}(\text{bulk}))\frac{D_{\text{I,aq}}}{\delta_{\text{aq}}} &= ([\text{IL}^+]'' - [\text{IL}^+]')\frac{D_{\text{IL,m}}}{\delta_{\text{m}}} \\ (c'_{\text{J}} - c_{\text{J}}(\text{bulk}))\frac{D_{\text{J,aq}}}{\delta_{\text{aq}}} &= ([\text{JL}^+]'' - [\text{JL}^+]')\frac{D_{\text{JL,m}}}{\delta_{\text{m}}} \end{aligned} \tag{2.29}$$

Although the steady-state model suggests several possible ways of improving the DL, its usefulness is limited because it disregards the time factor and is valid only for monovalent ions.

Another steady-state model, here called the **SDM2**, was presented around the same time as the SDM1 [67]. This model claims to be very general since it considers any number of ions of any charge. However, a closer inspection shows that only special cases are discussed in this paper. The possibility of time dependence is also hinted at, but only as an artificial addition.

Time-Dependent Diffusion Model The time-dependent diffusion model (here called the **TDM**) described by Morf et al. [68] takes into account only two monovalent ions, the preferred and the discriminated ion, and assumes that their diffusion coefficients in each phase are equal. The authors solve the diffusion problem inside each of the two phases by using the finite difference method. They use the explicit Euler time scheme for the space domain and Euler's method to solve the set of obtained ODEs.

At the left boundary of the first layer, the Dirichlet boundary conditions $(c_i^1(\lambda_0, t) = c_{iL} = \text{const})$ are assumed.

At each time step, the concentration of the ith ion in the kth discretization point of the jth layer $(t + \Delta t)$ is expressed as

$$c_i^{j,k}(t+\Delta t)=c_i^{j,k}(t)+\frac{D_i^j}{\delta_j^2}\left(c_i^{j,k-1}(t)-2c_i^{j,k}(t)+c_i^{j,k+1}(t)\right)\Delta t \tag{2.30}$$

where $\delta^j = (d_j/N_j - 1)$ represents the distance between two neighboring grid points.

At the membrane boundaries, the ion concentrations are given by the following equations for equilibrium:

$$c_1(t+\Delta t)=\frac{R_\mathrm{T}c_1'(t+\Delta t)}{c_1'(t+\Delta t)+K_\mathrm{exch}c_2'(t+\Delta t)},\quad c_2(t+\Delta t)=R_\mathrm{T}-c_1(t+\Delta t) \tag{2.31}$$

where c_i and c_i' are the concentrations inside and outside the membrane phase, R_T is the total concentration of anionic sites in the membrane, and K_exch is the exchange constant.

The potential of the ISE is calculated according to the well-known Nikolskii–Eisenman equation:

$$\varphi=\frac{RT}{F}\ln\frac{c_1^{1,N}+K_\mathrm{exch}c_2^{1,N}}{c_{1R}+K_\mathrm{exch}c_{2R}} \tag{2.32}$$

A second time-dependent model, here called the diffusion exchange model (**TDM-E**) [65], is an improvement of the TDM. It does not assume that the diffusion coefficients of each component in each phase are equal. In the diffusion layer and inside the membrane, the concentration changes are described by Equation (2.30).

On the right side of the diffusion layer, the time-dependent concentration is calculated using the equation

$$c_i^{1,N}(t+\Delta t)=c_i^{1,N}(t)+\frac{2D_i^1}{\delta_1(\delta_1+\delta_2)}\left(c_i^{1,N-1}(t)-c_i^{1,N}(t)\right)\Delta t$$
$$+\frac{2D_i^2}{\delta_2(\delta_1+\delta_2)}\left(c_i^{2,1}(t)-c_i^{2,0}(t)\right)\Delta t \tag{2.33}$$

At the boundary points of the membrane, the concentrations of the preferred and discriminated ions are given by the equations for equilibrium distribution:

$$c_1^{2,0}(t+\Delta t)=\frac{R_\mathrm{T}c_1^{1,N}(t+\Delta t)}{c_1^{1,N}(t+\Delta t)+K_\mathrm{exch}c_2^{1,N}(t+\Delta t)},\quad c_2^{2,0}(t+\Delta t)=R_\mathrm{T}-c_1^{2,0}(t+\Delta t)$$

$$c_1^{2,N}(t+\Delta t)=\frac{R_\mathrm{T}c_{1,R}}{c_{1,R}+K_\mathrm{exch}c_{2,R}},\quad c_2^{2,N}(t+\Delta t)=R_\mathrm{T}-c_1^{2,N}(t+\Delta t) \tag{2.34}$$

where R_T is the total concentration of anionic sites in the membrane and K_exch is the exchange constant.

The potential of the ISE is calculated according to the Nikolskii–Eisenman equation:

$$\varphi(t) = \frac{RT}{F}\ln\frac{c_1^{1,N}(t) + K_{\text{exch}} c_2^{1,N}(t)}{c_{1R} + K_{\text{exch}} c_{2R}} \tag{2.35}$$

A third time-dependent model, here described as the diffusion exchange coextraction model (**TDM-EC**), is a further extension of the TDM [65]. It considers both exchange and coextraction processes and takes into account three monovalent ions (preferred ion, discriminated ion, opposite ion). Equations (2.30) and (2.33) are used as in the case of the TDM-E model.

At the boundary points of the membrane, the concentrations of the preferred and discriminated ions are given by the following equations (instead of by (2.34)):

$$c_1^{2,0}(t) = \frac{R_{\text{T}} + \sqrt{\Delta_1}}{2 + 2K_{\text{exch}}\dfrac{c_2^{1,N}(t)}{c_1^{1,N}(t)}}, \quad c_2^{2,0}(t) = c_1^{2,0}(t) K_{\text{exch}} \frac{c_2^{1,N}(t)}{c_1^{1,N}(t)}$$

$$c_1^{2,N}(t) = \frac{R_{\text{T}} + \sqrt{\Delta_2}}{2 + 2K_{\text{exch}}\dfrac{c_{2,R}}{c_{1,R}}}, \quad c_2^{2,N}(t) = c_1^{2,N}(t) K_{\text{exch}} \frac{c_{2,R}}{c_{1,R}} \tag{2.36}$$

where

$$\Delta_1 = R_{\text{T}}^2 + 4\left(1 + K_{\text{exch}}\frac{c_2^{1,N}(t)}{c_1^{1,N}(t)}\right)\left(c_1^{1,N}(t) + c_2^{1,N}(t)\right) c_1^{1,N}(t) K_{\text{coex}}$$

$$\Delta_2 = R_{\text{T}}^2 + 4\left(1 + K_{\text{exch}}\frac{c_{2,R}}{c_{1,R}}\right)(c_{1,R} + c_{2,R}) c_{1,R} K_{\text{coex}} \tag{2.37}$$

2.4.3 Advanced Models Including Migration

The phase-boundary models and the models based on the DLM are very useful for the description of the behavior of ISEs. They do, however, have their limitations. A general model that takes into account all possible physicochemical processes that occur in an electrochemical system would be more relevant, both physically and mathematically, and would provide information about the behavior of sensors that cannot be deduced from the models mentioned earlier.

2.4.3.1 The Nernst–Planck–Poisson Model The Nernst–Planck–Poisson (NPP) model is an attempt at creating a general model. It considers both diffusion and migration; an unlimited number of ions, each with its own specific charge and its own specific diffusion coefficients in different phases; as well as the different dielectric permittivities of the different phases. The model describes the evolution of the electric field and of the concentrations of all ions in space and time. All these features provide the most accurate description ion sensors so far.

The NPP is, of course, only a model, which implies simplifications and assumptions. These, however, are much fewer and less unrealistic than those involved in previous models.

The first numerical simulation procedure for the time-dependent NPP problem using an explicit method was developed in 1965 [69]. Later, in 1975, a mixed implicit method (for an electric field) and an explicit method (for concentrations) were presented [70].

The application of the NPP model to membrane electrochemistry was presented in a seminal paper by Brumleve and Buck [71]. The authors developed an efficient finite difference scheme, totally implicit in time. The resulting set of nonlinear algebraic equations was solved using the Newton–Raphson method. Later, an approach based upon this idea, and dedicated to the general description of ISE behavior, was developed [72–75]. The first NPP model implementations using the method of lines (MOL) [76] were presented in [33, 77].

The first extension of the NPP model for a two-layer system was proposed in 2005 [78]. Later on, MOL extensions of the NPP model for an arbitrary number of layers were developed and implemented in C++ [79] or in MathCad [75] and MATLAB [80] scripts.

The NPP model is an initial-boundary value problem, which, for one dimension, is described by the set of equations given in the following. The system consists of n layers. Figure 2.11 shows the NPP system for two layers, one representing the diffusion layer of aqueous solution and another representing the membrane.

Each layer, with its own thickness, d^j, and dielectric permittivity, ε^j, is flat and isotropic. It can therefore be considered a continuous environment, in which the change of concentration of r components, c_i^j, and of the electric field, E^j, takes place in space (x) and time (t).

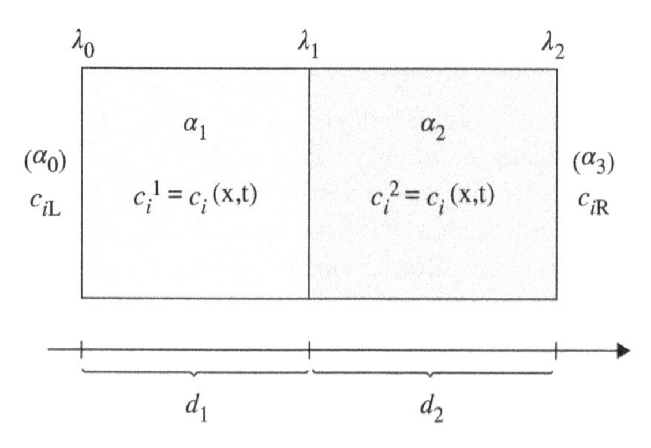

FIGURE 2.11 Scheme of a 2-layer system between two solutions. c_{iL} and c_{iR} are the concentrations of the ith component in the solution on the left and right side, respectively.

The ionic fluxes are expressed by the Nernst–Planck equation:

$$J_i^j(x,t) = -D_i^j\left[\frac{\partial c_i^j(x,t)}{\partial x} + \frac{F}{RT}z_i c_i^j(x,t)E^j(x,t)\right] \tag{2.38}$$

where $J_i^j(x,t)$ is the flux of the ith ion in the jth layer, D_i^j is the constant self-diffusion coefficient of the ith ion in the jth layer, $c_i^j(x,t)$ is the concentration of ith ion in the jth layer, z_i is the valence of ith ion, and $E^j(x, t)$ is the electric field in the jth layer.

The evolution of the electric field is represented by the Poisson equation:

$$\frac{\partial E^j(x,t)}{\partial x} = \frac{\rho^j(x,t)}{\varepsilon^j} \tag{2.39}$$

where the charge density is described by

$$\rho^j(x,t) = F\sum_i^n z_i c_i^j(x,t) \tag{2.40}$$

The mass conservation law describes the change of concentrations:

$$\frac{\partial c_i^j(x,t)}{\partial t} = -\frac{\partial J_i^j(x,t)}{\partial x} \tag{2.41}$$

The Poisson equation can be replaced by its equivalent form, the total current equation [69]:

$$I(t) = F\sum_i^n z_i J_i^j(x,t) + \varepsilon^j\frac{\partial E^j(x,t)}{\partial t} \tag{2.42}$$

The values of fluxes at the boundaries (the interface λ_j between layers α_j and α_{j+1}) are calculated using the modified Chang–Jaffe conditions [81] in the form

$$J_i^j(\lambda_j,t) = J_i^{j+1}(\lambda_j,t) = \overrightarrow{k_{i\lambda_j}}c_i^j(\lambda_j,t) - \overleftarrow{k_{i\lambda_j}}c_i^{j+1}(\lambda_j,t) \tag{2.43}$$

where $\overrightarrow{k_{i\lambda_j}}, \overleftarrow{k_{i\lambda_j}}$ are the first-order heterogeneous rate constants used to describe the interfacial kinetics. The $\overrightarrow{k_{i\lambda_j}}$ constant corresponds to the ion i, which moves from layer α_j to α_{j+1}, and $\overleftarrow{k_{i\lambda_j}}$ to the ion i, which moves from α_{j+1} to α_j.

Initial concentrations fulfill the electroneutrality condition, and consequently, there is no initial space charge in the membrane:

$$c_i^j(x,0) = c_{M_i}^j(x), \quad E^j(x,0) = 0 \quad \text{for} \quad x \in [0,d] \tag{2.44}$$

The membrane potential, $\phi(t)$, is given by

$$\phi(t) = -\int_{0}^{d_1} E^1(t,x)dx - \int_{d_1}^{d_2} E^2(t,x)dx \tag{2.45}$$

2.5 MODEL COMPARISON

It is well known, from abundant literature data [7, 8, 38, 66, 82], that a gradual decrease in the concentration of the preferred ion in the inner solution of a plastic membrane ISE leads to the improvement (decrease) of the DL. However, when the concentration of the preferred ion in the inner solution becomes too low, a super-Nernstian behavior, due to the overcompensation of the transmembrane fluxes of ions, is observed.

2.5.1 Steady-State Response Comparison

The results obtained using SDM1 and SDM2 [65] models give identical results, so for the sake of brevity, only the SDM1 results are discussed here.

Figure 2.12a shows that by decreasing the concentration of the preferred ion (I) in the IS, the DL can be improved by a few orders of magnitude toward the subnanomolar range. However, the decrease of [I] below a certain value (IS $< 10^{-5}$ M) induces a strong, apparently super-Nernstian response because the sample ions in the diffusion boundary layer are depleted.

Similar results for SDM1 were obtained using the TDM-EC model (Fig. 2.12c). Also, the TDM-E (Fig. 2.12b) model shows similar results for the curves that display super-Nernstian behavior but different ones for the curves that should have a higher than optimal DL. This is understandable, since the TDM-E model does not take coextraction processes into account and therefore cannot model the worsening of the DL caused by the leakage of the preferred ion into the diffusion layer.

Although SDM and TDM-EC are simplified models, which do not apply to complex practical cases (polyvalent ions, more than two ions, time influence), they compare satisfactorily with the general NPP model. The trends found when using the SDM and the TDM-EC are confirmed by the NPP and are of general validity for the construction of low DL ISEs.

2.5.2 Transient Response Comparison

Figure 2.13 illustrates the time-dependent response of an ISE (here called ISE10) with a 10^{-10} M concentration of the preferred ion in the IS calculated according to the TDM-EC and the NPP.

The calibration curves calculated according to these models show roughly the same tendencies. At first, with increasing measuring times (t), the DL decreases. A further increase of t causes the appearance of a super-Nernstian section, or jump, in the calibration curve. The longer the measuring time becomes, the more this jump moves toward higher concentrations, thus increasing the DL.

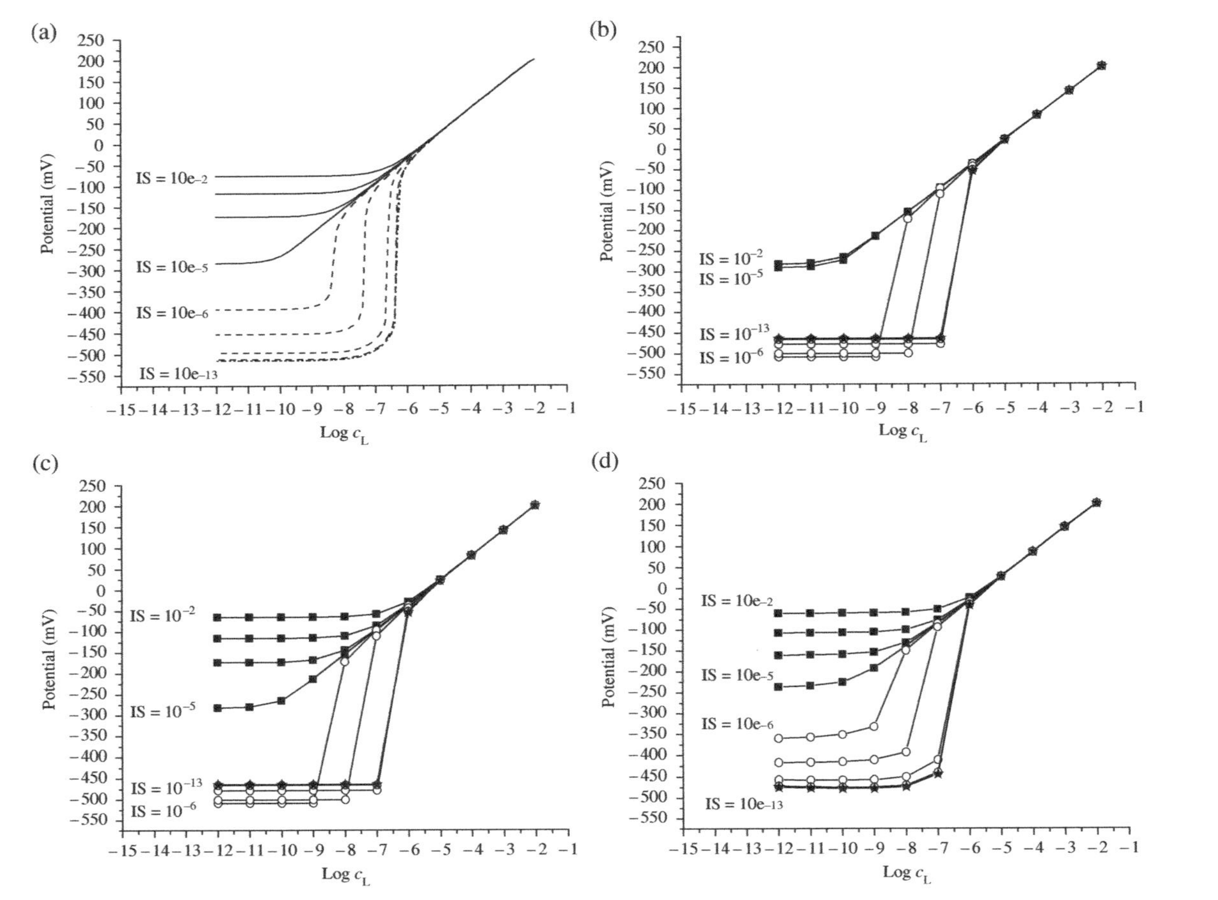

FIGURE 2.12 The influence of the concentration of preferred ion in the inner solution. Calibration curves obtained using the (a) SDM1, (b) TDM-E, (c) TDM-EC, and (d) NPP models. Reproduced from Ref. [65] by permission of Elsevier.

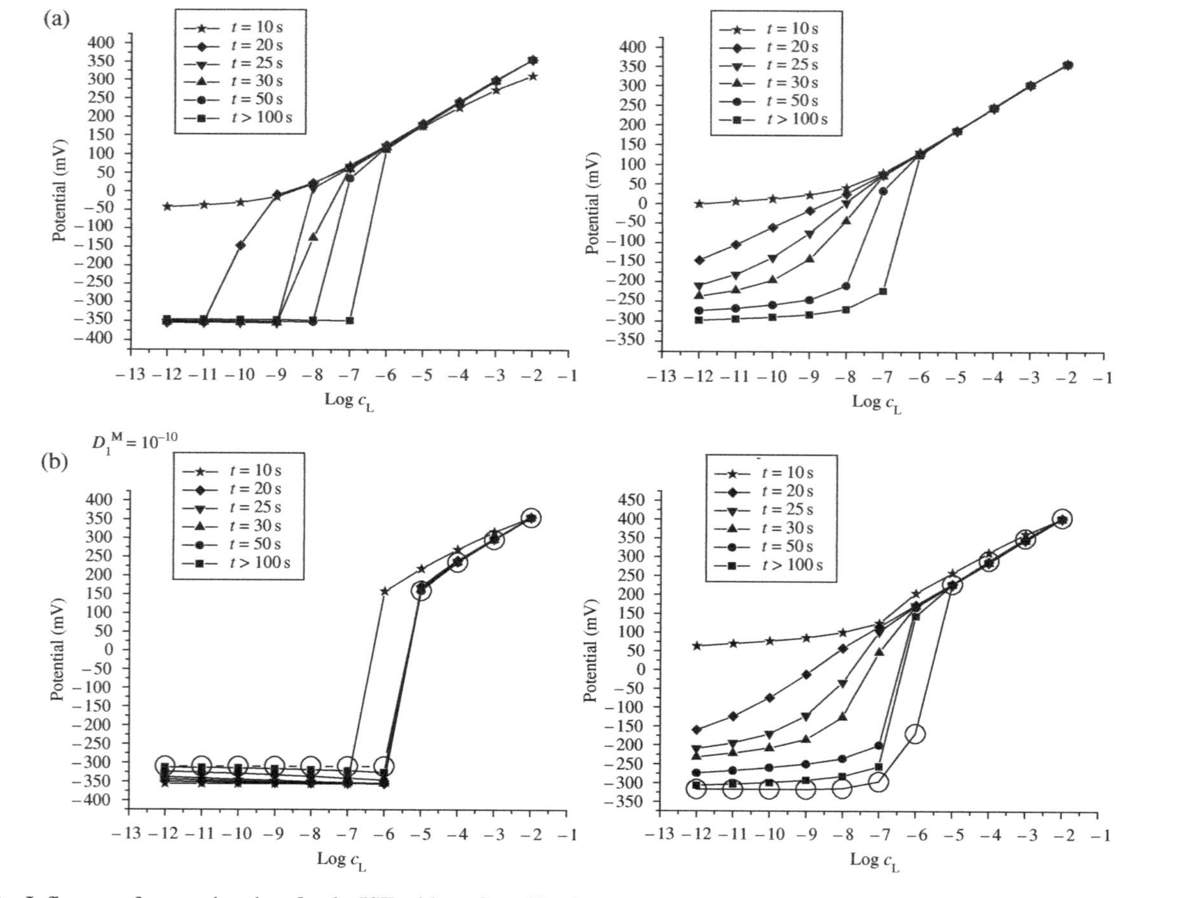

FIGURE 2.13 Influence of measuring time for the ISE with preferred ion inner solution concentration 10–10 obtained using the TDM-EC (left) and the NPP (right) model. (a) Equal diffusion coefficients (above), (b) different diffusion coefficients $D_I = 10 \cdot D_J$ (below). Steady-state curves marked with empty circles. Reproduced from Ref. [65] by permission of Elsevier.

A closer look at the calibration curves, however, reveals a difference between the models. The TDM-EC curves are much more angular, the NPP curves are "smoother" and more similar to the experimental ones.

The cases described earlier are of the "normal problem" type ("What is the value of the DL if we have a certain set of parameters?"). Equally or maybe more important is the so-called inverse problem ("Which set of parameters produces the best DL?").

The answer to this problem is illustrated in Figure 2.14a, which shows a three-dimensional contour plot of the DL versus the measurement time and the concentration of the preferred ion in the inner solution (DL-t-[I]).

The concentration of I is plotted on the x-axis, the measuring time on the y-axis, and the resulting value of the DL on the z-axis. The DL is depicted with the help of different color intensities; the darker the color, the lower the DL. The local/global minima can be read from the plot.

In order to obtain such contour plots, the measuring time was increased from 5 to 500 s with an interval of 5 s, and the ion concentration in the inner solution was varied over a range of 10 orders of magnitude (10^{-2} to 10^{-12} M) with an interval of 1 order of magnitude. For each calibration curve, nine points were calculated. Thus, $100 \times 12 = 1200$ calibration curves and altogether 10,800 points had to be calculated. This is a "brute-force" approach that requires a lot of computational effort, which is exponentially proportional to the number of investigated parameters. The process can be substantially speeded up by the use of more sophisticated techniques such as hierarchical genetic strategy (HGS) [83, 84].

A look at the contour plots in Figure 2.14a also reveals that while the models generally show the same tendency, their results substantially differ in detail. In practice, TDM-EC displays only one global minimum (IS5, steady state). The model suggests that the best value of the DL ($DL = 10^{-10.8}$) can be obtained with ISE5 using very long measuring times (steady state).

Simulations using NPP result in four local minima (of the value of DL) and one global. In simulations where the measuring time exceeds 5 min ($t > 300$ s)—and which are thus comparable to steady-state measurements—the lowest $DL = 10^{-9.1}$ is obtained for ISE5.

In transient-state simulations, an even better $DL = 10^{-10.6}$ can be noted for ISE6 ($t = 100$–200 s). A small, subnanomolar region $DL = 10^{-9.9}$ occurs with ISE7 at $45\ s < t < 50\ s$, and another $DL = 10^{-9}$ with ISE11 at $t = 25$ s. The latter could be very interesting in fast analysis, for example, with automated clinical analyzers.

The results presented in Figures 2.13a and 2.14a were obtained assuming that all species in water had the same diffusion coefficient and, similarly, that the diffusion coefficients of all species in the membrane were equal.

As we indicated earlier, the influence of different ionic diffusibility on the membrane formation process was ignored by all models. The NPP model is a good tool to visualize this fact. Figures 2.13b and 2.14b show the results obtained when assuming that the preferred ion has a 10 times higher diffusion coefficient in the membrane than the discriminated ion.

Now, the difference in the predictions of TDM-EC and NPP is striking. TEM-EC predicts that the steady state is reached almost immediately for $t > 10$ s. All curves,

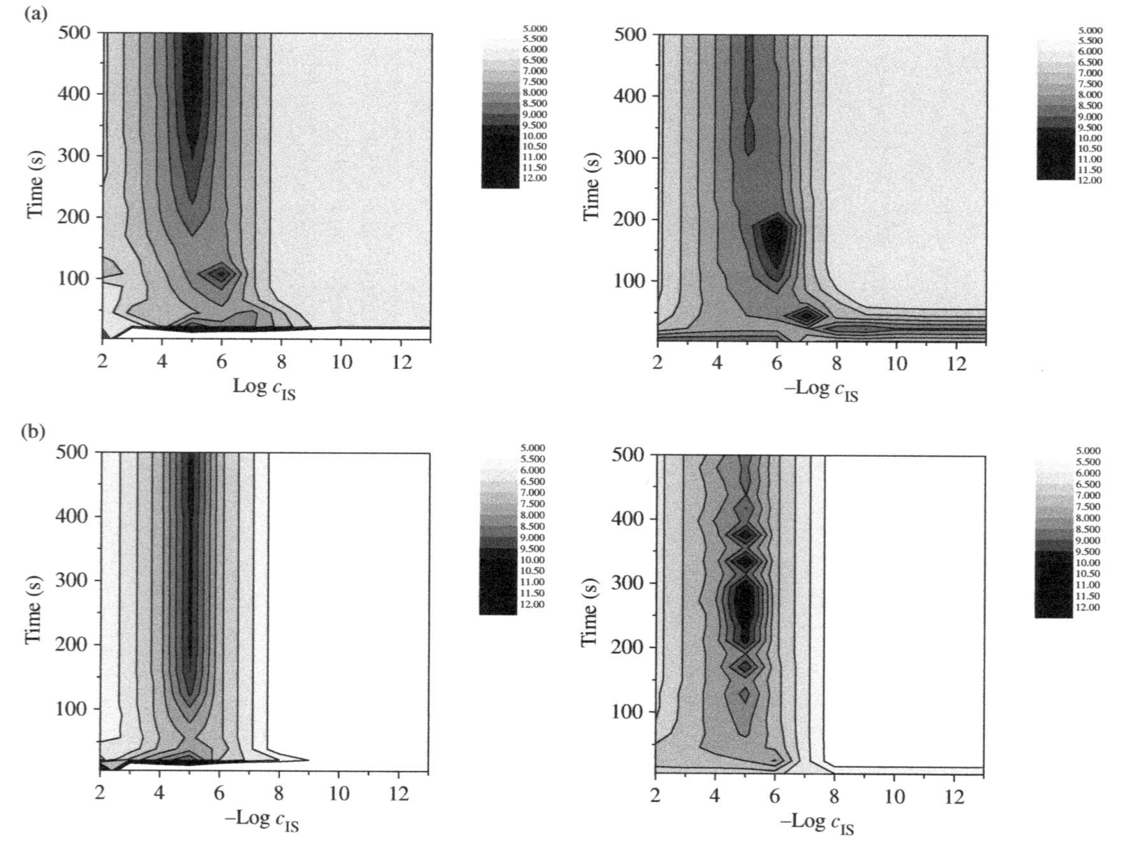

FIGURE 2.14 The time–concentration–detection limit maps obtained using the TDM-EC (left) and the NPP (right) model. (a) Equal diffusion coefficients (above), (b) different diffusion coefficients $D_I = 10 \cdot D_J$ (below). Reproduced from Ref. [65] by permission of Elsevier.

except those measured for 10 s, overlap each other and are the same as the steady-state curve (marked with empty circles). The NPP curves also show the influence of the faster transport of I, but this influence is considerably smaller. The curves for each measuring time length are still distinctly different from each other and different from the steady-state curve. This example nicely illustrates the effect of the potential formed inside the membrane (diffusion potential) on the evolution of the ISE signal. The electric field inside the membrane retards the transport of the faster moving ion and slows down the process of reaching steady state.

This phenomenon is also seen very clearly in the contour plot in Figure 2.14b. The map generated using the NPP model still resembles the one obtained for equal diffusion coefficients, although a faster transport effect is visible. The map obtained using the TDE-EC model shows that steady state is reached very quickly for all I concentrations in the inner solution (vertical stripes on the plot).

2.6 INVERSE PROBLEM

As mentioned earlier, the brute-force approach to solving the reverse problem is very (computationally) expensive. Very soon when considering more dimensions (physical parameters), it becomes unsolvable in acceptable time frame. However, this problem can be solved in an almost elegant or at least adequate fashion, using a hyphenated technique [85].

Figure 2.15 shows the time–concentration–DL map obtained using the NPP model (brute force) with overlaid points obtained with the NPP-HGS method. The NPP-HGS

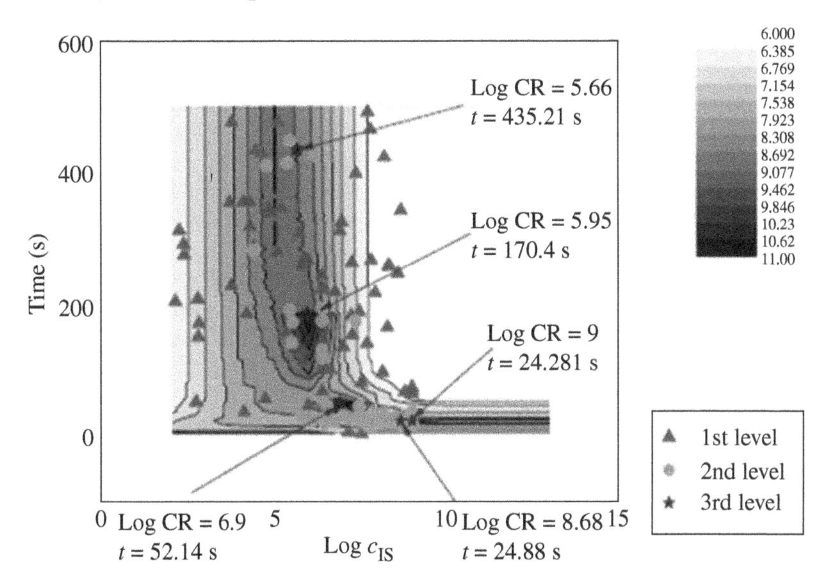

FIGURE 2.15 Time concentration map with all the individuals (points) of HGS. The symbols ▲, ●, and ★ denote the individuals of the first, second, and third populations, respectively. Reproduced from Ref. [85] by permission of ECS.

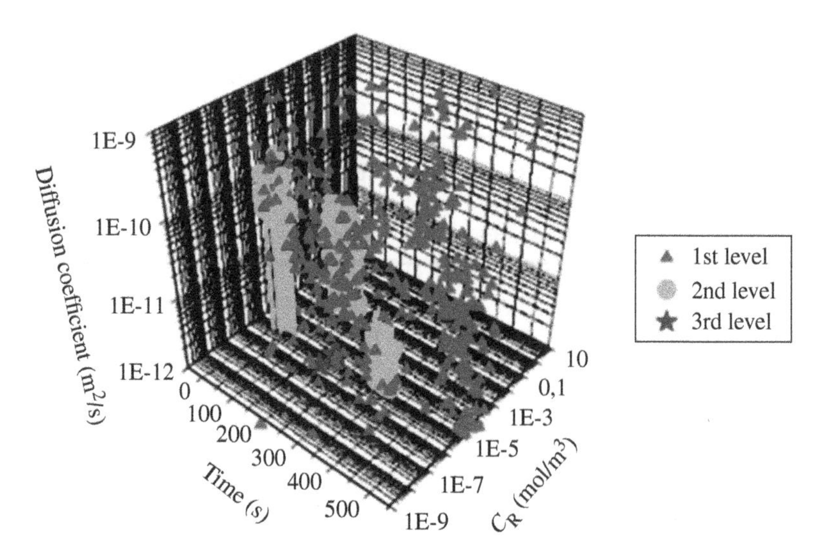

FIGURE 2.16 The individuals (points) of HGS in (D1, c1R, time) space. The symbols ▲, ●, and ★ denote the individuals of the first, second, and third populations, respectively. Reproduced from Ref. [65] by permission of ECS.

TABLE 2.1 Comparison of the dynamic models.

	Diffusion models	
NPP model	Time-dependent models	Steady-state models
Diffusion, migration (convection)	Diffusion	Diffusion
Exchange and coextraction described by heterogenous rate constants	Exchange and coextraction described by equilibrium constants	Exchange and coextraction described by equilibrium constants
Potential calculated from electrical field profile	Potential calculated from the phase boundary equation	Potential calculated from the phase boundary equation
Fluxes in all layers are codependent (concentrations-potential feedback)	Fluxes in both layers are independent	Fluxes are codependent (linear concentration profiles and mass balance)
Ions of any charge	Monovalent ions only	Monovalent ions only
Site distribution depends on electrical field distribution	Sites distribution constant	Sites distribution constant

algorithm was able to find all the minima on the map. It needed to calculate only 214 calibration curves in order to achieve this task. The computational effort was around six times smaller compared with the "brute-force" approach.

This result would be very unfeasible to obtain by the brute-force approach. We would have to calculate 12,000 calibration curves (132,000 points), which would take around 150 days. The computational effort using the NPP-HGS method was around 20 times smaller compared with the "brute-force" approach. The parallel version of HGS allows for further decrease of the computation time (Fig. 2.16).

2.7 IONS OF DIFFERENT CHARGES

In practical applications, ions of different charges are very often present in the solution. In such cases, the NPP is clearly superior to other models.

The experimental calibration curves of a calcium electrode, and the corresponding numerical simulations using the NPP model, are shown in Figure 2.17. The experiment was made and described in reference [66]. The simulations were done using

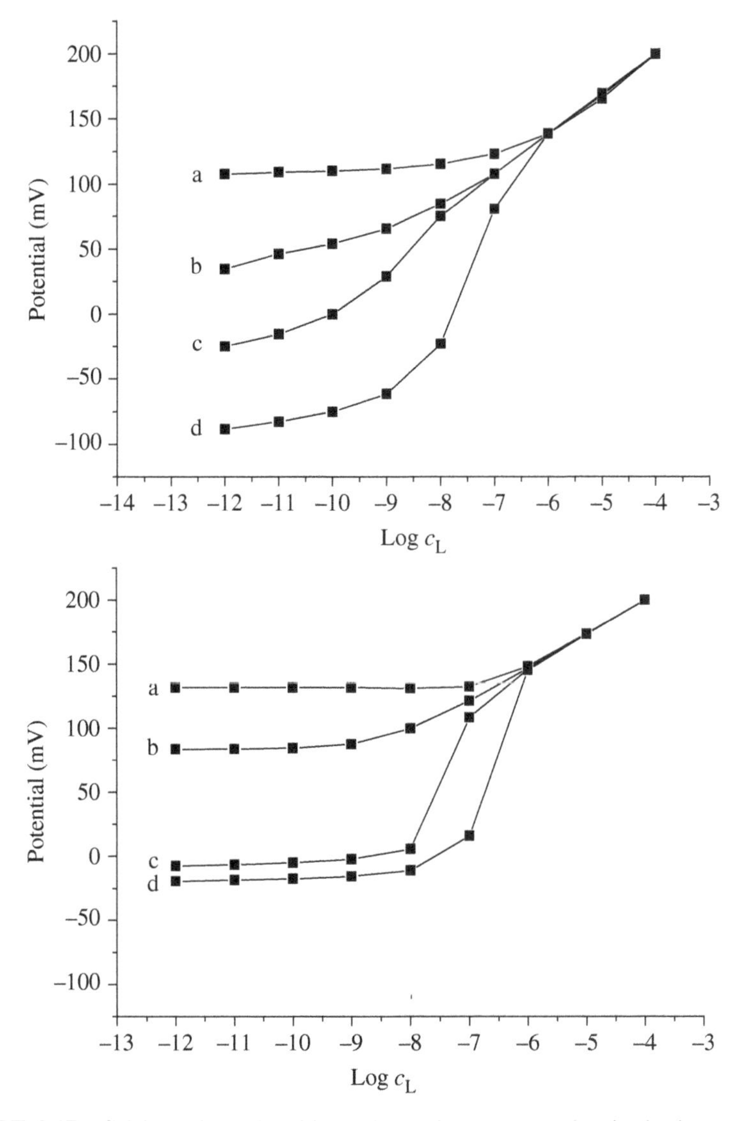

FIGURE 2.17 Calcium electrode with a primary ion concentration in the inner solution equal to (a) 10^{-2} M, (b) 3×10^{-8} M, (c) 1.3×10^{-10} M, and (d) 3×10^{-12} M. Experimental calibration curves (top) and theoretical curves obtained using the NPP model (bottom). Reproduced from Ref. [65] by permission of Elsevier.

the following parameters: two layers, $d_1 = 100$ µm, $d_2 = 200$ µm, $\varepsilon_1 = 7.07 \times 10^{-10}$, $\varepsilon_2 = 2.12 \times 10^{-10}$ (DOS+NPOE), $T = 298$ K. The time of each measurement was 1800 s. The NPP simulations were executed for a system containing six different ions. The inner solution concentrations for electrode A were $c_{Ca^{2+}} = 10^{-2}$ M, $c_{H^+} = 10^{-7}$ M, $c_{Na^+} = 0$; electrode B, $c_{Ca^{2+}} = 3 \times 10^{-8}$M, $c_{H^+} = 4.07 \times 10^{-6}$ M, $c_{Na^+} = 0.073$ M; electrode C, $c_{Ca^{2+}} = 1.3 \times 10^{-10}$ M, $c_{H^+} = 1.2 \times 10^{-7}$ M, $c_{Na^+} = 0.1$ M; and electrode D, $c_{Ca^{2+}} = 3 \times 10^{-12}$ M, $c_{H^+} = 2.5 \times 10^{-9}$ M, $c_{Na^+} = 0.12$ M. The number of discretization points was set to $N_1 = 60$ and $N_2 = 200$, and the first step of space discretization was set to 1×10^{-9}.

The resulting DL of the four electrodes used in the experiment were (i) $c_{1,L} = 10^{-6.8}$ M, (ii) $c_{1,L} = 10^{-8.5}$ M, (iii) $c_{1,L} = 10^{-8.2}$ M, and (iv) $c_{1,L} = 10^{-6.3}$ M. The DL obtained in the NPP simulations were (i) $c_{1,L} = 10^{-6.9}$ M, (ii) $c_{1,L} = 10^{-7.6}$ M, (iii) $c_{1,L} = 10^{-7.3}$ M, and (iv) $c_{1,L} = 10^{-6.3}$ M. The differences may be caused by the fact that the experimental values of the layer thicknesses and the diffusion coefficients inside the membrane were unknown. The presence of additional ions in the solution and the leaching of the ionic sites, drift, or other processes that were not considered in these particular simulations may also have had some effect. Nevertheless, this case is a very good example of a successful application of the model to the complicated real-world example of ISE behavior at low concentrations.

2.8 SUMMARY

This chapter is devoted to the description of the DL of ISEs with special attention to the role of transmembrane fluxes. Both the practical approaches and the theoretical models were discussed.

REFERENCES

1. D. Midgley, Detection limits of ion-selective electrodes. *Ion-Sel. Electrode Rev.* 1981, 3 (1), 43–104.
2. IUPAC, IUPAC Recommendations for nomenclature of ion-selective electrodes. *Pure Appl. Chem.* 1976, 48, 129.
3. R. P. Buck, E. Lindner, Recommendations for nomenclature of ion-selective electrodes—(IUPAC Recommendations 1994). *Pure Appl. Chem.* 1994, 66 (12), 2527–2536.
4. A. Lewenstam, Design and pitfalls of ion-selective electrodes. *Scand. J. Clin. Lab. Inv.* 1994, 54, 11–19.
5. IUPAC, *IUPAC Recommendations for nomenclature of ion-selective electrodes (Recommendations 1975).* Pergamon Press: Oxford, 1976.
6. D. Midgley, Limits of detection of ion-selective electrodes. *Anal. Proc.* 1984, 21 (8), 284–287.
7. T. Sokalski, A. Ceresa, T. Zwickl, E. Pretsch, Large improvement of the lower detection limit of ion-selective polymer membrane electrodes. *J. Am. Chem. Soc.* 1997, 119, 11347–11348.

8. T. Sokalski, T. Zwickl, E. Bakker, E. Pretsch, Lowering the detection limit of solvent polymeric ion-selective electrodes. 1. Modeling the influence of steady-state ion fluxes. *Anal. Chem.* 1999, 71 (6), 1204–1209.
9. Y. Umezawa, *Handbook of Ion-Selective Electrodes: Selectivity Coefficients*. CRC Press: Boca Raton, FL, 1990.
10. Y.M. Mi, S. Mathison, R. Goines, A. Logue, E. Bakker, Detection limit of polymeric membrane potentiometric ion sensors: how can we go down to trace levels? *Anal. Chim. Acta* 1999, 397 (1–3), 103–111.
11. R. Bereczki, B. Takacs, J. Langmaier, M. Neely, R.E. Gyurcsanyi, K. Toth, G. Nagy, E. Lindner, How to assess the limits of ion-selective electrodes: Method for the determination of the ultimate span, response range, and selectivity coefficients of neutral carrier-based cation selective electrodes. *Anal. Chem.* 2006, 78 (3), 942–950.
12. Z. Szigeti, T. Vigassy, E. Bakker, E. Pretsch, Approaches to improving the lower detection limit of polymeric membrane ion-selective electrodes. *Electroanalysis* 2006, 18 (13–14), 1254–1265.
13. T. Sokalski, M. Maj-Zurawska, A. Hulanicki, Determination of true selectivity coefficients of neutral carrier calcium selective electrode. *Mikrochim. Acta* 1991, I, 285–291.
14. E. Bakker, Determination of unbiased selectivity coefficients of neutral carrier-based cation-selective electrodes. *Anal. Chem.* 1997, 69 (6), 1061–1069.
15. W. Qin, T. Zwickl, E. Pretsch, Improved detection limits and unbiased selectivity coefficients obtained by using ion-exchange resins in the inner reference solution of ion selective polymeric membrane electrodes. *Anal. Chem.* 2000, 72 (14), 3236–3240.
16. A. Malon, A. Radu, W. Qin, Y. Qin, A. Ceresa, M. Maj-Zurawska, E. Bakker, E. Pretsch, Improving the detection limit of anion-selective electrodes: an iodide-selective membrane with a nanomolar detection limit. *Anal. Chem.* 2003, 75 (15), 3865–3871.
17. E. Bakker, M. Willer, E. Pretsch, Detection limit of ion-selective bulk optodes and corresponding electrodes. *Anal. Chim. Acta* 1993, 282 (2), 265–271.
18. R. E. Gyurcsanyi, E. Pergel, R. Nagy, I. Kapui, B. T. T. Lan, K. Toth, I. Bitter, E. Lindner, Direct evidence of ionic fluxes across ion selective membranes: a scanning electrochemical microscopic and potentiometric study. *Anal. Chem.* 2001, 73 (9), 2104–2111.
19. A. Michalska, J. Dumanska, K. Maksymiuk, Lowering the detection limit of ion-selective plastic membrane electrodes with conducting polymer solid contact and conducting polymer potentiometric sensors. *Anal. Chem.* 2003, 75 (19), 4964–4974.
20. A. Michalska, Improvement of analytical characteristic of calcium selective electrode with conducting polymer contact. The role of conducting polymer spontaneous charge transfer processes and their galvanostatic compensation. *Electroanalysis* 2005, 17 (5–6), 400–407.
21. K. Y. Chumbimuni-Torres, N. Rubinova, A. Radu, L. T. Kubota, E. Bakker, Solid contact potentiometric sensors for trace level measurements. *Anal. Chem.* 2006, 78 (4), 1318–1322.
22. T. Zwickl, T. Sokalski, E. Pretsch, Steady-state model calculations predicting the influence of key parameters on the lower detection limit and ruggedness of solvent polymeric membrane ion-selective electrodes. *Electroanalysis* 1999, 11 (10–11), 673–680.
23. A. Ceresa, T. Sokalski, E. Pretsch, Influence of key parameters on the lower detection limit and response function of solvent polymeric membrane ion-selective electrodes. *J. Electroanal. Chem.* 2001, 501 (1–2), 70–76.
24. T. Vigassy, C. G. Huber, R. Wintringer, E. Pretsch, Monolithic capillary-based ion-selective electrodes. *Anal. Chem.* 2005, 77 (13), 3966–3970.

25. M. Puntener, M. Fibbioli, E. Bakker, E. Pretsch, Response and diffusion behavior of mobile and covalently immobilized H^+-ionophores in polymeric membrane ion-selective electrodes. *Electroanalysis* 2002, 14 (19–20), 1329–1338.
26. T. Vigassy, R. E. Gyurcsanyi, E. Pretsch, Influence of incorporated lipophilic particles on ion fluxes through polymeric ion-selective membranes. *Electroanalysis* 2003, 15 (5–6), 375–382.
27. A. Radu, M. Telting-Diaz, E. Bakker, Rotating disk Potentiometry for inner solution optimization of low-detection-limit ion-selective electrodes. *Anal. Chem.* 2003, 75 (24), 6922–6931.
28. E. Lindner, R. E. Gyurcsanyi, R. P. Buck, Tailored transport through ion-selective membranes for improved detection limits and selectivity coefficients. *Electroanalysis* 1999, 11 (10-11), 695–702.
29. K. Pergel, R. E. Gyurcsanyi, K. Toth, E. Lindner, Picomolar detection limits with current-polarized Pb^{2+} ion-selective membranes. *Anal. Chem.* 2001, 73 (17), 4249–4253.
30. W. E. Morf, M. Badertscher, T. Zwickl, N. F. de Rooij, E. Pretsch, Effects of controlled current on the response behavior of polymeric membrane ion-selective electrodes. *J. Electroanal. Chem.* 2002, 526 (1–2), 19–28.
31. I. Bedlechowicz, T. Sokalski, A. Lewenstam, M. Maj-zurawska, Calcium ion-selective electrodes under galvanostatic current control. *Sensors Actuat. B-Chem.* 2005, 108 (1–2), 836–839.
32. I. Bedlechowicz-Sliwakowska, P. Lingenfelter, T. Sokalski, A. Lewenstam, M. Maj-Zurawska, Ion-selective electrode for measuring low Ca2+ concentrations in the presence of high K+, Na+ and Mg2+ background. *Anal. Bioanal. Chem.* 2006, 385 (8), 1477–1482.
33. W. Kucza, M. Danielewski, A. Lewenstam, EIS simulations for ion-selective site-based membranes by a numerical solution of the coupled Nernst–Planck–Poisson equations. *Electrochem. Commun.* 2006, 8 (3), 416–420.
34. J. M. Zook, R. P. Buck, R. E. Gyurcsanyi, E. Lindner, Mathematical model of current-polarized ionophore-based ion-selective membranes: Large current chronopotentiometry. *Electroanalysis* 2008, 20 (3), 259–269.
35. J. M. Zook, R. P. Buck, J. Langmaier, E. Lindner, Mathematical model of current-polarized ionophore-based ion-selective membranes. *J. Phys. Chem. B* 2008, 112 (7), 2008–2015.
36. M. A. Peshkova, T. Sokalski, K. N. Mikhelson, A. Lewenstam, Obtaining Nernstian response of a Ca(2+)-selective electrode in a broad concentration range by tuned galvanostatic polarization. *Anal. Chem.* 2008, 80 (23), 9181–9187.
37. G. Lisak, T. Sokalski, J. Bobacka, L. Harju, K. Mikhelson, A. Lewenstam, Tuned galvanostatic polarization of solid-state lead-selective electrodes for lowering of the detection limit. *Anal. Chim. Acta* 2011, 707 (1), 1–6.
38. J. Bobacka, A. Ivaska, A. Lewenstam, Potentiometric ion sensors. *Chem. Rev.* 2008, 108 (2), 329–351.
39. R. P. Buck, Theory of potential distribution and response of solid state membrane electrodes. I. Zero current. *Anal. Chem.* 1968, 40 (10), 1432–1439.
40. V. V. Bardin, Potentiometric determination of low concentrations of chloride ions. *Zavodskaya Laboratoriya* 1962, 28, 910–13.
41. F. G. K. Baucke, Potentials of electrodes of the second kind at low concentrations of common ion electrolyte. I. General discussion. *Electrochim. Acta* 1972, 17 (5), 845–9.

42. F. G. K. Baucke, Potentials of electrodes of the second kind at low concentrations of common ion electrolyte. II. Quantitative treatment of electrodes with salts with negligible complex formation. *Electrochim. Acta* 1972, 17 (5), 851–859.
43. J. Havas. In *IUPAC International Symposium on Selective Ion-Sensitive Electrodes*, Cardiff, 1973.
44. W. E. Morf, G. Kahr, W. Simon, Theoretical treatment of selectivity and detection limit of silver compound membrane electrodes. *Anal. Chem.* 1974, 46 (11), 1538–1543.
45. R. P. Buck, Electroanalytical chemistry of membranes. *CRC Crit. Rev. Anal. Chem.* 1975, 5 (4), 323–420.
46. R. P. Buck, Ion-selective electrodes. *Anal. Chem.* 1976, 48 (5), R23–R39.
47. N. Kamo, N. Hazemoto, Y. Kobatake, Limits of detection and selectivity coefficients of liquid membrane electrodes. *Talanta* 1977, 24 (2), 111–115.
48. N. Kamo, Y. Kobatake, K. Tsuda, Limits of detection and selectivity coefficients of a PVC-based anion-selective electrode. *Talanta* 1980, 27 (2), 205–208.
49. A. Hulanicki, A. Lewenstam, Interpretation of selectivity coefficients of solid-state ISEs by means of the diffusion layer model. *Talanta* 1977, 24 171–175.
50. A. Lewenstam, Construction and response mechanism of sulphide-based ion-selective membrane electrodes. Doctoral Thesis. Warsaw University, Poland 1977.
51. A. Hulanicki, R. Lewandowski, A. Lewenstam, Mechanism of the potential response of bromide-selective electrodes based on mercury salts. *Anal. Chim. Acta* 1979, 110 (2), 197–202.
52. A. Hulanicki, A. Lewenstam, M. Majzurawska, Behavior of iodide-selective electrodes at low concentrations of iodide. *Anal. Chim. Acta* 1979, 107 (JUN), 121–128.
53. A. Hulanicki, A. Lewenstam, Model for treatment of selectivity coefficients for solid-state ISEs. *Anal. Chem.* 1981, 53 (9), 1401–1405.
54. A. Hulanicki, A. Lewenstam, Variability of selectivity coefficients of solid-state ion-selective electrodes. *Talanta* 1982, 29 (8), 671–674.
55. W. E. Morf, Time-dependent selectivity behavior and dynamic response of silver halide membrane electrodes to interfering ions. *Anal. Chem.* 1983, 55 (7), 1165–1168.
56. A. Lewenstam, A. Hulanicki, T. Sokalski, Response mechanism of solid-state ISEs in the presence of interfering ions. *Anal. Chem.* 1987, 59 (11), 1539–1544.
57. M. Maj-Zurawska, T. Sokalski, A. Hulanicki, Interpretation of the selectivity and detection limit of liquid ion-exchanger electrodes. *Talanta* 1988, 35 (4), 281–286.
58. A. Lewenstam, A. Hulanicki, Selectivity coefficients of ion-sensing electrodes. *Sel. Electrode Rev.* 1990, 12 (2), 161–201.
59. A. Lewenstam, Correction. *Sel. Electrode Rev.* 1991, 13 (1), 129–131.
60. W. E. Morf, E. Pretsch, N. F. de Rooij, Theory and computer simulation of the time-dependent selectivity behavior of polymeric membrane ion-selective electrodes. *J. Electroanal. Chem.* 2008, 614 (1–2), 15–23.
61. J. Crank, *Mathematics of Diffusion*. Oxford University Press: Oxford, 1970.
62. M.E. Glicksman, *Diffusion in Solids: Field Theory, Solid-State Principles, and Applications*. John Wiley and Sons, Inc: New York 2000.
63. J.S. Kirkaldy, D.J. Young, *Diffusion in the Condensed State*. The Institute of Metals: London, 1985.
64. A. Jyo, N. Ishibashi, W. E. Morf, In *The Principles of Ion-Selective Electrodes and of Membrane Transport*, Akademiai Kiado: Budapest 1981: 1978.

65. J. J. Jasielec, T. Sokalski, R. Filipek, A. Lewenstam, Comparison of different approaches to the description of the detection limit of ion-selective electrodes. *Electrochim. Acta* 2010, 55 (22), 6836–6848.
66. T. Sokalski, A. Ceresa, M. Fibbioli, T. Zwickl, E. Bakker, E. Pretsch, Lowering the detection limit of solvent polymeric ion-selective membrane electrodes. 2. Influence of composition of sample and internal electrolyte solution. *Anal. Chem.* 1999, 71 (6), 1210–1214.
67. W. E. Morf, M. Badertscher, T. Zwickl, N. F. de Rooij, E. Pretsch, Effects of ion transport on the potential response of ionophore-based membrane electrodes: a theoretical approach. *J. Phys. Chem. B* 1999, 103 (51), 11346–11356.
68. W. E. Morf, E. Pretsch, N. F. de Rooij, Computer simulation of ion-selective membrane electrodes and related systems by finite-difference procedures. *J. Electroanal. Chem.* 2007, 602 (1), 43–54.
69. H. Cohen, J. W. Cooley, The numerical solution of the time-dependent Nernst–Planck equations. *Biophys. J.* 1965, 5, 145–162.
70. J. R. Sandifer, R. P. Buck, Algorithm for simulation of transient and alternating-current electrical properties of conducting membranes, junctions, and one-dimensional, Finite Galvanic Cells. *J. Phys. Chem.* 1975, 79 (4), 384–392.
71. T. R. Brumleve, R. P. Buck, Numerical solution of the Nernst–Planck and Poisson equation system with applications to membrane electrochemistry and solid state physics. *J. Electroanal. Chem.* 1978, 90 (1), 1–31.
72. T. Sokalski, A. Lewenstam, Application of Nernst–Planck and Poisson equations for interpretation of liquid-junction and membrane potentials in real-time and space domains. *Electrochem. Commun.* 2001, 3 (3), 107–112.
73. T. Sokalski, P. Lingenfelter, A. Lewenstam, Numerical solution of the coupled Nernst–Planck and Poisson equations for liquid junction and ion selective membrane potentials. *J. Phys. Chem. B* 2003, 107, 2443–2452.
74. P. Lingenfelter, I. Bedlechowicz-Sliwakowska, T. Sokalski, M. Maj-Zurawska, A. Lewenstam, Time-dependent phenomena in the potential response of ion-selective electrodes treated by the Nernst–Planck–Poisson model. 1. Intramembrane processes and selectivity. *Anal. Chem.* 2006, 78 (19), 6783–6791.
75. T. Sokalski, W. Kucza, M. Danielewski, A. Lewenstam, Time-dependent phenomena in the potential response of ion-selective electrodes treated by the Nernst–Planck–Poisson model. Part 2: Transmembrane processes and detection limit. *Anal. Chem.* 2009, 81 (12), 5016–5022.
76. W. E. Schiesser, *The Numerical Method of Lines: Integration of Partial Differential Equations*. Academic Press: San Diego, 1991.
77. R. Filipek, Modeling of diffusion in multicomponent systems. *Polish Ceramic Bull.* 2005, 90, 103–108.
78. P. Lingenfelter, T. Sokalski, A. Lewenstam, In *Unbiased Selectivity Coefficients: New Insights from the Nernst–Planck–Poisson Model*, Matrafured: Hungary, 2005.
79. J. Jasielec, Application of Nernst-Planck and Poisson equations for interpretation of liquid-junction and membrane potentials in real-time and space domains. Master Thesis, AGH-UST/AAU, 2008.
80. B. Grysakowski, B. Bozek, M. Danielewski, Electro-diffusion in multicomponent ion-selective membranes: numerical solution of the coupled Nernst–Planck–Poisson equations. *Diffusion Solids Liquids III* 2008, 273–276, 113–118.

81. H. C. Chang, G. Jaffe, Polarization in electrolytic solutions. 1. Theory. *J. Chem. Phys.* 1952, 20 (7), 1071–1077.

82. T. Sokalski, E. Pretsch, Low detection limit ion selective membrane electrodes. 9905515, July 23, 1998, 1998.

83. R. Schaefer, J. Kolodziej, R. Gwizdala, J. Wojtusiak, In *How simpletons can increase the community development - an attempt to hierarchical genetic computation*, 4th Conference on Evolutionary Algorithms and Global Optimization, Ladek Zdroj, 2000; pp 187–198.

84. B. Wierzba, A. Semczuk, J. Kolodziej, R. Schaefer, In *Hierarchical Genetic Strategy with real number encoding*, 6th Conference on Evolutionary Algorithms and Global Optimization, Lagow, 2003.

85. J. J. Jasielec, B. Wierzba, B. Grysakowski, T. Sokalski, M. Danielewski, A. Lewenstam, Novel strategy for finding the optimal parameters of ion selective electrodes. *ECS Transactions* 2011, 33 (26), 19–29.

3

ION TRANSPORT AND (SELECTED) ION CHANNELS IN BIOLOGICAL MEMBRANES IN HEALTH AND PATHOLOGY

Krzysztof Dołowy
Laboratory of Biophysics, Warsaw University of Life Sciences (SGGW), Warsaw, Poland

3.1 ION CHANNELS: STRUCTURE, FUNCTION, AND METHODS OF STUDY

3.1.1 Measurement of Ions Flowing through a Single Ion Channel

The idea of ion channels originates from experiments performed by Hodgkin and Huxley [1] on impulse conduction in nerve cells. They found that sodium and potassium ions cross the cell membrane via separate pathways. This idea was supported by the discovery of substances that specifically block the transport of either sodium or potassium ions. Whereas Hodgkin and Huxley measured the conduction of ions through thousands of channels simultaneously, the measurement of ions flowing through a single ion channel molecule required improved amplifiers and the development of new measurement techniques.

In the 1960s, the first method for measuring ion transport across artificial membranes was developed. The black lipid membrane (BLM) technique consists of forming a lipid bilayer on a small septum (50–250 μm) separating two solution

Electrochemical Processes in Biological Systems, First Edition. Edited by Andrzej Lewenstam and Lo Gorton.

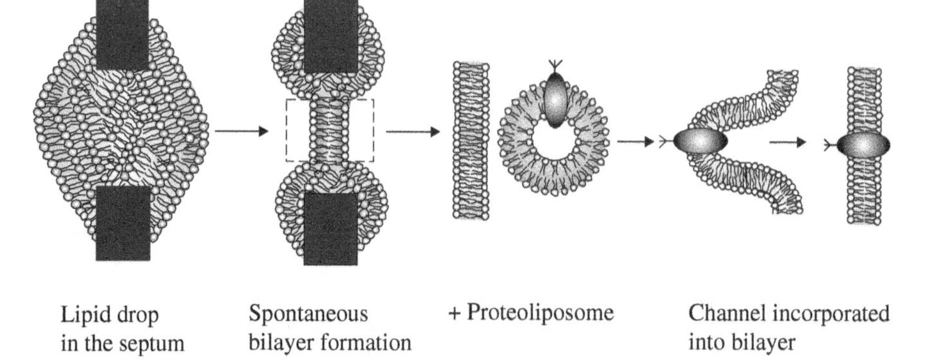

FIGURE 3.1 Black lipid membrane formation. Lipids dissolved in hydrocarbon are painted over the septum separating two electrode compartments. The lipids spontaneously form a lipid bilayer. If a proteoliposome is added to one side of the bilayer, it will incorporate into the bilayer and thereby allow the electrophysiological properties of the incorporated channel molecule to be measured.

compartments (Fig. 3.1). The lipids are either dissolved in the solvent (usually decane) and painted over the septum [2] or the septum is pushed through two separated lipid monolayers [3, 4]. It was shown that both methods produce a membrane of very high resistance ($10^3\ \Omega\ cm^2$) but that the membrane formed by the simpler painting method was twice the thickness of the cell membrane. Using the BLM method, it was possible to show that small peptides present in venoms and antibiotics (e.g., gramicidin) form pores in artificial membranes [5]. The pores had electrochemical properties expected of ion channel molecules (i.e., they open in response to a potential exceeding the critical value and switch between open and closed conformations). However, incorporating ion channel molecules into the artificial membrane proved to be difficult. Simple solubilization of cell membrane and separation of its proteins yielded inactive molecules. The method of separating membrane proteins requires a difficult and time-consuming procedure (Fig. 3.2). Much simpler is the use of native membrane vesicles obtained by fractionating cellular membranes [6]. The BLM method is burdensome and requires substantial patience. However, those who overcome these problems are left with an experimental system in which both surfaces are accessible for manipulation at the same time.

In the race to measure ion conduction through a single-channel molecule, the patch-clamp method succeeded [7]. The patch-clamp method [8] requires the pulling of a minute glass pipette (usually with a diameter of 1 μm). This pipette is filled with salt solution and provided with a silver chloride electrode. The measured cells are placed on an inverted microscope while the pipette is held by a micromanipulator. The pipette is then placed against the cell membrane and gentle suction is applied. It was shown that resistance of the seal between the pipette and the membrane is in the gigaohm range (typical resistance is $1\text{–}5 \cdot 10^{10}\ \Omega$). It is possible to obtain four different configurations of pipette attached to the membrane (Fig. 3.3). Each configuration has its respective advantages and drawbacks. The "cell-attached" configuration allows

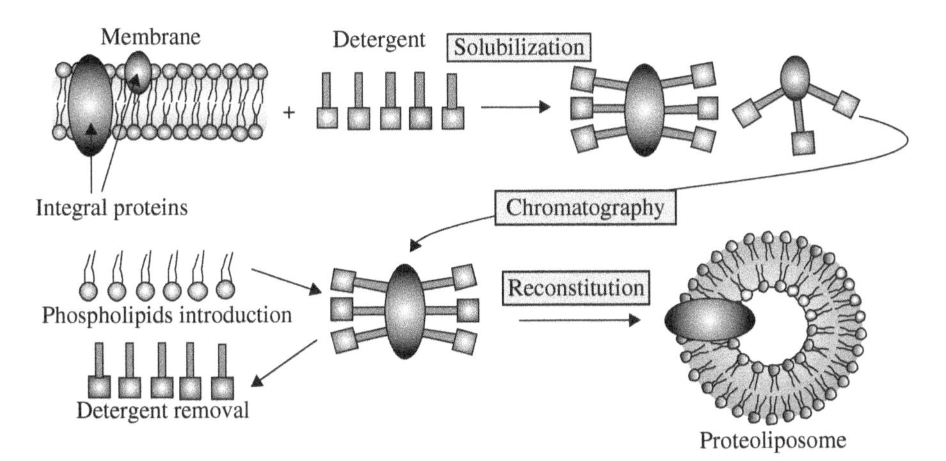

FIGURE 3.2 Purification of integral membrane proteins. To obtain an active channel molecule, one must dissolve the bilayer while not exposing the hydrophobic portion of the molecule to the aqueous environment. This is done by dissolving membrane proteins in long hydrophobic chain detergents that substitute for lipid molecules. Then, the dissolved proteins are separated by chromatography. When pure protein is obtained, detergent molecules are slowly replaced by lipids during dialysis. At the end of the purification, proteoliposomes are obtained.

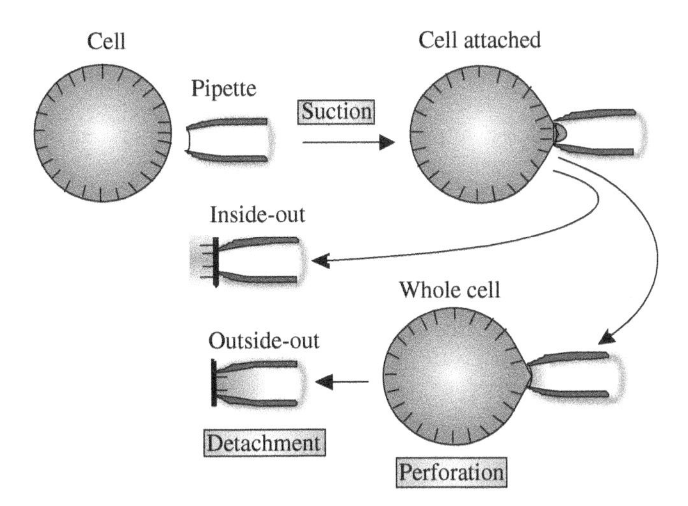

FIGURE 3.3 Four patch-clamp configurations. A micropipette supplied with a silver chloride electrode is brought near the surface of a cell on an inverted microscope. Mild suction is then applied. A piece of membrane is aspirated into the pipette, often forming a gigaseal (i.e., a seal with gigaohm resistance). This configuration is called "cell attached," and it allows channels in the small area beneath the pipette to be studied. If a short electric impulse of high amplitude is then applied, the membrane under the pipette perforates and the "whole-cell" configuration is obtained. The "whole-cell" configuration allows simultaneous measurements of ions flowing through the whole-cell membrane, while internal solution is replaced with the pipette filling solution. By applying mild suction to the "whole-cell" configuration, it is possible to achieve the "outside-out" configuration. By taking the pipette with the cell in the "cell-attached" configuration out from the solution to the air and back to the solution, the "inside-out" configuration is achieved.

one to measure ions flux through a single channel of the native membrane and leaves the intracellular medium unchanged, although no manipulation of either medium composition is possible. The "whole-cell" configuration allows simultaneous measurement of multiple ion channels (e.g., of very small conductivity); however, the intracellular medium is replaced by pipette filling solution. The "inside-out" and "outside-out" configurations allow measurements of the effects of different substances applied to one side of the membrane, while the other has the composition of the pipette filling solution. The patch clamp is easier and much more controllable than the BLM technique. Most data concerning the properties of single channels present in membranes were obtained using the patch-clamp technique.

3.1.2 Intrinsic Limitations of the Patch-Clamp and BLM Methods

The current flowing through ion channels is small and channels open for a very short period of time. Thus, current fluctuation (Johnson noise) is the limiting factor for ion channel detection. To detect a channel, the noise σ must be considerably lower than the current flowing through the channel molecule. Also, one must sample the current with high frequency to separate between opened and closed states of the channel and not average the current between these two states (this condition is described by f, the upper cutoff frequency). The other important factor affecting the measurement is the internal resistance of the signal source R. At present, the internal resistance of the meter is usually very high, on the order of $10^{13}\ \Omega$; thus, it is the resistance of the lipid bilayer that is the limiting factor of the measurement. The area of the cell membrane under the patch-clamp pipette is on the order of 1 μm^2 and has a resistance of $R = 10^{11}\ \Omega$. With the BLM method, the area of the bilayer is on the order of 1000 μm^2 and the resistance is $R = 10^8\ \Omega$. The relationship between the noise and parameters of the measured systems are given by the following equation:

$$\sigma^2 = \frac{4kTf}{R}$$

where k is the Boltzmann constant and T is the absolute temperature. The dependence between the amplitude and the opening time is shown in Figure 3.4. There is a detection limit for both patch-clamp and BLM methods. It is likely that there are ion channels with amplitudes (or open times) beyond the reach of single-channel measurement methods [9]. There are methods by which one can estimate single-channel conductance from nonstationary fluctuation analysis of independent openings of thousands of very-low-conductance channels [10]. There may also be channels with conductivity beyond the reach of any present-day technique—microchannels. Although the existence of such channels has been postulated and their putative properties predicted, they are still hypothetical [11]. A completely new technique is required that will enable measurement of the minute currents flowing through single microchannels and two other well-known types of molecules for transporting ions: transporters (exchangers) and pumps. Large channels allow the transport of 10^{6-7} elementary charges per second, microchannels allow 10^{4-5}, transporters allow 10^{3-4}, and pumps allow no more

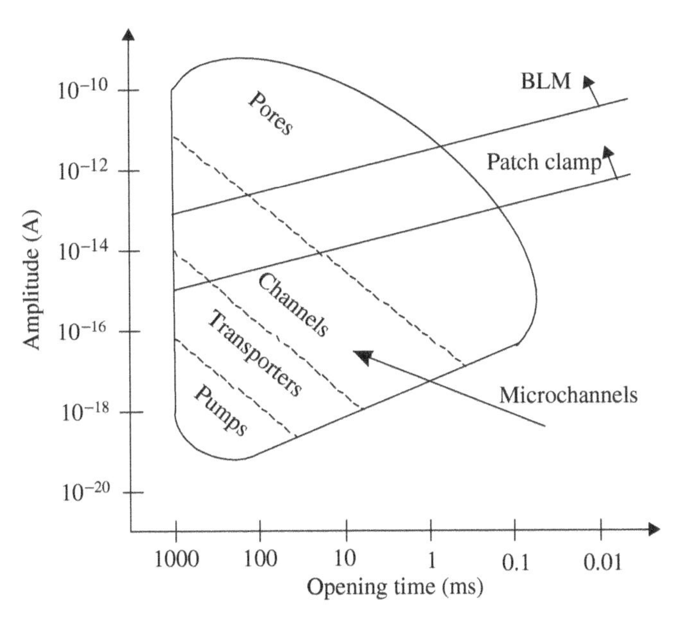

FIGURE 3.4 Amplitude as a function of opening time for different types of molecules transporting ions across the cell membrane. The two lines represent the detection limits of the BLM and patch-clamp methods for single-channel events. Below the limit, other methods allowing simultaneous measurements of multiple microchannels, transporters, or ion pumps must be used.

than 10^2 (Fig. 3.4). Thus, to study the transport of ions through a transporter molecule or a pump, many transporters are measured simultaneously using cells or vesicles made from cell membranes, and the change in ion concentration is recorded by means of radioactive isotopes, fluorescent probes, or ion-selective electrodes.

3.1.3 Ion Channel Structure and Functions

There are numerous textbooks describing ion channels [9, 12]. Essentially, the lipid bilayer is spanned by a peptide forming an α-helix consisting of 20–25 mostly hydrophobic amino acids. If one side of the helix spanning the bilayer is made of 4–5 hydrophilic amino acids, then when 4–6 α-helices are brought together, they can form the pore portion of an ion channel. The hydrophobic portion of the peptide faces the hydrophobic lipid bilayer, while the hydrophilic portion facilitates the translocation of ions. Such a simple structure is characteristic for some venoms and antibiotics. The channel usually consists of multiple membrane-spanning domains, and only 4–6 of them form the hydrophilic pore.

In contrast to the pore-forming peptide, the channel molecule has a fixed geometry and allows a constant number of ions to flow through it in a given unit of time. The channel fluctuates between closed and open states. The probability of opening increases for some channels with a change of membrane potential, for others with

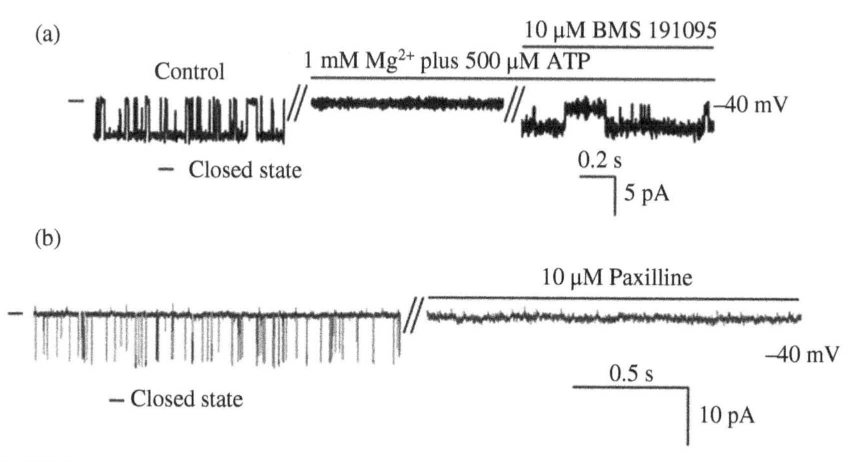

FIGURE 3.5 Single-channel recordings. (a) The effect of complex ATP/Mg (blocker) and BMS 191096 (opener) on the mito-KATP channel from a bovine heart. (b) The effect of paxilline (blocker) on the mito-BKCa channel from astrocytoma cells (Piotr Bednarczyk unpublished results).

the addition or removal of a specific molecule, and for others by application of mechanical stress. The search for specific openers or blockers of channels is an important field of pharmacology (Fig. 3.5).

In living organisms, ion channels are responsible for numerous functions (e.g., nerve conduction, muscle contraction, hormone secretion, and water transport). They perform these functions by changing membrane potentials, increasing calcium ion concentrations, and transporting both cations and anions, thereby inducing osmotic water flux.

There are few ion channels or transporters whose amino acid sequence has been identified. Very few channels have been cloned and crystallized. The three-dimensional structure has been determined for even fewer. In the genome, ion channel proteins are encoded not as a single molecule but in separate parts that are then assembled together. Some of these parts are present in each ion channel of a given type. The other parts are usually in the beginning and end of the amino acid chain, and they have a few exchangeable variants. Thus, "the same" ion channel might have slightly different properties when it is present in different tissues or different organelles—so-called split variants [13]. For a small number of ion channels and transporters, specific activators and blockers have been identified. The commonly used activators have many side effects when used in cells [14]. Very few ion channels or transporters have been eliminated from living animals. The results obtained with these KO animals are often ambiguous because the organisms have many defect-compensating mechanisms.

The ionic and solute composition of body fluid varies greatly. In Table 3.1, typical values are shown. There is an electrochemical gradient of certain ions and solutes across the plasma membrane and the membranes of organelles.

TABLE 3.1 Solute composition of extracellular fluid, cytoplasm and mitochondria.

Solute	Plasma	Interstitium	Cell cytoplasm	Mitochondrion
Na^+ (mM)	142	145	15	~5
K^+ (mM)	4.4	4.5	120	~40
Ca^{2+} (mM) ionized	1.2	1.2	10–100 nM	?
Ca^{2+} (mM) total	2.5	?	?	>1000
Cl^- (mM)	102	116	20	Low
HCO_3^- (mM)	22	25	15	?
$H_2PO_4^- + HPO_4^{2-}$ (mM) ionized	0.7	0.8	0.7	?
pH	7.4	7.4	7.2	7.2–8.2
Proteins (g/dl)	7	1	30	500

?, values not known.

3.2 ION CHANNELS IN HEALTH AND PATHOLOGY

Ion channels serve multiple functions in organisms. Defective ion channel molecules lead to hundreds of different diseases. The best known are neuromuscular disorders caused by mutations in voltage-gated sodium, potassium, calcium, and chloride channels, as well as acetylcholine-gated channels. These mutations lead to symptoms such as weakness or stiffness of muscles, periodic muscle paralysis, epilepsy, and migraines. Numerous reviews regarding ion channel-related diseases have been published [15–23].

In this paper, the mechanisms of the three most common channel-related diseases will be described: diabetes, stroke, and cystic fibrosis (CF). In neonatal and type 2 diabetes, both the channel and the mechanism of action are known. In CF, the channel responsible for the disease is known, but the mechanism remains obscure. Ion channels are not directly responsible for a stroke, but they are suspected to play a role in stroke prevention. A hypothetical mechanism of ischemic preconditioning will be presented.

3.2.1 Potassium Channels of the Plasma Membrane and Their Defects in Diabetes and Hyperinsulinemia

Ashcroft et al. [24] discovered that potassium channels are blocked by ATP in β-cells of the pancreas. These cells are responsible for insulin secretion when the glucose concentration in blood rises. This channel (K_{ATP}) is also present in many other cell membranes and organelles. Three or four other channels are also involved in the process of insulin secretion: the voltage-dependent calcium channel (Ca_v), the calcium-activated calcium channel (Ca_{Ca}), the calcium-activated big conductance potassium channel (BK_{Ca}), and the voltage-dependent potassium channel (K_v). The sequence of events leading to insulin secretion is presented in Figure 3.6. When the glucose concentration rises, ATP is produced by mitochondria. A high concentration of ATP closes the K_{ATP}

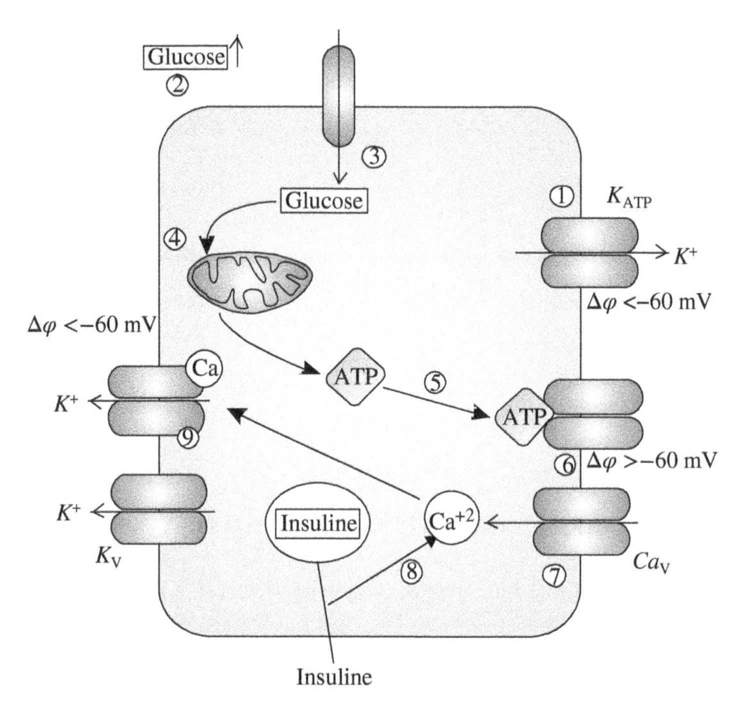

FIGURE 3.6 The events leading to insulin secretion from pancreatic β-cells. 1. The ATP concentration in β-cells is low. Plasma membrane K_{ATP} channels are open. K^+ ions flow out from the cytoplasm leading to the formation of a high negative membrane potential $\Delta\psi$. 2. Glucose concentration outside the cell rises. 3. Glucose is transported into the cell. 4. Glucose reaches mitochondria where it is used to produce ATP. 5. The concentration of ATP inside the cell rises and blocks the K_{ATP} channel. 6. K^+ ions no longer flow out from the cell and the membrane potential becomes less negative. 7. The rising membrane potential opens Ca_v channels and Ca^{2+} ions flow into the cell. The increased calcium concentration in the cytoplasm opens calcium-dependent calcium channels, which release more calcium ions from internal cellular stores (not shown). 8. Via a complex series of events (not shown), calcium ions lead to the secretion of insulin stored in intracellular vesicles. 9. Calcium ions open BK channels (and/or voltage-dependent potassium channels K_v). K^+ ions flow out from the cell, restoring the high negative membrane potential and terminating the insulin secretion.

channel. Potassium ions are unable to flow out from the cell and the membrane potential becomes less negative. The less negative potential opens the Ca_v channel, which allows calcium ions to flow into the cytoplasm. Calcium ions flowing into the cytoplasm also open the Ca_{Ca} channels, which in turn release even more calcium from internal calcium stores. In a series of reactions, calcium ions lead to the secretion of insulin stored in intracellular vesicles. The process of secretion is terminated by the opening of the BK_{Ca} channel or the K_v channel present in the plasma membrane. Potassium ions flow out from the cytoplasm through the BK_{Ca} or K_v channel, restoring the high negative membrane potential terminating calcium influx through the Ca_v channel and insulin secretion.

Diabetes is an illness in which insulin is not secreted or is secreted in an insufficient amount. The opposite property—too much insulin secretion (hyperinsulinemia)—leads to a deficiency of glucose in the blood. The major cause of both diseases is a genetic defect in the K_{ATP} channel—especially in its ATP-sensing domain [25–30]. The manifestation of the disease is especially severe if defective genes are acquired from both parents (the homozygous condition). The severity of the disease is proportional to the ATP sensitivity of the channel (i.e., if the channel is more open, then less insulin is secreted, leading to diabetes). Transient neonatal diabetes mellitus (TNDM) is caused by heterozygous mutations (e.g., G53S, G53R, I182V) reducing the fraction of K_{ATP} channels that are not blocked by MgATP from 6 to 12%. More severe permanent neonatal diabetes mellitus (PNMD) is caused by different mutations (e.g., R201H) that reduce the fraction of unblocked K_{ATP} channels to 22%. The most serious V59G and I296D mutations reduce the fraction of unblocked K_{ATP} channels to 36%, leading to DEND syndrome (i.e., developmental delay, muscle weakness, epilepsy, and neonatal diabetes). On the other hand, the Y12X mutation is the cause of congenital hyperinsulinism of infancy (HI) leading to hypoglycemia, which without therapy may cause irreversible brain damage. Mild forms of the disease such as type 2 diabetes mellitus are caused by different mutations (e.g., E23K) and may be controlled by diet and K_{ATP} channel blockers like glibenclamide [31]. Mild forms of hyperinsulinemia may be controlled by diazoxide, a K_{ATP} channel opener.

3.2.2 Ion Channels of the Inner Mitochondrial Membrane and Their Role in Protection from Ischemic Heart Injury

The ion channels of the cell membrane have been studied since the 1970s. Access to inner membrane channels is much more difficult. It is practically impossible to reach the membranes of organelles by puncturing the plasma membrane. Thus, organelles must be first separated from the cell and then studied by means of the patch-clamp or BLM methods. One of the most extensively studied organelles is the mitochondrion. In mitochondria, there are many ion channels. Among them, there are channels similar to the K_{ATP} and BK_{Ca} channels of the cell membrane. The mito-K_{ATP} channel has been found in many different organisms including amoebae, plants, and many different human and animal tissues [32–37]. The mito-BK_{Ca} channel has also been found in many different tissues [38–40]. There are some differences between plasma membrane channels and mitochondrial channels. There are also differences between mito-K_{ATP} channels of different tissues. Among other channels present in mitochondria, there is the calcium channel mito-Ca [41]. Mitochondria produce ATP on the expense of energy stored in protons (Fig. 3.7). The energy stored in protons can be described as a classic of electrochemical potential difference of protons $\Delta\overline{\mu_H}$ between the cytoplasm and the matrix (inside the mitochondrion):

$$\Delta p = \Delta\overline{\mu_H} \cdot F = 63[\text{mV}] \cdot \Delta\text{pH} + \Delta\psi$$

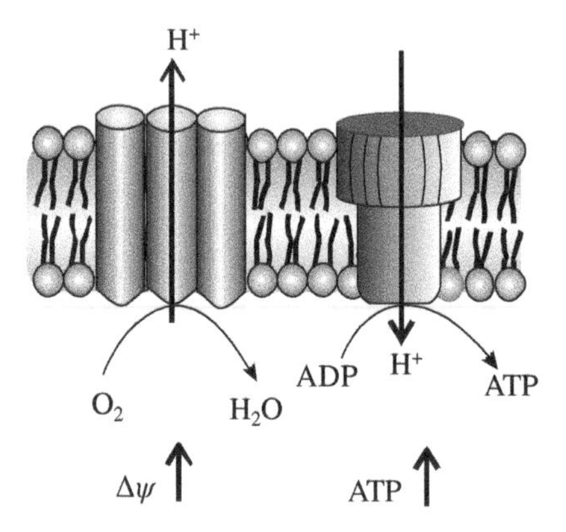

FIGURE 3.7 Mitochondria produce ATP on the expense of energy stored in protons. The oxidative chain expels protons that return via ATP synthase.

where Δp is the so-called proton-motive force used in bioenergetics instead of the electrochemical potential difference, $\Delta\psi$ is the electrical potential difference, and the factor 63 mV is for 37°C. Typical values of Δp are on the order of 220 mV. The electrical potential difference $\Delta\psi$ equals to 60–90% of the Δp. The physiological role of mitochondrial channels is not clear. It is obvious, however, that the influx of cations compensating for the efflux of protons via the oxidative chain leads to an increase in ΔpH [42].

Murry et al. [43] discovered that successive periods of ischemia (no oxygen) and reperfusion (oxygen supply restoration) protect the heart muscle from injury caused by a consecutive longer period of ischemia. Protection of the heart by preconditioning was accompanied by a higher concentration of ATP in preconditioned hearts compared to nonpreconditioned ones. A similar phenomenon was discovered with hearts preconditioned by changing the calcium concentration—a phenomenon called calcium preconditioning [44–46]. Calcium preconditioning is also accompanied by an increase in ATP concentration. Preconditioning can also be achieved by opening mito-K_{ATP} [47–50] and mito-BK_{Ca} channels [51]. Mito-K_{ATP} and mito-BK_{Ca} channel blockers abolish the protective effects of the channel openers. These findings may lead to the development of new stroke-protective drugs. Mitochondria are likely the organelles that are involved in preconditioning because they use oxygen, produce ATP, store calcium, and are involved in programmed cell death (apoptosis). There are many hypotheses on mitochondrial involvement in cell death during ischemia and during reperfusion. The major factors leading to heart injury (according to different hypotheses) are reactive oxygen species production, reduced ATP concentration, increased calcium concentration in the cytoplasm, the formation of permeability transition pore, and the release of cytochrome C leading to apoptosis (e.g., [52–54]). There is, however, no hypothesis explaining how these factors lead to preconditioning.

Although it is possible to associate the heart-protective effect of calcium preconditioning and potassium channel openers with the increase of ΔpH in mitochondria,

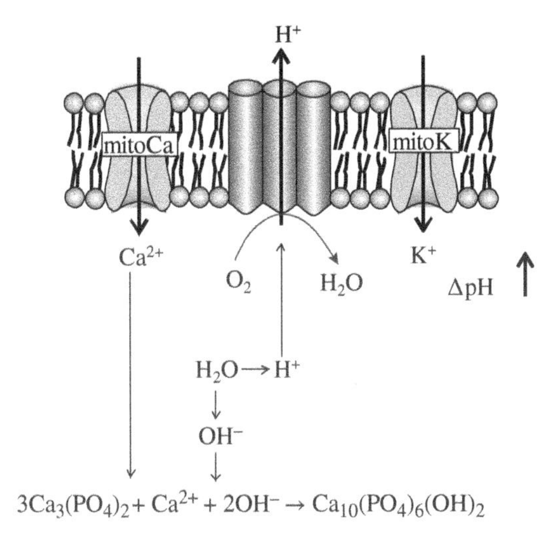

FIGURE 3.8 As the ATP concentration in the cytoplasm is not high, mito-K_{ATP} channels are partly open. Potassium influx into the mitochondrial matrix increases ΔpH. If the calcium concentration in the cytoplasm becomes high, mito-BK_{Ca} channels open leading to the production of ΔpH. Calcium ion influx via mito-Ca channels also increases ΔpH (calcium preconditioning). It is likely that hydroxyapatite is formed from apatite, calcium, and hydroxyl ions. Formation of hydroxyapatite produces large pH buffering capacity of the mitochondrial matrix.

two questions still remain, namely, whether ischemic preconditioning also increases ΔpH and how increased ΔpH protects heart cells from injury caused by ischemia.

From an electrochemical point of view, the mechanism of heart protection might involve ion channels. In the case of sufficient nutrient and oxygen supply, ATP is produced and its concentration rises. The increased membrane potential $\Delta\psi$ sucks available calcium ions from the cytoplasm. It is likely that hydroxyapatite is formed when calcium ions are moved into the mitochondrial matrix (Fig. 3.8). Hydroxyapatite is likely to be a pH buffer of mitochondria. Potassium channel openers and external calcium rise lead to increased ΔpH value in the matrix. Then, both mito-K_{ATP} (high ATP concentration) and mito-BK_{Ca} (no calcium in cytoplasm) channels close. Further expulsion of protons through the oxidative chain increases Δp and also increases the ATP concentration until it reaches the maximum energetic efficiency of the oxidative chain. Blocking the oxygen supply to the heart causes a gradual decrease of oxygen concentration in the vicinity of mitochondria and decreases the speed of proton expulsion from the matrix by the oxidative chain. $\Delta\psi$ decreases, both mito-K_{ATP} and mito-BK_{Ca} channels are closed, but ATP can still be produced at the expense of ΔpH and calcium ion efflux via the mito-Ca channel maintains the electroneutrality of the process. Transformation of hydroxyapatite to apatite prolongs the process of ischemic ATP production (Fig. 3.9). Thus, the higher the ΔpH and the higher hydroxyapatite concentration, the longer mitochondria will produce ATP during ischemia and

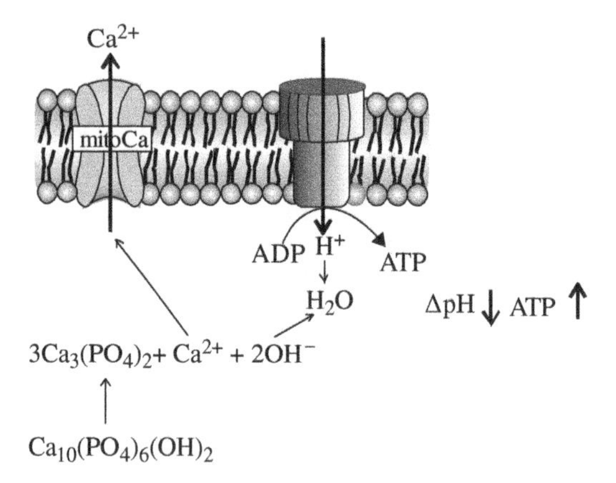

FIGURE 3.9 During ischemia, the oxidative chain is stopped, but ATP synthase is still working to produce ATP at the beginning due transformation of hydroxyapatite to apatite and then at the expense of ΔpH. The electroneutrality of the later process is maintained by an outward flux of calcium ions via mito-Ca channels. The higher the amount of hydroxyapatite and ΔpH, the longer mitochondria produces ATP during ischemia.

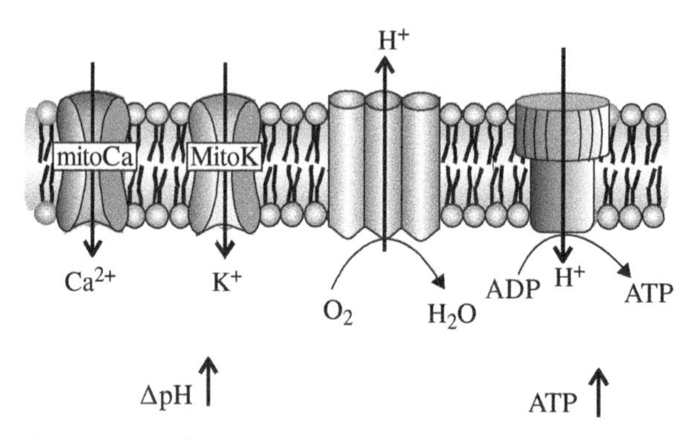

FIGURE 3.10 Reperfusion (i.e., resumption of the oxygen supply) restarts the oxidative chain, which expels protons from the matrix. As both potassium channels and the calcium channel are open, the expulsion of protons increases ΔpH rather than $\Delta\psi$.

thereby prevent damage to heart muscle cells. When ΔpH is used up, ATP is no longer produced, calcium ions appear in the cytoplasm, $\Delta\psi$ decreases, and both mito-K_{ATP} and mito-BK_{Ca} channels open. Reperfusion (i.e., resumption of the oxygen supply) restarts the oxidative chain, which expels protons from the matrix (Fig. 3.10). As both potassium channels and the calcium channel are open, the expulsion of protons increases ΔpH rather than $\Delta\psi$. A short period of ischemia followed by reperfusion leads to the formation of a high value of ΔpH. In the case of the next longer ischemic event, the high value of ΔpH allows ATP production at the expense of ΔpH for a longer time, protecting the heart muscle from ischemic injury.

3.2.3 The CFTR Channel, Its Role in Water Transport across the Epithelial Cell Layer, and Its Defects in CF

CF is the most common fatal genetic disorder among people of northern and central European descent, affecting 1:3000 newborns. It affects especially the lungs and pancreases. In the lungs, very viscous mucus that cannot be cleared from the airways is present. Viscous mucus leads to opportunistic infections. Multiple incidents of inflammation deteriorate the function of the lungs, leading to premature death. Also, pancreatic ducts become blocked by dense mucus secretion leading to poor digestion and malnutrition. CF is caused by the defect in a single gene that encodes a chloride channel present in the epithelial cell layer called the CF transmembrane conductance regulator (CFTR) [55, 56]. Since in CF chloride transport is also affected in the gut and sweat glands, it was sought that it is the defective water transport across the epithelial cell layer that is responsible for producing too dense mucus.

The epithelial cell layer is watertight. Water is transported across epithelial cell layer by means of osmosis. Electroneutral transport of a single cation accompanied by a single anion causes the osmotic flow of 450 water molecules. If the chloride transport route is blocked by defective CFTR channel molecule, there are no electroneutral transport and no water flow across the epithelial cell layer. Epithelial cell layer is polarized, that is, its apical and basolateral faces have different sets of ion-transporting proteins and is able to transport water in both directions. For years, scientists argue whether it is the decreased secretion or enhanced absorption of water from the apical that is responsible for viscous apical mucus (e.g., [57]). Obviously, both sides of the debate cannot be right in the same time. Either defective CFTR channel decreases water secretion as should be expected from the simple osmotic model or enhanced water transport via other not obvious mechanism involving enhanced sodium absorption. Apparently, while chloride transport is decreased in CF, the sodium conductance is not affected [58]. There is still another hypothesis concerning the dense mucus production in CF. Quinton [59] suggested that in CF bicarbonate secretion is affected. His hypothesis is based on experimental results [60] showing condensation of mucus in the absence of bicarbonate anions. From an electrochemical point of view, while CFTR is called the chloride channel, it is in fact an anion channel, with low selectivity coefficient $Cl^-/HCO_3^- = 4$ [61]. Is it possible that CFTR channel is responsible for chloride, bicarbonate, and water transport in the same time?

The study of epithelial cell layer goes back to Ussing's experiments on the water transport across the frog skin [62]. The sealed layer was placed between the two compartments, and the ions moving across the skin were measured by radioisotopes. The other possibility was the use of voltage clamp or current clamp technique (Fig. 3.11a). The most common is the use of voltage clamped at zero (short-circuited) between apical and basolateral faces of the tissue system. The current measured in such configuration is called the short-circuit current. The method has obvious limitation—one cannot recognize between cations flowing inward from anions flowing outward. Moreover, only the total current of anions and cations flowing through both basolateral and apical membranes can be measured. The activators and/or blockers of channels, pumps, and transporters are used to get an insight into what is being measured and

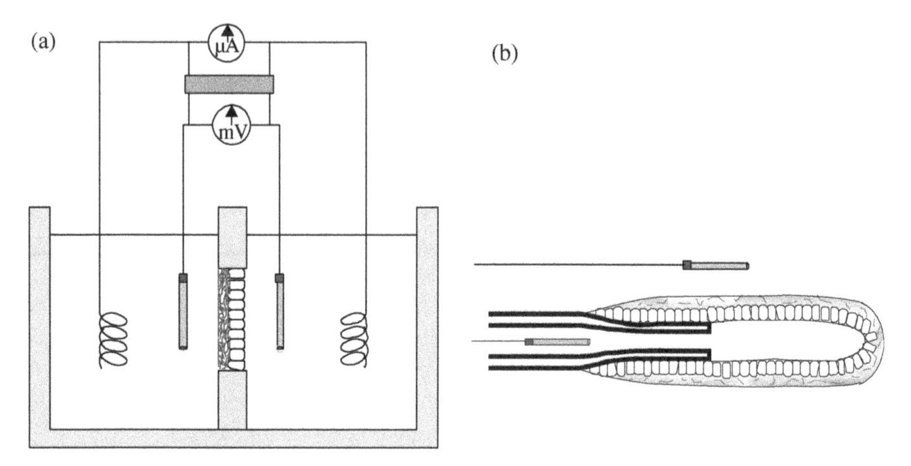

FIGURE 3.11 (a) The sealed tissue layer is placed between the two compartments, and ions moving across can be measured by radioisotopes or by means of voltage clamp or current clamp techniques [62]. (b) The isolated sweat gland impaled on microcapillary is used to measure ion transport during sweat production [77].

which transporters are involved. However, neither activators nor blockers are completely specific and often have side effects influencing cell physiology [14].

Studying epithelial transport is difficult. Firstly, there are many different transporting molecules involved in a process. Secondly, there are many different processes, that is, water transport, cell volume control, and pH homeostasis, that are using the same transporting molecules. Thirdly, we have limited means of studying epithelial cells. We can study intact skin or gut, which is composed of many layers of different cells. We can also study some cell types grown in vitro on porous support. But cell monolayers might not reveal a whole mechanism of transport. Fourthly, there are different epithelial tissues, for example, sweat gland, pancreatic duct, and bronchial lining, which might employ different transport mechanisms.

There are different channels, transporters, exchangers, and pumps in epithelia [63]. On the apical face of the bronchial epithelium, there are the CFTR anion channels, CaCC (calcium-activated anion channel), and ENaC (amiloride-sensitive sodium channel), which play a critical role in regulating water secretion and absorption. All three channels have low conductivity: CFTR 9–11pS [64], CaCC 4–8pS [65], and ENaC 4–5pS [66]. While ENaC channel is very specific for sodium ion, two anionic channels have low selectivity ratio conducting chloride over bicarbonate $Cl^-/HCO_3^- = 4$ for CFTR and $Cl^-/HCO_3^- = 2$ for CaCC. The other molecules involved in ion transport in bronchial epithelium are Na-K-ATPase, the most abundant pump, which expels $3Na^+$ ions from the cytoplasm in exchange for $2K^+$ ions using one ATP molecule and is present exclusively on the basolateral face of epithelia. The NaK2Cl transporter (regulating NaCl and KCl influx into the cell) and $Na2HCO_3$ transporter are present on the basolateral face of the bronchial epithelium. On both sides of the bronchial membrane, there are Na^+/H^+ and Cl^-/HCO_3^- exchangers. There are many other ion channels that are present in epithelial cells such as potassium channels [67], volume-sensitive outward rectifier (VSOR) channel [68], and outward rectifier

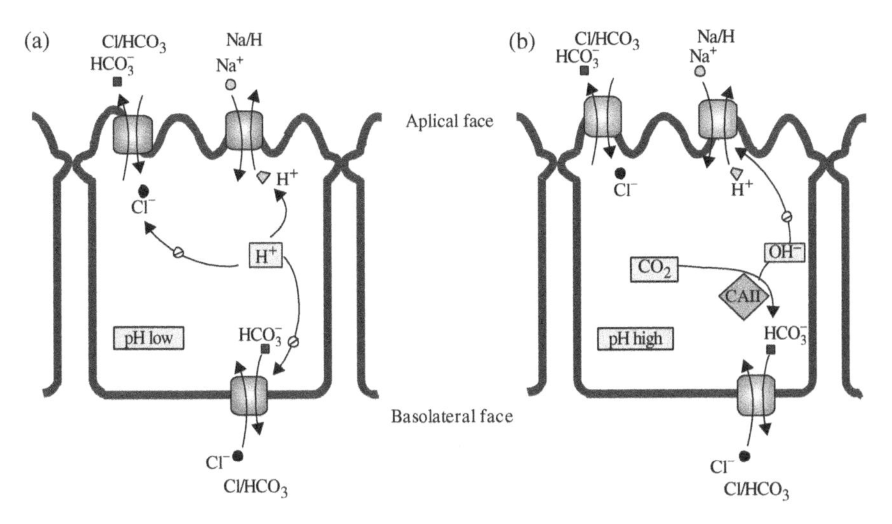

FIGURE 3.12 Ion transporters involved in cell pH homeostasis. (a) pH too high. (b) pH too low. (CAII—carbonic anhydrase).

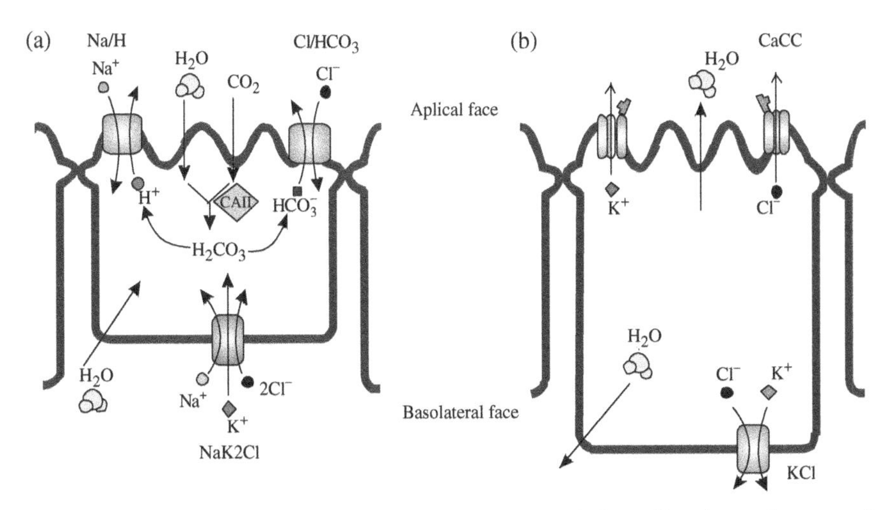

FIGURE 3.13 Ion channels and transporters involved in cell volume homeostasis. (a) Volume too small. (b) Volume too large.

chloride channel (ORCC) [69, 70], the latter two seem to be regulated by the CFTR channel. Other channels and transporters also seem to be regulated by CFTR [71, 72]. The activities of some of these molecules involving ion transport are affected by cAMP or elevated calcium concentrations in the cell [73], and they are activated by P2Y receptors [74]. Some of these molecules are involved in cell pH homeostasis [75, 76] (Fig. 3.12) and volume control processes (Fig. 3.13).

The study of the sweat gland [77] gives the best insight into the mechanisms of water and ion transport in CF. The salty taste of sweat was the first discovered manifestation of CF since hypotonic sweat chloride concentrations in CF patients exceed

60 mM (and the value for non-CF people is around 20 mM). The sweat gland can be isolated from the skin and put on the tip of a capillary holding reference electrode (Fig. 3.11b). Thus, not only the short-circuit current could be measured but also the ionic composition of the sweat can be determined. The data obtained suggest that isotonic sweat is produced in the secretory coil of the sweat gland by CFTR-independent mechanism and when it passes through the reabsorptive duct, the sodium chloride is removed by concerted action of CFTR and ENaC channels transporting electroneutrally NaCl.

It is not certain whether the mechanism of sweat secretion can be extended on the bronchial epithelium. But the sweat gland measurements suggest that ENaC and CFTR channels are responsible for absorption of apical fluid and CaCC and paracellular routes of sodium are responsible for secretion of the fluid. Secretion in bronchial epithelium might involve calcium signaling that activates NaK2Cl and $Na2HCO_3$ transporters on the basolateral face and CaCC and one of the potassium channels on the apical face, and sodium is transported by paracellular route. The process seems to be electroneutral, and the typical potential difference of nearly −60 mV should be

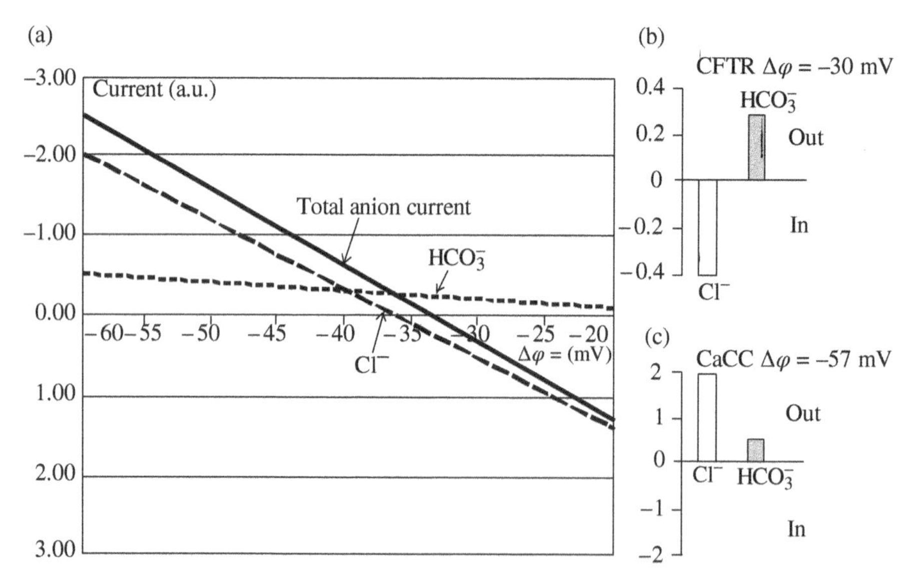

FIGURE 3.14 Electrochemical considerations. (a) Chloride and bicarbonate currents flowing via CFTR channels for different potential difference between apical fluid and cytoplasm. (b) Chloride and bicarbonate currents flowing via CaCC secretion phase of human bronchial apical membrane (four chloride anions are accompanied by one bicarbonate anion). (c) During the absorptive phase, chloride anions flowing into the cell are exchanged for bicarbonate anions flowing out from the cell via CFTR channel. For calculations, cytoplasm chloride and bicarbonate concentrations $[Cl]_{cyt} = 20$ mM and $[HCO_3]_{cyt} = 15$ mM were accepted and $[Cl]_{asl} = 116$ mM and $[HCO_3]_{asl} = 25$ mM for apical surface liquid. Selectivity ratio $Cl^-/HCO_3^- = 4$ for CFTR and $Cl^-/HCO_3^- = 2$ for CaCC were adopted. The potential difference is between apical fluid and cytoplasm.

present across the apical face of the bronchial epithelium membrane. Electrochemical considerations (Fig. 3.14) suggested that during secretion four chloride anions are accompanied by one bicarbonate anion. The mechanism of secretion and molecules involved are shown in Figure 3.15. During the absorption phase, opening of ENaC channel causes influx of sodium cations via the apical face of the bronchial cell, and electric potential difference across the apical face drops what is manifested as change of overall potential difference between apical and basolateral fluid. The change of transapical potential leads to equilibrium between chloride absorption and bicarbonate secretion (Fig. 3.14). Thus, in non-CF people, the absorption phase leads to enrichment of apical fluid in bicarbonate ions (Fig. 3.16) striking agreement with Quinton's [59] "bicarbonate" hypothesis of CF. In CF patients, there is no bicarbonate secretion, and influx of sodium ions is accompanied by secretion of potassium ions and transport of chloride anions across the basolateral face of the bronchial cells, which triggers osmotic cell volume increase and subsequent regulatory processes (Fig. 3.13).

While electrochemical considerations supports Quinton's [59] bicarbonate hypothesis of CF, we still lack hard experimental evidence of how ions flow across bronchial cell layer. The developments of completely new techniques are required to get an insight into the transport process and give potential hope for patients of this incurable disease.

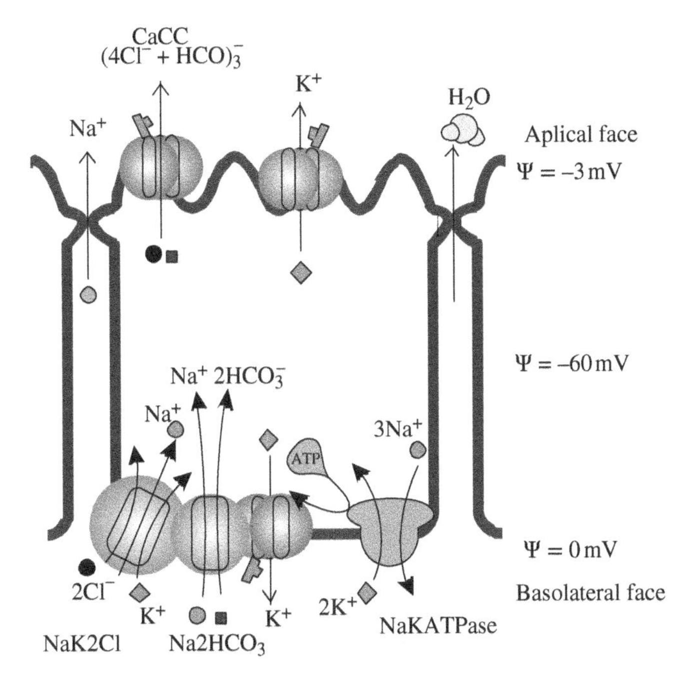

FIGURE 3.15 The mechanism of water secretion in human bronchial epithelium. Fluid transported to the apical face is rich in chloride ions flowing via CaCC.

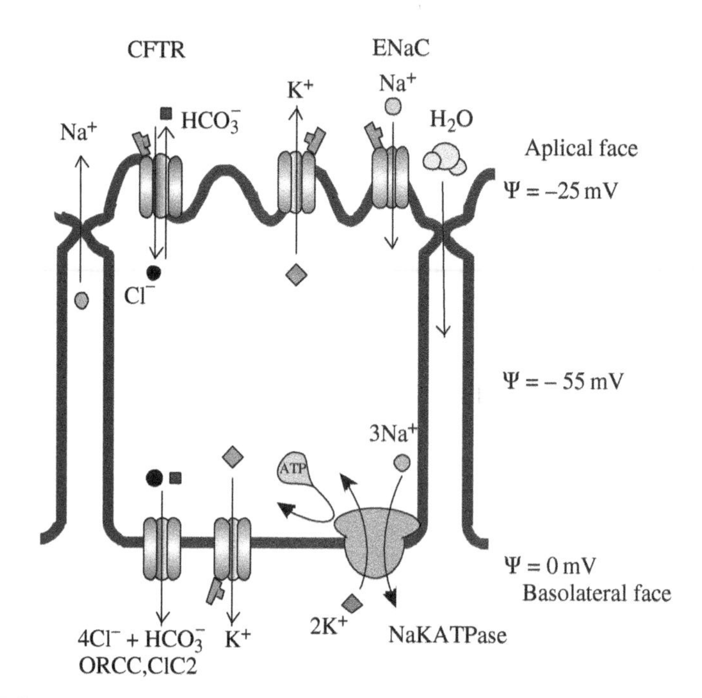

FIGURE 3.16 The mechanism of water absorption in human bronchial epithelium involves both ENaC and CFTR channels. During the absorptive phase, there is an exchange of chloride ions moving into the cell and bicarbonate ions moving out via CFTR channel. The defective CFTR channel in cystic fibrosis (CF) leads to low bicarbonate concentration in bronchial fluid causing condensation of mucus in the lungs—striking agreement with Quinton's [59] "bicarbonate" hypothesis of mucus condensation in CF.

ACKNOWLEDGMENTS

The author is indebted to Mrs. Wanda Jarzabek for drawing figures. The paper was supported by grants from the Polish Ministry of Science and Higher Education N N401 182839 and MitoNet.Pl.

REFERENCES

1. Hodgkin AL, Huxley AF. 1952. A quantitative description of membrane current and its application to conduction and excitation in nerve. *J Physiol* 117: 500–544.
2. Mueller P, Rudin DO, Tien HT, Wescott WC. 1962. Reconstitution of cell membrane structure in vitro and its transformation into an excitable system. *Nature* 194: 979–980.
3. Montal M, Mueller P. 1972. Formation of bimolecular membranes from the lipid monolayers and a study of their electric properties. *Proc Natl Acad Sci USA* 69: 3561–3566.

4. Miller C. 1983. Integral membrane channels: studies in model membranes. *Physiol Rev* 63: 1209–1242.
5. Hladky SB, Haydon DA. 1970. Discreteness of conductance change in bimolecular lipid membranes in the presence of certain antibiotics. *Nature* 225: 451–453.
6. Schindler H. 1980. Formation of planar bilayers from artificial or native membrane vesicles. *FEBS Lett* 122: 77–79.
7. Neher E, Sackmann B. 1976. Single-channel currents recorded from membrane of denervated frog muscle fibres. *Nature* 260: 799–802.
8. Sackmann B, Neher E eds. 1995. *Single Channel Recordings*. Second ed. Plenum Press, New York.
9. Hille B. 1992, 2001. *Ionic Channels of Excitable Membranes*. Sinauer, Sunderland.
10. Sigworth FJ. 1980. The variance of sodium channels current fluctuations at the node of Ranvier. *J Physiol* 307: 97–129.
11. Dolowy K. 2003. A new double-chamber model of ion channels. Beyond the Hodgkin and Huxley model. *Cell Mol Biol Lett* 8: 749–775.
12. Aidley DJ, Stanfield PR. 1996. *Ion Channels. Molecules in Action*. Cambridge University Press, New York.
13. Pongs O. 1992. Molecular biology of voltage-dependent potassium channels. *Physiol Rev* 72: S69–S88.
14. Szewczyk A, Kajma A, Malinska D, Wrzosek A, Bednarczyk P, Zabłocka B, Dołowy K. 2010. Pharmacology of mitochondrial potassium channels: dark side of the field. *FEBS Lett* 584: 2063–2069.
15. Ashcroft, FM. 2000. *Ion Channels and Disease*. Academic Press, New York.
16. Ashcroft FM. 2006. From molecule to malady. *Nature* 440: 440–447.
17. Bernard G, Shevell MI. 2008. Channelopathies: a review. *Pediatr Neurol* 38: 73–85.
18. Catterall WA. 2010. Ion channel voltage sensors: structure, function, and pathophysiology. *Neuron* 67: 915–928.
19. Dworakowska B. Dolowy K. 2000. Ion channels-related diseases. *Acta Biochim Pol* 47: 685–703.
20. Jurkat-Rott K, Holzherr B, Fauler M, Lehmann-Horn F. 2010. Sodium channelopathies of skeletal muscle result from gain or loss of function. *Pflugers Arch* 460: 239–248.
21. Kullmann DM, Waxman SG. 2010. Neurological channelopathies: new insights into disease mechanisms and ion channel function. *J Physiol* 588.11: 1823–1827.
22. Planells-Cases R, Jentsch TJ. 2009. Chloride channelopathies. *Biochim Biophys Acta* 1792: 173–189.
23. Striessnig J, Bolz HJ, Koschak A. 2010. Channelopathies in Cav1.1, Cav1.3, and Cav1.4 voltage-gated L-type Ca^{2+} channels. *Pflugers Arch* 460: 361–374.
24. Ashcroft FM, Harrison DE, Ashcroft SJH. 1984. Glucose induces closure of single potassium channels in isolated rat pancreatic β-cells. *Nature* 312: 446–448.
25. Ashcroft FM. 2005. ATP-sensitive potassium channelopathies: focus on insulin secretion. *J Clin Invest* 115: 2047–2058.
26. Aittoniemi J, Fotinou C, Craig TJ, de Wet H, Proks P, Ashcroft FM. 2009. SUR1: a unique ATP-binding cassette protein that functions as an ion channel regulator. *Phil Trans R Soc B* 364: 257–267.

27. Akrouh A, Halcomb SE, Nichols CG, Sala-Rabanal M. 2009. Molecular biology of KATP channels and implications for health and disease. *IUBMB Life* 61: 971–978.
28. Flanagan SE, Patch AM, Mackay DJG, Edghill EL, Gloyn AL, Robinson D, Shield JPH, Temple K, Ellard S, Hattersley AT. 2007. Mutations in ATP-sensitive K- channel genes cause transient neonatal diabetes and permanent diabetes in childhood or adulthood. *Diabetes* 56: 1930–1937.
29. Flechtner I, de Lonlay P, Polak M. 2006. Diabetes and hypoglycaemia in young children and mutations in the Kir6.2 subunit of the potassium channel. *Diabetes Metab* 32: 569–580.
30. Pinney SE, MacMullen C, Becker S, Lin YW, Hanna C, Thorton P, Ganguly A, Shyng SL, Stanley CA. 2008. Clinical characteristics and biochemical mechanisms of congenital hyperinsulinism associated with dominant K_{ATP} channel mutations. *J Clin Invest* 118: 2877–2886.
31. Drews G, Dufer M. 2012. Role of K_{ATP} channels in β-cell resistance to oxidative stress. *Diabetes Obes Metab* 14: 120–128.
32. Bednarczyk P, Dolowy K, Szewczyk A. 2008. New properties of mitochondrial ATP-regulated potassium channels. *J Bioenerg Biomembr* 40: 325–335.
33. Choma K, Bednarczyk P, Koszela-Piotrowska I, Kulawiak B, Kudin A, Kunz W. Dolowy K, Szewczyk A. 2009. Single channel studies of the ATP-regulated potassium channel in brain mitochondria *J Bioenerg Biomembr* 41: 323–334.
34. Inoue I, Nagase H, Kishi K, Higuti T. 1991. ATP-sensitive K^+ channel in the mitochondrial inner membrane. *Nature* 352: 244–247.
35. Kicinska A, Swida A, Bednarczyk P, Koszela-Piotrowska I, Choma K, Dolowy K, Szewczyk A, Jarnuszkiewicz W. 2007. ATP-sensitive potassium channel in mitochondria of the eukaryotic microorganism Acanthamoeba castellanii. *J Biol Chem* 282: 17433–17441.
36. Pastore D, Stopelli MC, Di Fonzo N, Passarella S. 1999. The existence of the K^+ channel in plant mitochondria. *J Biol Chem* 274: 26683–26690.
37. Yarov-Yarovoy V, Paucek P, Jaburek M, Garlid KD. 1997. The nucleotide regulatory sites on mitochondrial KATP channel face the cytosol. *Biochim Biophys Acta* 1321: 128–136.
38. Sato T, Saito T, Saegusa N, Nakaya H. 2005. Mitochondrial Ca^{2+}-activated K^+ channels in cardiac myocytes: a mechanism of the cardioprotective effect and modulation by protein kinase A. *Circulation* 111: 198–203.
39. Siemen D, Loupatatzis C, Borecky J, Gulbins E, Lang F. 1999. Ca^{2+}—activated K channel of the BK-type in the inner mitochondrial membrane of the human glioma cell line. *Biochem Biophys Res Com* 257: 549–554.
40. Skalska J, Bednarczyk P, Piwońska M, Kulawiak B, Wilczyński G, Dołowy K, Kunin AP, Kunz WS, Szewczyk A. 2009. Calcium ions regulate K^+ uptake into brain mitochondria: the evidence for a novel potassium channel. *Int J Mol Sci* 10: 1104–1120.
41. Kirichok Y, Krapivinsky G, Clapham DE. 2004. The mitochondrial calcium uniporter is a highly selective ion channel. *Nature* 427: 360–364.
42. Nicholls DG, Ferguson SJ. 2002. In *Bioenergetics 3*, Academic Press, Amsterdam.
43. Murry CE, Jennings RB, Reimer KA. 1986. Preconditioning with ischemia: a delay of lethal cell injury in ischemic myocardium. *Circulation* 74: 1124–1136.
44. Meldrum DR, Cleveland JCJ, Sheridan BC, Rowland RT, Banerjee A, Harken AH. 1996a. Cardiac preconditioning with calcium: clinically accessible myocardial protection. *J Thorac Cardiovasc Surg* 112: 778–786.

45. Meldrum DR, Cleveland JCJ, Mitchell MB, Sheridan BC, Gamboni-Robertson F, Harken AH, Banerjee A. 1996b. Protein kinase C mediates Ca2(+)-induced cardioadaptation to ischemia-reperfusion injury. *Am J Physiol* 271: R718–R726.
46. Miyawaki H, Zhou X, Ashraf M. 1996. Calcium preconditioning elicits strong protection against ischemic injury via protein kinase C signaling pathway. *Circ Res* 79: 137–146.
47. Costa ADT, Quinlan CL, Andrukhiv A, West IC, Jaburek M, Garlid KD. 2006. The direct physiological effects of mitoK_{ATP} opening on heart mitochondria. *Am J Physiol Heart Circ Physiol* 290: H406–H415.
48. Grover GJ, McCullough JR, Henry DE, Conder ML, Sleph PG. 1989. Anti-ischemic effect of the potassium channel activators pinacidil and cromakalim and the reversal of these effects with the potassium channel blocker glyburide. *J Pharmacol Exp Ther* 251: 98–104.
49. Grover GJ, Dzwonczyk S, Parham CS, Sleph PG. 1990. The protective effects of cromakalim and pinacidil on reperfusion function and infarct size in isolated perfused rat hearts and anesthetized dogs. *Cardiovasc Drugs Ther* 4: 465–474.
50. McPherson CD, Pierce GN, Cole WC. 1993. Ischemic cardioprotection by ATP-sensitive K^+ channels involves high-energy phosphate preservation. *Am J Physiol* 265: H1809–H1818.
51. Xu W, Liu Y, Wang S, McDonald T, Van Eyk JE, Sidor A, O'Rourke B. 2002. Cytoprotective role of Ca^{2+}—activated K^+ channels in the cardiac inner mitochondrial membrane. *Science* 298: 1029–1033.
52. Dzeja PP, Bast P, Ozcan C, Valverde A, Holmuhamedov EL, Van Wylen DG, Terzic A. 2003. Targeting nucleotide-requiring enzymes: implications for diazoxide-induced cardioprotection. *Am J Physiol Heart Circ Physiol* 284: H1048–H1056.
53. Garlid KD, Costa ADT, Quinlan CL, Pierre SV, Dos Santos P. 2009. Cardioprotective signaling to mitochondria. *J Mol Cell Cardiol* 46: 858–866.
54. Nicholls DG. 2008. Oxidative stress and energy crises in neuronal dysfunction. *Ann NY Acad Sci* 1147: 53–60.
55. Riordan JR. 1993. The cystic fibrosis transmembrane conductance regulator. *Annu Rev Physiol* 55: 609–630.
56. Welsh MJ, Tsui LC, Boat TF, Baudet AL. 1995. Cystic fibrosis. In *The Metabolic and Molecular Bases of Inherited Disease*. Scriver CR, Baudet AL, Sly WS, Valle D. eds. vol. 3. 3799–3876. McGraw Hill: New York.
57. Kunzelmann K, Schreiber R. 2012. Airway epithelial cells—hyperabsorption in CF? *Int J Biochem Cell Biol* 44: 1232–1235.
58. Itani OA, Chen JH, Karp PH, Ernst S, Keshavjee S, Parekh K, Klesney-Tait J, Zabner J, Welsh MJ. 2011. Human cystic fibrosis airway epithelia have reduced Cl^- conductance but not increase Na^+ conductance. *Proc Natl Acad Sci U S A* 108: 10260–10265.
59. Quinton PM. 2010. Role of epithelial HCO_3^- transport in mucin secretion: lesson from cystic fibrosis. *Am J Physiol Cell Physiol* 299: C1222–C1233.
60. Chen YET, Yang N, Quinton PM, Chin WC. 2010. A new role for bicarbonate in mucus formation. *Am J Physiol Lung Cell Mol Physiol* 299: L542–L549.
61. Kim D, Steward MC. 2009. The role of CFTR in bicarbonate secretion by pancreatic duct and airway epithelia. *J Med Invest* 56: 336–342.
62. Ussing HH, Zerahn K. 1951. Active transport of sodium as the source of electric current in the short-circuited isolated frog skin *Acta Physiol Scand* 23: 110–127.

63. Toczylowska-Maminska R, Dolowy K. 2012. Ion transporting proteins of human bronchial epithelium. *J Cell Biochem* 113: 426–432.
64. Cai Z, Chen JH, Hughes LK, Li H, Sheppard DN. 2006. The physiology and pharmacology of the CFTR Cl channel. *Adv Mol Cell Biol* 38: 109–143.
65. Wei L, Vankeerberghen A, Cuppens H, Eggermont J, Cassiman JJ, Droogmans G, Nilius B. 1999. Interaction between calcium-activated chloride channels and the cystic fibrosis transmembrane conductance regulator. *Pflugers Arch* 438:635–641.
66. Gaillard EA, Kota P, Gentzsch M, Dokholyan NV, Stutts MJ, Tarran R. 2010. Regulation of the epithelial Naţ channel and airway surface liquid volume by serine proteases. *Eur J Physiol* 460: 1–17.
67. Bardou O, Trinh NTN, Brochiero E. 2009. Molecular diversity and function of K- channels in airway and alveolar epithelial cells. *Am J Physiol Lung Cell Mol Physiol* 296: 145–155.
68. Ando-Akatsuka Y, Abdullaev IF, Lee EL, Okada Y, Sabirov RZ. 2002. Down-regulation of volume-sensitive Cl- channels by CFTR is mediated by the second nucleotide-binding domain. *Pflugers Arch* 445: 177–186.
69. Hryciw DH, Guggino WB. 2000. Cystic fibrosis transmembrane conductance regulator and the outwardly rectifying chloride channel: a relationship between two chloride channels expressed in epithelial cells. *Clin Exp Pharmacol Physiol* 27: 892–895.
70. Kloch M, Milewski M, Nurowska E, Dworakowska B, Cutting GR, Dolowy K. 2010. The H-loop in the second nucleotide-binding domain of the cystic fibrosis transmembrane conductance regulator is required for efficient chloride channel closing. *Cell Physiol Biochem* 25: 169–180.
71. Kunzelmann K. 2001. CFTR: Interacting with everything? *News Physiol Sci* 16: 167–170.
72. Li C, Naren AP. 2005. Macromolecular complexes of cystic fibrosis transmembrane conductance regulator and its interacting partners. *Pharmacol Therap* 108: 208–223.
73. Kunzelmann K, Milenkovic VM, Spitzner M, Soria RB, Schreiber R. 2007. Calcium-dependent chloride conductance in epithelia: is there a contribution by Bestrophin? *Pflugers Arch* 454: 879–889.
74. Bucheimer RE, Linden J. 2003 Purinergic regulation of epithelial transport. *J Physiol* 5552: 311–321.
75. Fischer H, Widdicombe JH. 2006. Mechanisms of acid and base secretion by the airway epithelium. *J Membr Biol* 211: 139–150.
76. Haggie PM, Verkman AS. 2009. Defective organellar acidification as a cause of cystic fibrosis lung disease: reexamination of a recurring hypothesis. *Am J Physiol Lung Cell Mol Physiol* 296: 859–867.
77. Quinton PM. 2007. Cystic fibrosis: lessons from the sweat gland. *Physiology* 22: 212–225.

4

ELECTRICAL COUPLING THROUGH GAP JUNCTIONS BETWEEN ELECTRICALLY EXCITABLE CELLS

Yaara Lefler and Marylka Yoe Uusisaari

Department of Neurobiology, The Institute of Life Sciences and Edmond and Lily Safra Center for Brain Sciences (ELSC), The Hebrew University, Jerusalem, Israel

4.1 MOLECULAR CHARACTERISTICS OF GAP JUNCTIONS

Gap junctions (GJs) enable direct connection between neighboring cells and are formed of clusters of hundreds to thousands of transcellular hydrophilic channels arranged in a plaque-like formation (Fig. 4.1a). The channels are made of two hemichannels (called connexons [1]), each residing on the opposing sides of connected cells. Connexons are composed of six subunits (connexins) that are a family of highly evolutionarily conserved integral transmembrane proteins with at least 20 isoforms in humans and mice [2], 19 of which can be considered as orthologue pairs on the basis of their amino acid sequence [3]. Connexins are most commonly named according to their molecular weights (e.g., Cx36 has a mass of approximately 36 kDa [4]).

Within the cell membrane, the six connexins are arranged in a circular assembly forming a large pore (16–20 Å of diameter in mammalian cells) permitting bidirectional intercellular diffusion of small (<1000 Da or <16 Å) molecules such as nutrients, ions, amino acids or short peptides, and second messengers (e.g., K^+, Ca^{2+}, IP_3, cAMP; for reviews, see [5–8]). Larger biomolecules, such as nucleic acids and proteins, are precluded from passing through GJ channels.

Electrochemical Processes in Biological Systems, First Edition. Edited by Andrzej Lewenstam and Lo Gorton.

(a) Structure of gap junctions

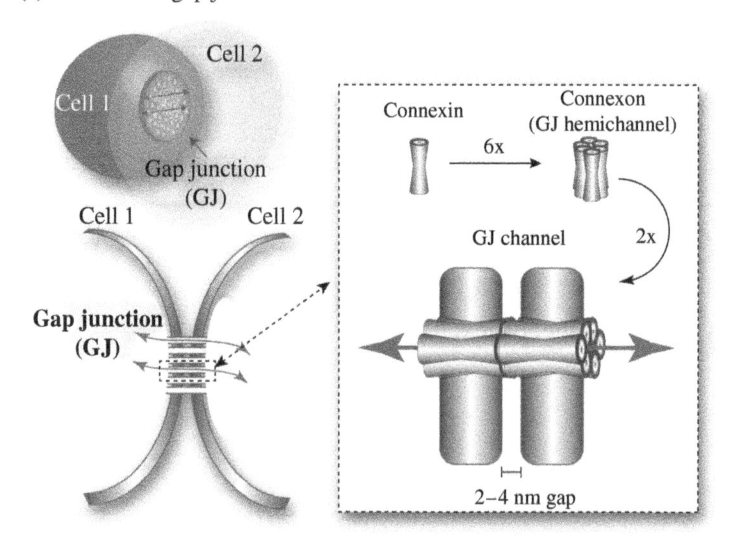

(b) Measuring GJ coupling in neurons

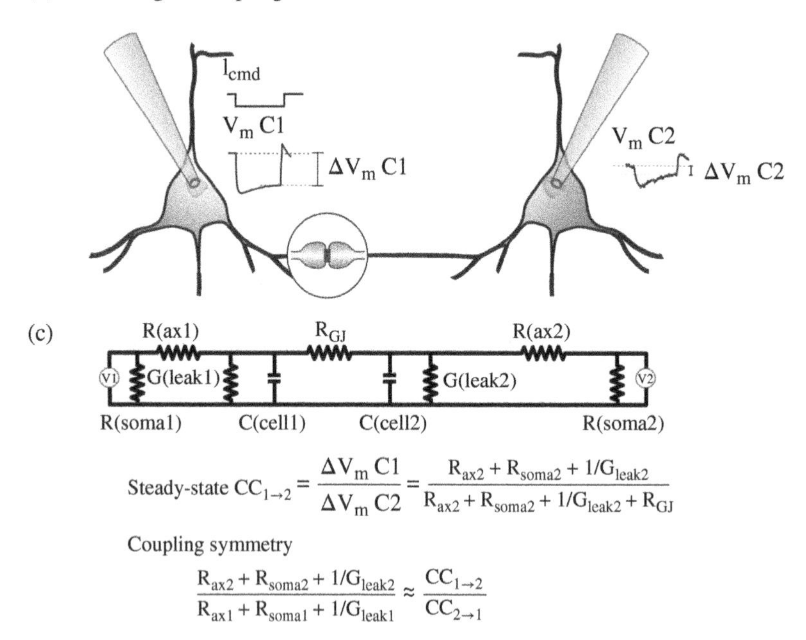

FIGURE 4.1 *(Continued)*

(d) Low-pass filtering in gap junctional transmission

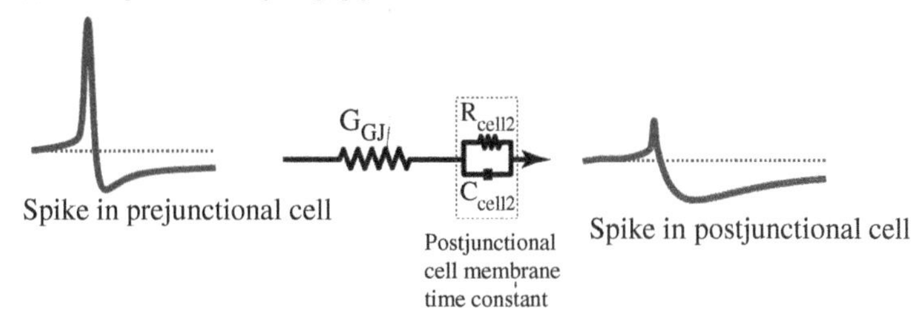

(e) Neurotransmitter-dependent shunt uncoupling a GJ

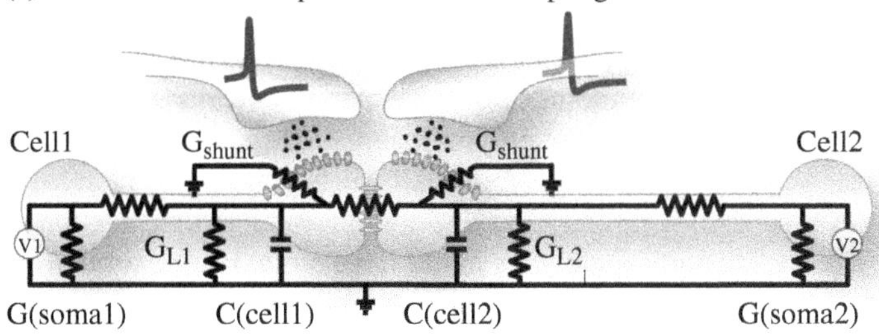

FIGURE 4.1 (a) A schematic representation of the general gap junction (GJ) structure. The GJ plaque connecting two closely apposed cells (left panels) allows bidirectional flux of ions and small biomolecules through up to thousands of individual GJ channels arranged in the plaque. Each GJ channel is composed of two GJ hemichannels or connexons, each one crossing the plasma membrane of one connected cell (right); the membranes are separated from each other by a 2–4 nm gap of extracellular space. The channels are, in turn, composed of six connexin proteins, arranged in a circular formation with a pore in the middle. (b) Empirical determination of GJ conductance between two coupled neurons, where a GJ is present at a distal dendritic location. Voltage changes (ΔV) in two coupled cells are compared during injection of a negative command current (I_{cmd}) step into cell 1, and the steady-state coupling coefficient (CC) is defined as the ratio of voltage changes in cell 2 and cell 1. (c) Equivalent circuit roughly approximating the electrical elements contributing to the measured steady-state CC. In addition to the somatic (input) resistances (R_{soma}), the anatomical location of the GJ on dendrites and the dendritic cable properties (axial resistivity (R_{ax}) and leak conductances (G_{leak})) are a major determinant of the coupling efficacy. Notably, if a GJ is residing on a tip of a long dendrite, the leak conductance through dendritic channels will be a major factor. The differences in relative positions of GJs on dendrites will result in coupling asymmetry, due to the mismatches in total resistances in the coupled cells; in general, the coupling will be stronger from a lower- to higher-resistance cell. (d) The neuronal capacitance (C_{cell} in panel B) together with the resistances determines the time constant of the coupled neurons and results in the GJ communication acting as first-order low-pass filter for signals. As a result, neuronal electric signals are differentially transmitted based on their temporal features; a strongly depolarizing and fast action potential in prejunctional cell (left) may result in a slow hyperpolarization in the postjunctional cell (right). (e) A schematic drawing depicting the mechanism recently shown to modulate the GJ coupling among inferior olivary neurons. The sites of GJ connections are targeted by axonal terminals, which, upon activation, release neurotransmitters that activate shunting conductances on both sides of the GJ, thereby greatly decreasing intersomatic signal transmission.

The arrangement of the six connexin subunits into a connexon hemichannel [9–11] takes place in the cytosol of each cell, and they are then translocated to the plasma membrane. The junctional channels connecting the intracellular spaces of the two cells form when hemichannels residing in the cell membrane of one cell dock with their connexon counterparts in the opposing cell [12, 13].

Most cells can express multiple subtypes of connexins, and either homomeric (composed of identical subunits) or heteromeric (with various subunits) connexon hemichannels can be formed. The two connexon hemichannels forming a transcellular channel can be either of identical (homotypic) or differing (heterotypic) connexin composition. Each subunit contributes to the fine-tuning of resultant GJ properties; as an example, the flux of molecules passing through GJ channels formed of C × 43 is approximately 3.3 times higher than that of C × 26 (6000 molecules(s) vs. 1800 molecules(s) [14]). The multitude of known connexin subunits and the possibility of formation of heteromeric and heterotypic hemichannels would seem to allow a large functional variety of GJ compositions. However, mice with targeted deletion of many connexin genes still exhibit relatively normal structure and function, suggesting that many of the individual subtypes might have redundant or similar properties and thus could be interchangeable [15, 16]. Furthermore, despite the large number of theoretically possible connexin and hemichannel combinations, only a limited number of them are known to form transcellular channels [17–19].

During development, the composition of connexin subtypes changes, indicating that their functional roles are likely shifting as the animal matures. For instance, C × 45 is strongly expressed in the brain for the first 2 weeks of mouse postnatal development and is largely absent in the adult; similarly, C × 36 expression first increases during early postnatal age and subsequently diminishes, even though in certain brain regions it remains strongly expressed throughout adulthood. In general, electrical coupling is stronger (i.e., connexin expression level is higher and GJs are found more abundantly between cells) in developing tissues. Unlike in the adult brain (discussed in the following), the GJ coupling in immature neural system is involved in coordination of the arrangement of cellular groups via direct diffusion of signaling molecules [20–24].

4.2 DISTRIBUTION OF GJS IN THE BRAIN

Even though originally thought to be restricted to only some tissue types or developing structures, it is now known that connexin proteins are widely expressed in the animal body with various functions. The full list of regions with GJs is beyond the scope of the present review; the interested reader is kindly directed to several broad reviews on the subject [8, 9, 25]. Here, we will focus on GJs and their functions in cells of the central nervous system (CNS (glial cells and neurons)).

4.2.1 Glia

Glial cells are small, nonneuronal cells of the nervous system that outnumber the neurons severalfold. Their functions were originally thought to be restricted to homeostatic management of neuronal tissue and support of neuronal cells; however, their crucial roles in interneuronal signaling and modulation of network activity are becoming increasingly clear. Importantly, they are known to express a variety of voltage-sensitive ion channels and thereby responsive to electrical signals [26]; furthermore, they are capable of releasing transmitters in response to electrical transients [27].

Astrocytes are the most abundant type of glial cells that are crucial for the correct functioning of the neuronal network. They are extensively coupled via GJs comprising of C × 43 and C × 30 as well as others [28]; GJ-coupled astrocytes form an interconnected mesh of individual glial cells, a glial syncytium. This syncytium operates as a "sink" for by-products of action potential firing (such as K^+ or glutamate) among the nearby neurons. The interconnected astrocytes are thus able to redistribute these by-products, thereby maintaining homeostatic conditions. In addition, they regulate the extracellular pH levels as well as provide neurons with sufficient amounts of nutrients and replenish their neurotransmitter stores. This network is also known to be involved in propagation of calcium waves [29] and modulation of chemical synaptic transmission between neurons [30], and thereby play a significant role in coordinating neuronal network activity as well in processes related to memory formation in the brain.

Oligodendrocytes and Schwann cells are best known as the glial cells that form the insulating sheets (myelin) around neuronal axons, immensely improving electrical signal conduction [31]. They form the myelin by growing multiple, spiral layers of lipid membranes around the axons. Curiously, in these cells, the GJs are found not only between individual glial cells [32, 33] but also between nearby membranes of the same cell within the spiraling structure, improving diffusion of proteins, signaling molecules, and calcium throughout the cell cytoplasm. Furthermore, as an important exception to the rule of GJs coupling only cells of the same type in the adult animal, oligodendrocytes are electrically coupled to the astrocytic syncytium, allowing siphoning of excess potassium released from spiking axons into the extracellular space and thereby ensuring reliable action potential transmission.

4.2.2 Neurons

The notion that neurons communicate electrically is almost as old as the idea of bioelectricity itself [34], as it was originally thought that neurons form directly connected networks (syncytium) within which the signals would propagate as electrical transients [35]. This concept was forgotten along subsequent rise and triumph of the "neuron doctrine" [36, 37] revolving around the idea of neurons as separate units that communicate via discrete, chemically transmitted synapses. Although the presence of electrical synapses between neurons was demonstrated already in the 1950s in crayfish, shrimp, and fish nervous systems [38–40], they were long dismissed as minor contributors to neural function in higher vertebrates most of the twentieth century [5]. This was to a large part due to methodological constraints and difficulties in demonstrating GJs in mammalian neurons. Technological advances in detection and visualization of specific proteins, however, have now revealed that connexins and GJs are abundant in the CNS of all animals and that the most prevalent connexin type (the C × 36 [41]) is expressed in a multitude of neuronal types in various brain regions (reviewed in [11, 42]).

4.3 ELECTRICAL SIGNALING THROUGH GJS

Even though the direct diffusion of metabolites and ions has significant functions (as discussed earlier) for the glial cell, to date, strong evidence for such diffusion through GJs having physiological significance between neurons in the adult brain is lacking. GJs in the mature CNS exhibit strong cationic selectivity, which makes them nearly impermeable for many negatively charged signaling molecules and metabolites. In contrast, cations such as K^+ (which is a major charge carrier in neurons) are highly permeable, suggesting that neuronal GJs are specialized for electrical rather than metabolic communication [43].

4.3.1 Electrical Conductance

The main GJs in the CNS (based on Cx36) have the smallest single-channel conductance of any type described (~10–15 pS [44]; compared with 310 pS conductance of Cx37 found in motor neurons). Also, compared with larger GJs found in the retina [45] or in Mauthner cells of fish auditory system [46] with thousands of connexin channels per GJ plaque, the Cx36-based GJs of many mature brain regions have only approximately 150–380 connexin channels each [47]. In addition, the probability of individual connexon channels being open seems to be very low (in the order of 4–9%), resulting in a total mean junctional conductance of approximately 0.2 nS [19, 23, 48–50]. This rough estimate (hundreds of pS of conductance per single GJ plaque) is comparable to what can be observed as a conductance for single chemical synapses [51] (even though much larger conductance (over 2 nS) chemical synapses have been described [52]). While considering this relatively small GJ conductance, one should bear in mind that there are usually numerous GJs connecting two individual neurons; the parallel sum of all the junctional conductances may reach up to 50% of the cell's input conductance [48, 53, 54]. In all, it is clear that the gap junctional currents can be comparable to chemical synapses in their power to affect neuronal electrical activity.

The conductance of a subpopulation of GJs shows rectification (i.e., the efficacy of electrical signaling in one direction is greater than in the other), partly due to their connexin subtype-dependent ionic selectivity [55, 56]. Furthermore, many if not all GJs show subtype-specific transjunctional or transmembrane voltage gating at least when expressed in oocytes [6, 57]. This, combined with the possibility of heterotypic channel assembly, has been proposed to contribute to rectification of GJ signaling. However, the physiological significance of voltage gating or rectification in the case of the GJs containing Cx36 [18, 44] is unclear, at least when it comes to physiologically relevant voltage ranges in mammalian CNS [57–59].

The difficulty in quantifying the voltage gating (as well as GJ channel rectification) in neurons stems in part from the fact that unlike oocytes in which such studies are commonly performed, neurons are very small in size and precise measurement of the electrical properties of individual GJs becomes a challenging task. The common way of determining neuronal gap junctional coupling (or coupling coefficient, CC) is by simultaneous voltage and current recording from both cell bodies of the coupled pair (Fig. 4.1b). A current pulse is injected into one side of the pair, and the resulting voltage drops in both coupled cells are measured. The CC is defined as the ratio between post- and prejunctional voltages and typically ranges from several to tens of percent [60–62]. This rather straightforward and simple measurement becomes further complicated in neurons for two reasons: first, the very small size of neurons (~1–2 orders of magnitude smaller than oocytes) makes recordings technically difficult and, second, because GJs are most commonly found on electrotonically distal neuronal processes (dendrites and axons) [63, 64]. The latter issue complicates the measurements because the neuronal GJ CC observed in the aforementioned manner reflects not only the conductance through the GJ channels but to a large extent also the access resistance and leakage through the dendrites connecting the soma to the location of the GJ(s) (Fig. 4.1c).

Furthermore, dendrites are not simple uniform cables but are rather nonlinear in their intrinsic electrical properties [65, 66]. Therefore, interplay between GJ currents and intrinsic voltage-gated Ca^{2+} or Na^{+} channels may act to depress or amplify the electrical transmission between coupled neurons [67, 68]. The diameter of the dendrites also affects the propagation of electrical signals [64] and thereby contributes to the observed GJ coupling efficacy. As these dendritic parameters are rarely known to the experimenter, the neuronal GJ CC need to be always taken as descriptive, rather than exact physical measures.

4.3.2 Asymmetry of GJ Coupling

Since ionic current flow can occur freely between the two cells through connexin channels, electrical transmission via GJs can be bidirectional [69]. However, even if the GJ channels would behave in an entirely ohmic manner (i.e., without any rectification), the resultant coupling between neurons is likely to be asymmetrical, simply because the coupled cells can have different conductances (Fig. 4.1c). Even with simple linear ionic conductances, the CC will be lower from a high- to a low-conductance cell than in the other direction. An effective conductance difference can be due to the random distribution of GJs on dendrites at different electrotonic distances [42].

Indeed, examples of asymmetry can be found in numerous brain structures, and it has been proposed to play a significant role in determining the preferred directionality of information flow within these networks [60, 61, 67].

4.3.3 Temporal Features of GJ Communication

Another consequence of GJ localization on nonlinear dendrites at electrotonically distant sites is the low-pass filter-like behavior of the electrical synapse, as the transjunctional current needs to charge the parallel capacitance of the postjunctional cell (Fig. 4.1c, d). Thus, postjunctional voltage responses are delayed as well as attenuated relative to the prejunctional potential. The delay is most evident for fast signals such as neuronal action potentials. Consequently, relatively small but slow signals, such as subthreshold voltage oscillations, are communicated more effectively.

An important consequence of the low-pass filtering properties of GJs is the possibility of selective transmission of action potential components. The neuronal action potentials have prominent biphasic (fast depolarizing—slow hyperpolarizing) waveforms; due to the low-pass filtering, the fast phases are greatly attenuated (resulting in a small postjunctional "spikelet" instead of a full-blown action potential). In contrast, the smaller but much slower hyperpolarization following the depolarizing spike will be less attenuated and therefore, paradoxically, more likely to influence the postsynaptic cell. This hyperpolarizing prejunctional signal may in fact dominate the GJ-mediated signaling and result in relatively long-lasting postjunctional hyperpolarization. Thus, depending on the characteristics of the action potential waveform in the prejunctional cell, electrical synapses can either excite or inhibit the postsynaptic cell [70]. Furthermore, the differential transmission of the depolarizing and hyperpolarizing phases of the action potential can result in a phenomenon where the postjunctional spikelets show faster decays than expected for a waveform spreading through a dendrite with a given membrane time constant. This is the result of the hyperpolarization following the prejunctional spike that counteracts the depolarizing current and thus a much shortened or even biphasic spikelet will result.

Finally, the low-pass filtering will reduce noise with very-high-frequency components and, thus, may improve the signal-to-noise ratio in certain communication pathways [71].

4.4 ROLES OF GJ-MEDIATED ELECTRICAL SIGNALING IN BRAIN FUNCTION

As follows from what was discussed before, the physiological effects of GJ signaling depend on the specific subunit composition as well as on the intrinsic properties of the cells that they connect. Still, even in very different neuronal structures (retina, respiratory nuclei, inferior olive (IO), etc.) that operate with different time scales, it seems that GJs are often involved in synchronization of network activity as the bidirectional current flow endows them with the capability to synchronize discharges of interconnected cells [42].

The significance of synchronous, rhythmic brain activity has been recognized for more than half a century [72] and is thought to reflect temporal coordination of neural activity across brain networks required for many behavioral and cognitive processes [73]. The idea that GJs are involved in synchronization of neuronal firing was brought forward with first reports demonstrating neuronal electrotonic interactions [39, 56], and presently, their role in the generation, synchronization, and shaping of neuronal network activity is widely accepted [74–76]. When two or more spiking neurons are coupled via fast and reciprocal GJs, they may rapidly synchronize their firing. The coupling does not need to be strong for synchronization to occur; in the mammalian brain, neuronal action potentials and subthreshold oscillations are known to synchronize even with weak or moderate coupling strengths [54, 59, 67, 77–80]. Again, the specific outcome of GJ coupling depends on the details of the connection and of the properties of the connected cells, and instead of synchronization, modeling works show that in certain cases weak GJ coupling can lead to antiphasic or asynchronous firing [74, 81–83]. This counterintuitive result can be understood in terms of the aforementioned temporal properties of GJ signaling: with suitable prejunctional action potential frequencies and waveforms, prejunctional action potential will in effect inhibit spiking in the electrically coupled neighbors.

Even though the present review focuses on these properties among networks of neurons, analogous synchronizing roles for GJ communication are found in other body structures with synchronous or oscillatory activity (such as the heart muscle or digestive system [84–87]). Even in developing neural tissue, where the GJ-mediated passage of metabolites and second messengers between the cells plays a central role in the coordination of neuronal circuit formation, the electrical GJ communication is also essential for the generation of rhythmic patterns required for the formation of adult neuronal networks [88–90]. Therefore, when GJs are present among electrically responsive cells, at least some level of synchronous and/or oscillatory network behavior is expected to emerge.

4.4.1 GJs in an Oscillating Brain Structure: The IO

A remarkable example of a brain structure where gap junctional signaling plays a major role is the IO. The IO was one of the first sites in the mammalian brain where electrical coupling was demonstrated [60, 80, 91, 92] and high levels of Cx36 expression was documented [19, 22, 93, 94]. The importance of electrical signaling through GJs is further stressed by the fact that they are the only communication pathway between IO neurons [92, 95].

The physiological significance of IO becomes apparent when examining its anatomical arrangement within the olivocerebellar networks responsible for motor coordination and cognitive functions [96–98]. The IO neuron axons (called climbing fibers) give rise to one of the two major cerebellar input pathways [99] and form one of the largest chemical synaptic junctions in the nervous system onto the cerebellar Purkinje neurons (PNs) in the cerebellar cortex [100, 101]. PNs then project to the cerebellar nuclei, which in turn send axons to the IO. Notably, all of these connections are topographically arranged so that one can define discrete groups of IO neurons, PNs, and cerebellar nuclear neurons that are linked by synaptic connections and thereby form a feedback loop [102].

Functionally, two hypotheses are being discussed in regard to IO function. The first posits that climbing fibers originating in IO are related to motor learning and function as an "error signal" that would modify synaptic strengths within the cerebellar cortex [103–106]. The second hypothesis sees the climbing fiber signaling as a provider of precise timing signal for correct execution of movements [97, 107].

The biophysical correlate of these timing-related functions can be found in the IO, where many neurons have an unusual propensity to generate large, spontaneous, synchronous, subthreshold fluctuations of membrane potential at 1–10 Hz [60, 80, 108–112]. The propagation of these remarkably stable and precise oscillations, as well as the phase differences emerging between oscillating units, can be in fact one of the fundamental "timing signals" of the brain [113, 114].

The synchronicity of rhythmic IO activity is dependent on the GJ coupling [54, 92, 109, 115, 116], but it has been also proposed that the electrical coupling not only mediates synchrony but is, in fact, essential for generation of the oscillations themselves [81, 117–121]. The experimental and theoretical evidence that support both views are fueling the fierce debate that is ongoing at the time of writing this review. The role of GJs and especially the modulation of GJ coupling between the IO neurons is a target of vigorous research, as it is among the key questions to be resolved regarding this debate as well as the aforementioned controversy between the role of IO in either motor learning or execution.

4.5 PLASTICITY AND MODULATION OF GAP JUNCTIONAL COUPLING

A fundamental aspect of GJ coupling is that it can be modulated rapidly, and thus, the sensitivity of cells to signals from their coupled counterparts can be adjusted. Extreme suppression of the coupling strength can effectively disconnect cells from each other and thus reorganize subnetworks of synchronously active neurons according to behavioral needs.

It has been long known that intracellular Ca^{2+} concentration and pH can rapidly open or close GJ channels, as demonstrated in well-controlled systems such as oocytes or embryonic cells [122, 123]. The GJ conductance is reduced with increases of intracellular $[H^+]_i$ or $[Ca^{2+}]_i$, but whether these constitute a physiological mechanism of channel regulation has long been debated [123, 124]. In most coupled systems of cells, pH is a stronger regulator of GJ coupling than Ca^{2+}, and in general, $[Ca^{2+}]_i$ must rise to pathologically high concentrations (hundreds of micromolars; compared with peak $[Ca^{2+}]_i$ of hundreds of nanomolars during strong physiological calcium waves [125]), for GJs to close [126]. Thus, it has been suggested that the GJ closure with excessively high intracellular calcium concentration mainly acts during cell death as a protective measure to prevent calcium-dependent cellular damage [127] in the surrounding cells.

Neural activity can either acidify or alkalinize the intracellular pH of central mammalian neurons by several tenths of a pH unit [128], and thus, neuronal state can up- or downregulate its GJ coupling. In the case of GJs composed of certain connexin subtypes, the relationship between pH and coupling strength is very steep and centered on the normal resting pH_i [129, 130], suggesting relevance in physiological or pathological situations (discussed in the following). The pH changes are proposed to modulate GJ conductance via shifts in the connexin channel open probabilities as well as possibly by modification of voltage gating [129] and thus can effectuate very rapidly (within milliseconds).

A much slower but an equally important form of activity-dependent modulation of GJ conductance involves G-protein-mediated signaling cascades and phosphorylation. As all connexin subtypes have been found to contain multiple phosphorylation, this kind of regulation has recently received increasing attention, especially related to the possibility of aberrations in phosphorylation state being involved in neuropathologies.

Phosphorylation can directly influence connexin channel opening probability or regulate the assembly, trafficking, and turnover of GJ channels [131–133]. Via this pathway, impermeable extracellular agents such as neurotransmitters can modify GJ properties. Indeed, a large variety of neurotransmitters (such as dopamine, serotonin, histamine, and glutamate) are known to affect neuronal GJs via such signaling cascades (for review, see [134]). It appears that there is no simple rule by which to predict what action a given neurotransmitter will have on coupling; indeed, activation of one neurotransmitter receptor can have opposite effects on GJ coupling when examined in different neurons. Except for the immature neocortex where all neurotransmitters thus far tested reduce GJ coupling, the only common feature for GJ modulation based on neurotransmitter-dependent intracellular molecular pathways is the time scale of the effects (ranging from several minutes to hours [135, 136]).

In addition to neurotransmitter release-dependent modulation of GJs, it has recently been reported [61] that intense spiking activity in electrically coupled neurons themselves can have a significant and long-lasting downregulating effect on the coupling. Such activity in coupled neurons can also modify symmetry of the GJ transmission, thus shifting the preferred direction of electrical signaling among neurons. The exact mechanism for these modifications that develop over the course of several minutes is presently unclear but likely involves phosphorylation-dependent closing or undocking of individual connexons from the GJ.

A much faster (in the order of tens of milliseconds) neurotransmitter-dependent modulatory effect can be found in the IO. As discussed earlier, the electrotonic coupling is thought to play a crucial role in synchronizing IO oscillations and in shaping the spatial arrangement of groups of concurrently oscillating neurons [109, 115, 137]. Indeed, it has been shown that manipulations that affect GJ coupling in IO do result in changes in the spatiotemporal features of IO output (manifested as rearrangement of the synchronous activity in the cerebellar cortex [138–142]).

The precise mechanism of this GJ-mediated modulation of olivary synchrony has been a key question in the cerebellar field for decades. Curiously, it was found that each and every location in the IO where GJs couple neighboring neurons is specifically targeted by axon terminals originating from the cerebellar nuclei [143–147]. This observation led to the hypothesis that these afferents may reduce passive electrotonic propagation along the dendrites (i.e., change the dendritic conductance) and instantaneously modify the efficacy of electrical coupling (in other words, the CC) between the cell somata Fig. 4.1e [92, 148, 149]. This idea, as powerful as it was, could not be demonstrated directly due to technical limitations until very recently, when the advent of viral optogenetic methods allowed specific activation of defined axonal terminals in a network. It was determined experimentally that activation of the cerebellar axons strategically terminating in the vicinity of olivary GJs decreases the CC and blocks the subthreshold oscillations of IO neurons (150). Thus, the spatial arrangement of electrotonically coupled neuronal groups and their dynamic behavior in the IO can be determined by signals originating in the cerebellum [98]. This, in return, will bestow the cerebellum with an ability to essentially "choose" which of its regions are acting in synchrony.

4.6 CLINICAL RELEVANCE

Even though the role of abnormal chemical synaptic transmission in ontogeny of various neurological diseases and injuries (dopamine and Parkinson's disease, GABA and epilepsy, glutamate and excitotoxicity after trauma [151–156]) has stood in the spotlight of clinical research for decades, the role of GJs in both healthy and pathological neuronal functions has been ignored until rather recently. Emerging evidence has, however, progressively contributed to reassessment of the GJ roles, and presently, electrical transmission can be viewed as a form of communication that is complementary to chemical signaling with which it interacts. Therefore, it should be expected that disturbances in chemical transmission can also manifest in aberrant electrical coupling and thereby profoundly alter network-level activity and behavior.

A prominent group of neurological disorders where pathological network-level activity plays a central role are those categorized as epilepsies. This diverse group of disorders is characterized by periodic and rather unpredictable occurrence of seizures, which are defined as transient behavioral or sensory changes attributable to the synchronous and rhythmic firing of large populations of neurons in the CNS.

As GJs are essential for network synchronization, it is not surprising that many experimentally induced states of epileptic-like activity as well as epileptiform discharges in human brain tissue slices resected from patients with intractable epilepsy depend on GJs [157–161]. Even though the causal relation between GJs and clinical epileptogenesis is often not clear, intriguing links are being found that bind abnormal neuronal activity and GJ modulation together suggesting that modification of synaptic and electric transmission can synergistically lead to epileptogenesis.

First, as discussed earlier, changes in intracellular pH driven by intense neuronal activity [128] can modify GJ-coupled neuronal communication [162]; specifically, intracellular alkalinization enhances coupling through certain GJ types and thus could boost neuronal network activity. This can build up to the pathological, hypersynchronous activity and thereby emergence of epileptic seizures (see, e.g., [159]). Indeed, it has been suggested that brain alkalinization, perhaps induced by hyperventilation, contributes to seizure initiation in certain clinical and experimental contexts [163–165]. Second, alterations in connexin expression levels are known to occur during epileptogenic processes or seizure activity [166, 167].

GJ expression also increases during and after neuronal injuries, such as ischemia, traumatic brain injury, and inflammation [168–172]. The time window of enhanced connexin expression (~2 h post-injury) coincides with the period of massive glutamate release from injured cells suggesting a possible causal relationship between glutamate release and GJ expression. Glutamate-dependent neuronal death or excitotoxicity [173] plays a critical role in the delayed effects of neuronal injury [153] via well-characterized mechanisms that include hyperactivation of glutamate receptors, massive influx of Ca^{2+} ions, and thereby overactivation of Ca^{2+}-dependent signaling pathways. However, it has been recently postulated that the universal mechanism for massive glutamate-dependent neuronal death during neuronal injury involves increase in GJ expression triggered by glutamate [167]. While overactivation of glutamate receptors results in neurodegenerative processes, they will be limited to a small group of neurons near the injury. However, it has been demonstrated that the neuronal death spreads to much larger volume of neurons in the presence of GJs, likely because of propagation of GJ-permeable neurodegenerative signals between the coupled neurons [174, 175].

The molecular and cellular mechanisms of modulation of GJ coupling during injury are only recently being understood, but the research holds potential for development of novel therapeutic approaches for posttraumatic neuroprotection. In addition, aberrations of the GJ coupling are also implicated in a multitude of other disorders (such as multiple sclerosis, Charcot–Marie–Tooth disease (CMTX), Pelizaeus–Merzbacher-like disease (PMLD), HIV-associated neurologic disorders, cystic fibrosis, migraine, Parkinson's disease, Alzheimer's disease, cardiac dysfunction, cancer, and skin diseases [31, 176–183]). However, there are presently no drugs in clinical use that selectively target GJ-mediated intercellular communication [9]. This is in part due to the difficulty of designing pharmacological agents that can affect the connexin modulatory sites, which usually reside within the intracellular space of the coupled cells and thus are well insulated from external pharmacological agents.

4.7 CONCLUDING REMARKS

GJs have been long overshadowed by the chemical synapses in terms of perceived significance in both healthy and pathological brain functions, perhaps because of the relative simplicity of their underlying functional mechanism and the general view that they lack "interesting" plastic properties. Emerging evidence has, however, progressively revised the minimalist view of GJs as channels for diffusive distribution of metabolites and simple supporters of network synchronization. As a result, electrical synapses are now known to be a dynamic and modifiable form of interneuronal communication that may modify concurrent chemical synaptic transmission. Electrical transmission should be viewed, therefore, as a complementary form of communication, instead of an alternative to chemical synaptic transmission.

ACKNOWLEDGMENTS

The authors wish to thank Prof. Yosef Yarom for kind support for the work and Ms. Hermina Nedelescu and Ms. Lea Ankri for help and inspiration during manuscript preparation.

REFERENCES

1. Goodenough, D.A., The structure of cell membranes involved in intercellular communication. *Am J Clin Pathol*, 1975. 63(5): 636–45.
2. Willecke, K., et al., Structural and functional diversity of connexin genes in the mouse and human genome. *Biol Chem*, 2002. 383(5): 725–37.

3. Sohl, G. and K. Willecke, An update on connexin genes and their nomenclature in mouse and man. *Cell Commun Adhes*, 2003. 10(4–6): 173–80.
4. Beyer, E.C., D.L. Paul, and D.A. Goodenough, Connexin43: a protein from rat heart homologous to a gap junction protein from liver. *J Cell Biol*, 1987. 105(6 Pt 1): 2621–9.
5. Loewenstein, W.R., Junctional intercellular communication: the cell-to-cell membrane channel. *Physiol Rev*, 1981. 61(4): 829–913.
6. Harris, A.L., Emerging issues of connexin channels: biophysics fills the gap. *Q Rev Biophys*, 2001. 34(3): 325–472.
7. Veenstra, R.D., Size and selectivity of gap junction channels formed from different connexins. *J Bioenerg Biomembr*, 1996. 28(4): 327–37.
8. Saez, J.C., et al., Plasma membrane channels formed by connexins: their regulation and functions. *Physiol Rev*, 2003. 83(4): 1359–400.
9. Nielsen, M.S., et al., Gap junctions. *Compr Physiol*, 2012. 2(3): 1981–2035.
10. Makowski, L., et al., Gap junction structures. II. Analysis of the x-ray diffraction data. *J Cell Biol*, 1977. 74(2): 629–45.
11. Sohl, G., S. Maxeiner, and K. Willecke, Expression and functions of neuronal gap junctions. *Nat Rev Neurosci*, 2005. 6(3): 191–200.
12. Yeager, M., V.M. Unger, and M.M. Falk, Synthesis, assembly and structure of gap junction intercellular channels. *Curr Opin Struct Biol*, 1998. 8(4): 517–24.
13. Segretain, D. and M.M. Falk, Regulation of connexin biosynthesis, assembly, gap junction formation, and removal. *Biochim Biophys Acta*, 2004. 23: 1–2.
14. Valiunas, V., Cyclic nucleotide permeability through unopposed connexin hemichannels. *Front Pharmacol*, 2013. 4(75): 00075.
15. Guldenagel, M., et al., Visual transmission deficits in mice with targeted disruption of the gap junction gene connexin36. *J Neurosci*, 2001. 21(16): 6036–44.
16. Buhl, D.L., et al., Selective impairment of hippocampal gamma oscillations in connexin-36 knock-out mouse in vivo. *J Neurosci*, 2003. 23(3): 1013–8.
17. Bruzzone, R., T.W. White, and D.L. Paul, Connections with connexins: the molecular basis of direct intercellular signaling. *Eur J Biochem*, 1996. 238(1): 1–27.
18. Al-Ubaidi, M.R., et al., Functional properties, developmental regulation, and chromosomal localization of murine connexin36, a gap-junctional protein expressed preferentially in retina and brain. *J Neurosci Res*, 2000. 59(6): 813–26.
19. Teubner, B., et al., Functional expression of the murine connexin 36 gene coding for a neuron-specific gap junctional protein. *J Membr Biol*, 2000. 176(3): 249–62.
20. Condorelli, D.F., et al., Cellular expression of connexins in the rat brain: neuronal localization, effects of kainate-induced seizures and expression in apoptotic neuronal cells. *Eur J Neurosci*, 2003. 18(7): 1807–27.
21. Maxeiner, S., et al., Spatiotemporal transcription of connexin45 during brain development results in neuronal expression in adult mice. *Neuroscience*, 2003. 119(3): 689–700.
22. Belluardo, N., et al., Expression of connexin36 in the adult and developing rat brain. *Brain Res*, 2000. 865(1): 121–38.
23. Galarreta, M. and S. Hestrin, Electrical and chemical synapses among parvalbumin fast-spiking GABAergic interneurons in adult mouse neocortex. *Proc Natl Acad Sci U S A*, 2002. 99(19): 12438–43.
24. Belousov, A.B. and J.D. Fontes, Neuronal gap junctions: making and breaking connections during development and injury. *Trends Neurosci*, 2013. 36(4): 227–36.

25. Evans, W.H. and P.E. Martin, Gap junctions: structure and function (Review). *Mol Membr Biol*, 2002. 19(2): 121–36.
26. Baker, M.D., Electrophysiology of mammalian Schwann cells. *Prog Biophys Mol Biol*, 2002. 78(2–3): 83–103.
27. Parpura, V. and R. Zorec, Gliotransmission: Exocytotic release from astrocytes. *Brain Res Rev*, 2010. 63(1–2): 83–92.
28. Nagy, J.I., et al., Connexin26 in adult rodent central nervous system: demonstration at astrocytic gap junctions and colocalization with connexin30 and connexin43. *J Comp Neurol*, 2001. 441(4): 302–23.
29. Orthmann-Murphy, J.L., C.K. Abrams, and S.S. Scherer, Gap junctions couple astrocytes and oligodendrocytes. *J Mol Neurosci*, 2008. 35(1): 101–16.
30. Perea, G., M. Navarrete, and A. Araque, Tripartite synapses: astrocytes process and control synaptic information. *Trends Neurosci*, 2009. 32(8): 421–31.
31. Nualart-Marti, A., C. Solsona, and R.D. Fields, Gap junction communication in myelinating glia. *Biochim Biophys Acta*, 2013. 1: 69–78.
32. Menichella, D.M., et al., Connexins are critical for normal myelination in the CNS. *J Neurosci*, 2003. 23(13): 5963–73.
33. Odermatt, B., et al., Connexin 47 (Cx47)-deficient mice with enhanced green fluorescent protein reporter gene reveal predominant oligodendrocytic expressionof Cx47 and display vacuolized myelin in the CNS. *J Neurosci*, 2003. 23(11): 4549–59.
34. Piccolino, M. and N.J. Wade, The frog's dancing master: science, seances, and the transmission of myths. *J Hist Neurosci*, 2013. 22(1): 79–95.
35. Eccles, J.C., The synapse: from electrical to chemical transmission. *Annu Rev Neurosci*, 1982. 5: 325–39.
36. Bock, O., Cajal, Golgi, Nansen, Schafer and the neuron doctrine. *Endeavour*, 2013. 17(13): 00042–2.
37. Glickstein, M., Golgi and Cajal: the neuron doctrine and the 100th anniversary of the 1906 Nobel Prize. *Curr Biol*, 2006. 16(5): R147–51.
38. Furshpan, E.J. and D.D. Potter, Mechanism of nerve-impulse transmission at a crayfish synapse. *Nature*, 1957. 180(4581): 342–3.
39. Watanabe, A., The interaction of electrical activity among neurons of lobster cardiac ganglion. *Jpn J Physiol*, 1958. 8(4): 305–18.
40. Bennett, M.V., S.M. Crain, and H. Grundfest, Electrophysiology of supramedullary neurons in *Spheroides maculatus*. II. Properties of the electrically excitable membrane. *J Gen Physiol*, 1959. 43: 189–219.
41. Condorelli, D.F., et al., Expression of Cx36 in mammalian neurons. *Brain Res Brain Res Rev*, 2000. 32(1): 72–85.
42. Bennett, M.V. and R.S. Zukin, Electrical coupling and neuronal synchronization in the Mammalian brain. *Neuron*, 2004. 41(4): 495–511.
43. Bukauskas, F.F., Neurons and beta-cells of the pancreas express connexin36, forming gap junction channels that exhibit strong cationic selectivity. *J Membr Biol*, 2012. 245(5–6): 243–53.
44. Srinivas, M., et al., Functional properties of channels formed by the neuronal gap junction protein connexin36. *J Neurosci*, 1999. 19(22): 9848–55.
45. Mansour, H., et al., Connexin 30 expression and frequency of connexin heterogeneity in astrocyte gap junction plaques increase with age in the rat retina. *PLoS One*, 2013. 8(3): 14.

46. Tuttle, R., S. Masuko, and Y. Nakajima, Freeze-fracture study of the large myelinated club ending synapse on the goldfish Mauthner cell: special reference to the quantitative analysis of gap junctions. *J Comp Neurol*, 1986. 246(2): 202–11.
47. Fukuda, T. and T. Kosaka, Ultrastructural study of gap junctions between dendrites of parvalbumin-containing GABAergic neurons in various neocortical areas of the adult rat. *Neuroscience*, 2003. 120(1): 5–20.
48. Deans, M.R., et al., Synchronous activity of inhibitory networks in neocortex requires electrical synapses containing connexin36. *Neuron*, 2001. 31(3): 477–85.
49. Connors, B.W. and M.A. Long, Electrical synapses in the mammalian brain. *Annu Rev Neurosci*, 2004. 27: 393–418.
50. Lin, J.W. and D.S. Faber, Synaptic transmission mediated by single club endings on the goldfish Mauthner cell. I. Characteristics of electrotonic and chemical postsynaptic potentials. *J Neurosci*, 1988. 8(4): 1302–12.
51. Alle, H. and J.R. Geiger, GABAergic spill-over transmission onto hippocampal mossy fiber boutons. *J Neurosci*, 2007. 27(4): 942–50.
52. Kirischuk, S., N. Veselovsky, and R. Grantyn, Relationship between presynaptic calcium transients and postsynaptic currents at single gamma-aminobutyric acid (GABA)ergic boutons. *Proc Natl Acad Sci U S A*, 1999. 96(13): 7520–5.
53. Amitai, Y., et al., The spatial dimensions of electrically coupled networks of interneurons in the neocortex. *J Neurosci*, 2002. 22(10): 4142–52.
54. Long, M.A., et al., Rhythmicity without synchrony in the electrically uncoupled inferior olive. *J Neurosci*, 2002. 22(24): 10898–905.
55. Phelan, P., et al., Molecular mechanism of rectification at identified electrical synapses in the Drosophila giant fiber system. *Curr Biol*, 2008. 18(24): 1955–60.
56. Furshpan, E.J. and D.D. Potter, Transmission at the giant motor synapses of the crayfish. *J Physiol*, 1959. 145(2): 289–325.
57. Gonzalez, D., J.M. Gomez-Hernandez, and L.C. Barrio, Molecular basis of voltage dependence of connexin channels: an integrative appraisal. *Prog Biophys Mol Biol*, 2007. 94(1–2): 66–106.
58. Rash, J.E., et al., Molecular and functional asymmetry at a vertebrate electrical synapse. *Neuron*, 2013. 79(5): 957–69.
59. Gibson, J.R., M. Beierlein, and B.W. Connors, Two networks of electrically coupled inhibitory neurons in neocortex. *Nature*, 1999. 402(6757): 75–9.
60. Devor, A. and Y. Yarom, Electrotonic coupling in the inferior olivary nucleus revealed by simultaneous double patch recordings. *J Neurophysiol*, 2002. 87(6): 3048–58.
61. Haas, J.S., B. Zavala, and C.E. Landisman, Activity-dependent long-term depression of electrical synapses. *Science*, 2011. 334(6054): 389–93.
62. Iball, J. and A.B. Ali, Endocannabinoid release modulates electrical coupling between CCK cells connected via chemical and electrical synapses in CA1. *Front Neural Circuits*, 2011. 5(17): 00017.
63. Fukuda, T., Structural organization of the gap junction network in the cerebral cortex. *Neuroscientist*, 2007. 13(3): 199–207.
64. Nadim, F. and J. Golowasch, Signal transmission between gap-junctionally coupled passive cables is most effective at an optimal diameter. *J Neurophysiol*, 2006. 95(6): 3831–43.
65. Silver, R.A., Neuronal arithmetic. *Nat Rev Neurosci*, 2010. 11(7): 474–89.

66. London, M. and M. Hausser, Dendritic computation. *Annu Rev Neurosci*, 2005. 28: 503–32.
67. Mann-Metzer, P. and Y. Yarom, Electrotonic coupling interacts with intrinsic properties to generate synchronized activity in cerebellar networks of inhibitory interneurons. *J Neurosci*, 1999. 19(9): 3298–306.
68. Haas, J.S. and C.E. Landisman, State-dependent modulation of gap junction signaling by the persistent sodium current. *Front Cell Neurosci*, 2011. 5(31): 00031.
69. Hormuzdi, S.G., et al., Electrical synapses: a dynamic signaling system that shapes the activity of neuronal networks. *Biochim Biophys Acta*, 2004. 23: 1–2.
70. Pereda, A.E., et al., Gap junction-mediated electrical transmission: regulatory mechanisms and plasticity. *Biochim Biophys Acta*, 2013. 1: 134–46.
71. Medvedev, G.S. and S. Zhuravytska, Shaping bursting by electrical coupling and noise. *Biol Cybern*, 2012. 106(2): 67–88.
72. Bremer, F., Cerebral and cerebellar potentials. *Physiol Rev*, 1958. 38(3): 357–88.
73. Wang, X.J., Neurophysiological and computational principles of cortical rhythms in cognition. *Physiol Rev*, 2010. 90(3): 1195–268.
74. Chow, C.C. and N. Kopell, Dynamics of spiking neurons with electrical coupling. *Neural Comput*, 2000. 12(7): 1643–78.
75. Migliore, M., M.L. Hines, and G.M. Shepherd, The role of distal dendritic gap junctions in synchronization of mitral cell axonal output. *J Comput Neurosci*, 2005. 18(2): 151–61.
76. Traub, R.D., et al., Contrasting roles of axonal (pyramidal cell) and dendritic (interneuron) electrical coupling in the generation of neuronal network oscillations. *Proc Natl Acad Sci U S A*, 2003. 100(3): 1370–4.
77. Korn, H., C. Sotelo, and F. Crepel, Electronic coupling between neurons in the rat lateral vestibular nucleus. *Exp Brain Res*, 1973. 16(3): 255–75.
78. Landisman, C.E., et al., Electrical synapses in the thalamic reticular nucleus. *J Neurosci*, 2002. 22(3): 1002–9.
79. Beierlein, M., J.R. Gibson, and B.W. Connors, A network of electrically coupled interneurons drives synchronized inhibition in neocortex. *Nat Neurosci*, 2000. 3(9): 904–10.
80. Benardo, L.S. and R.E. Foster, Oscillatory behavior in inferior olive neurons: mechanism, modulation, cell aggregates. *Brain Res Bull*, 1986. 17(6): 773–84.
81. Sherman, A. and J. Rinzel, Rhythmogenic effects of weak electrotonic coupling in neuronal models. *Proc Natl Acad Sci USA*, 1992. 89(6): 2471–4.
82. Lewis, T.J. and J. Rinzel, Dynamics of spiking neurons connected by both inhibitory and electrical coupling. *J Comput Neurosci*, 2003. 14(3): 283–309.
83. Pfeuty, B., et al., Electrical synapses and synchrony: the role of intrinsic currents. *J Neurosci*, 2003. 23(15): 6280–94.
84. Hanani, M., G. Farrugia, and T. Komuro, Intercellular coupling of interstitial cells of Cajal in the digestive tract. *Int Rev Cytol*, 2005. 242: 249–82.
85. Young, R.C., Myocytes, myometrium, and uterine contractions. *Ann N Y Acad Sci*, 2007: 18.
86. Rohr, S., Role of gap junctions in the propagation of the cardiac action potential. *Cardiovasc Res*, 2004. 62(2): 309–22.
87. Kanczuga-Koda, L., Gap junctions and their role in physiology and pathology of the digestive tract. *Postepy Hig Med Dosw*, 2004. 58: 158–65.

88. Rekling, J.C., K.H. Jensen, and H. Jahnsen, Spontaneous cluster activity in the inferior olivary nucleus in brainstem slices from postnatal mice. *J Physiol*, 2012. 590(Pt 7): 1547–62.
89. Saint-Amant, L. and P. Drapeau, Synchronization of an embryonic network of identified spinal interneurons solely by electrical coupling. *Neuron*, 2001. 31(6): 1035–46.
90. Minlebaev, M., Y. Ben-Ari, and R. Khazipov, Network mechanisms of spindle-burst oscillations in the neonatal rat barrel cortex in vivo. *J Neurophysiol*, 2007. 97(1): 692–700.
91. Llinas, R., R. Baker, and C. Sotelo, Electrotonic coupling between neurons in cat inferior olive. *J Neurophysiol*, 1974. 37(3): 560–71.
92. Sotelo, C., R. Llinas, and R. Baker, Structural study of inferior olivary nucleus of the cat: morphological correlates of electrotonic coupling. *J Neurophysiol*, 1974. 37(3): 541–59.
93. Rash, J.E., et al., Immunogold evidence that neuronal gap junctions in adult rat brain and spinal cord contain connexin-36 but not connexin-32 or connexin-43. *Proc Natl Acad Sci U S A*, 2000. 97(13): 7573–8.
94. Condorelli, D.F., et al., Cloning of a new gap junction gene (Cx36) highly expressed in mammalian brain neurons. *Eur J Neurosci*, 1998. 10(3): 1202–8.
95. De Zeeuw, C.I., et al., Morphological correlates of bilateral synchrony in the rat cerebellar cortex. *J Neurosci*, 1996. 16(10): 3412–26.
96. Yarom, Y. and D. Cohen, The olivocerebellar system as a generator of temporal patterns. *Ann N Y Acad Sci*, 2002. 978: 122–34.
97. Llinas, R.R., Cerebellar motor learning versus cerebellar motor timing: the climbing fibre story. *J Physiol*, 2011. 589(Pt 14): 3423–32.
98. De Zeeuw, C.I., et al., Spatiotemporal firing patterns in the cerebellum. *Nat Rev Neurosci*, 2011. 12(6): 327–44.
99. Desclin, J.C., Histological evidence supporting the inferior olive as the major source of cerebellar climbing fibers in the rat. *Brain Res*, 1974. 77(3): 365–84.
100. Eccles, J.C., R. Llinas, and K. Sasaki, The excitatory synaptic action of climbing fibres on the Purkinje cells of the cerebellum. *J Physiol*, 1966. 182(2): 268–96.
101. Uusisaari, M. and E. De Schutter, The mysterious microcircuitry of the cerebellar nuclei. *J Physiol*, 2011. 589(Pt 14): 3441–57.
102. Ruigrok, T.J., Ins and outs of cerebellar modules. *Cerebellum*, 2011. 10(3): 464–74.
103. Ito, M. and M. Kano, Long-lasting depression of parallel fiber-Purkinje cell transmission induced by conjunctive stimulation of parallel fibers and climbing fibers in the cerebellar cortex. *Neurosci Lett*, 1982. 33(3): 253–8.
104. Ito, M., Cerebellar long-term depression: characterization, signal transduction, and functional roles. *Physiol Rev*, 2001. 81(3): 1143–95.
105. Raymond, J.L., S.G. Lisberger, and M.D. Mauk, The cerebellum: a neuronal learning machine? *Science*, 1996. 272(5265): 1126–31.
106. Marr, D., A theory of cerebellar cortex. *J Physiol*, 1969. 202(2): 437–70.
107. Jacobson, G.A., D. Rokni, and Y. Yarom, A model of the olivo-cerebellar system as a temporal pattern generator. *Trends Neurosci*, 2008. 31(12): 617–25.
108. Crill, W.E., Unitary multiple-spiked responses in cat inferior olive nucleus. *J Neurophysiol*, 1970. 33(2): 199–209.
109. Llinas, R. and Y. Yarom, Oscillatory properties of guinea-pig inferior olivary neurones and their pharmacological modulation: an in vitro study. *J Physiol*, 1986. 376: 163–82.

110. Chorev, E., Y. Yarom, and I. Lampl, Rhythmic episodes of subthreshold membrane potential oscillations in the rat inferior olive nuclei in vivo. *J Neurosci*, 2007. 27(19): 5043–52.
111. Khosrovani, S., et al., In vivo mouse inferior olive neurons exhibit heterogeneous subthreshold oscillations and spiking patterns. *Proc Natl Acad Sci U S A*, 2007. 104(40): 15911–6.
112. Bazzigaluppi, P., et al., Properties of the nucleo-olivary pathway: an in vivo whole-cell patch clamp study. *PLoS One*, 2012. 7(9): 27.
113. Lampl, I. and Y. Yarom, Subthreshold oscillations of the membrane potential: a functional synchronizing and timing device. *J Neurophysiol*, 1993. 70(5): 2181–6.
114. Mathy, A., et al., Encoding of oscillations by axonal bursts in inferior olive neurons. *Neuron*, 2009. 62(3): 388–99.
115. Leznik, E. and R. Llinas, Role of gap junctions in synchronized neuronal oscillations in the inferior olive. *J Neurophysiol*, 2005. 94(4): 2447–56.
116. Makarenko, V. and R. Llinas, Experimentally determined chaotic phase synchronization in a neuronal system. *Proc Natl Acad Sci U S A*, 1998. 95(26): 15747–52.
117. Bleasel, A.F. and A.G. Pettigrew, Development and properties of spontaneous oscillations of the membrane potential in inferior olivary neurons in the rat. *Brain Res Dev Brain Res*, 1992. 65(1): 43–50.
118. Lampl, I. and Y. Yarom, Subthreshold oscillations and resonant behavior: two manifestations of the same mechanism. *Neuroscience*, 1997. 78(2): 325–41.
119. Yarom, Y., Rhythmogenesis in a hybrid system—interconnecting an olivary neuron to an analog network of coupled oscillators. *Neuroscience*, 1991. 44(2): 263–75.
120. Manor, Y., et al., To beat or not to beat: a decision taken at the network level. *J Physiol Paris*, 2000. 94(5–6): 375–90.
121. Marshall, S.P., et al., Altered olivocerebellar activity patterns in the connexin36 knockout mouse. *Cerebellum*, 2007. 6(4): 287–99.
122. Loewenstein, W.R. and B. Rose, Calcium in (junctional) intercellular communication and a thought on its behavior in intracellular communication. *Ann N Y Acad Sci*, 1978. 307: 285–307.
123. Rose, B. and R. Rick, Intracellular pH, intracellular free Ca, and junctional cell-cell coupling. *J Membr Biol*, 1978. 44(3–4): 377–415.
124. Rozental, R., M. Srinivas, and D.C. Spray, How to close a gap junction channel. Efficacies and potencies of uncoupling agents. *Methods Mol Biol*, 2001. 154: 447–76.
125. Ross, W.N., Understanding calcium waves and sparks in central neurons. *Nat Rev Neurosci*, 2012. 13(3): 157–68.
126. Spray, D.C., et al., Gap junctional conductance: comparison of sensitivities to H and Ca ions. *Proc Natl Acad Sci U S A*, 1982. 79(2): 441–5.
127. Decrock, E., et al., Calcium and connexin-based intercellular communication, a deadly catch? *Cell Calcium*, 2011. 50(3): 310–21.
128. Chesler, M. and K. Kaila, Modulation of pH by neuronal activity. *Trends Neurosci*, 1992. 15(10): 396–402.
129. Palacios-Prado, N., et al., pH-dependent modulation of voltage gating in connexin45 homotypic and connexin45/connexin43 heterotypic gap junctions. *Proc Natl Acad Sci U S A*, 2010. 107(21): 9897–902.

130. Spray, D.C., A.L. Harris, and M.V. Bennett, Gap junctional conductance is a simple and sensitive function of intracellular pH. *Science*, 1981. 211(4483): 712–5.
131. Marquez-Rosado, L., et al., Connexin43 phosphorylation in brain, cardiac, endothelial and epithelial tissues. *Biochim Biophys Acta*, 2012. 8(92): 26.
132. Laird, D.W., Connexin phosphorylation as a regulatory event linked to gap junction internalization and degradation. *Biochim Biophys Acta*, 2005. 10(2): 172–82.
133. Lampe, P.D. and A.F. Lau, Regulation of gap junctions by phosphorylation of connexins. *Arch Biochem Biophys*, 2000. 384(2): 205–15.
134. Hatton, G.I., Synaptic modulation of neuronal coupling. *Cell Biol Int*, 1998. 22(11–12): 765–80.
135. Landisman, C.E. and B.W. Connors, Long-term modulation of electrical synapses in the mammalian thalamus. *Science*, 2005. 310(5755): 1809–13.
136. Pereda, A.E., et al., Ca^{2+}/calmodulin-dependent kinase II mediates simultaneous enhancement of gap-junctional conductance and glutamatergic transmission. *Proc Natl Acad Sci U S A*, 1998. 95(22): 13272–7.
137. Leznik, E., V. Makarenko, and R. Llinas, Electrotonically mediated oscillatory patterns in neuronal ensembles: an in vitro voltage-dependent dye-imaging study in the inferior olive. *J Neurosci*, 2002. 22(7): 2804–15.
138. Lang, E.J., I. Sugihara, and R. Llinas, GABAergic modulation of complex spike activity by the cerebellar nucleoolivary pathway in rat. *J Neurophysiol*, 1996. 76(1): 255–75.
139. Lang, E.J., Organization of olivocerebellar activity in the absence of excitatory glutamatergic input. *J Neurosci*, 2001. 21(5): 1663–75.
140. Lang, E.J., GABAergic and glutamatergic modulation of spontaneous and motor-cortex-evoked complex spike activity. *J Neurophysiol*, 2002. 87(4): 1993–2008.
141. Bell, C.C. and T. Kawasaki, Relations among climbing fiber responses of nearby Purkinje cells. *J Neurophysiol*, 1972. 35(2): 155–69.
142. Llinas, R. and K. Sasaki, The functional organization of the olivo-cerebellar system as examined by multiple Purkinje cell recordings. *Eur J Neurosci*, 1989. 1(6): 587–602.
143. Sotelo, C., T. Gotow, and M. Wassef, Localization of glutamic-acid-decarboxylase-immunoreactive axon terminals in the inferior olive of the rat, with special emphasis on anatomical relations between GABAergic synapses and dendrodendritic gap junctions. *J Comp Neurol*, 1986. 252(1): 32–50.
144. de Zeeuw, C.I., et al., A new combination of WGA-HRP anterograde tracing and GABA immunocytochemistry applied to afferents of the cat inferior olive at the ultrastructural level. *Brain Res*, 1988. 447(2): 369–75.
145. de Zeeuw, C.I., et al., Ultrastructural study of the GABAergic, cerebellar, and mesodiencephalic innervation of the cat medial accessory olive: anterograde tracing combined with immunocytochemistry. *J Comp Neurol*, 1989. 284(1): 12–35.
146. De Zeeuw, C.I., et al., Microcircuitry and function of the inferior olive. *Trends Neurosci*, 1998. 21(9): 391–400.
147. Fredette, B.J. and E. Mugnaini, The GABAergic cerebello-olivary projection in the rat. *Anat Embryol*, 1991. 184(3): 225–43.
148. Onizuka, M., et al., Solution to the inverse problem of estimating gap-junctional and inhibitory conductance in inferior olive neurons from spike trains by network model simulation. *Neural Netw*, 2013. 47: 51–63.

149. Llinas, R., Eighteenth Bowditch lecture. Motor aspects of cerebellar control. *Physiologist*, 1974. 17(1): 19–46.
150. Lefler, Y., et al., Cerebellar Inhibitory Input to the Inferior Olive Decreases Electrical Coupling and Blocks Subthreshold Oscillations. *Neuron*, 2014. 81(6): 1389–400.
151. Bisaglia, M., et al., Dysfunction of dopamine homeostasis: clues in the hunt for novel Parkinson's disease therapies. *Faseb J*, 2013. 27(6): 2101–10.
152. Werner, F.M. and R. Covenas, Classical neurotransmitters and neuropeptides involved in generalized epilepsy: a focus on antiepileptic drugs. *Curr Med Chem*, 2011. 18(32): 4933–48.
153. Choi, D.W., Glutamate neurotoxicity and diseases of the nervous system. *Neuron*, 1988. 1(8): 623–34.
154. Goldberg, E.M. and D.A. Coulter, Mechanisms of epileptogenesis: a convergence on neural circuit dysfunction. *Nat Rev Neurosci*, 2013. 14(5): 337–49.
155. Rakhade, S.N. and F.E. Jensen, Epileptogenesis in the immature brain: emerging mechanisms. *Nat Rev Neurol*, 2009. 5(7): 380–91.
156. Pavlov, I., et al., Cortical inhibition, pH and cell excitability in epilepsy: what are optimal targets for antiepileptic interventions? *J Physiol*, 2013. 591(Pt 4): 765–74.
157. Carlen, P.L., Curious and contradictory roles of glial connexins and pannexins in epilepsy. *Brain Res*, 2012. 3: 54–60.
158. Uusisaari, M., et al., Spontaneous epileptiform activity mediated by GABA(A) receptors and gap junctions in the rat hippocampal slice following long-term exposure to GABA(B) antagonists. *Neuropharmacology*, 2002. 43(4): 563–72.
159. Kohling, R., et al., Prolonged epileptiform bursting induced by 0-Mg(2+) in rat hippocampal slices depends on gap junctional coupling. *Neuroscience*, 2001. 105(3): 579–87.
160. Traub, R.D., et al., Axonal gap junctions between principal neurons: a novel source of network oscillations, and perhaps epileptogenesis. *Rev Neurosci*, 2002. 13(1): 1–30.
161. Kraglund, N., M. Andreasen, and S. Nedergaard, Differential influence of non-synaptic mechanisms in two in vitro models of epileptic field bursts. *Brain Res*, 2010. 9: 85–95.
162. Gonzalez-Nieto, D., et al., Regulation of neuronal connexin-36 channels by pH. *Proc Natl Acad Sci U S A*, 2008. 105(44): 17169–74.
163. Cunningham, M.O., et al., Glissandi: transient fast electrocorticographic oscillations of steadily increasing frequency, explained by temporally increasing gap junction conductance. *Epilepsia*, 2012. 53(7): 1205–14.
164. Schuchmann, S., et al., Experimental febrile seizures are precipitated by a hyperthermia-induced respiratory alkalosis. *Nat Med*, 2006. 12(7): 817–23.
165. Schuchmann, S., et al., Respiratory alkalosis in children with febrile seizures. *Epilepsia*, 2011. 52(11): 1949–55.
166. Collignon, F., et al., Altered expression of connexin subtypes in mesial temporal lobe epilepsy in humans. *J Neurosurg*, 2006. 105(1): 77–87.
167. Belousov, A.B., Novel model for the mechanisms of glutamate-dependent excitotoxicity: role of neuronal gap junctions. *Brain Res*, 2012. 3: 123–30.
168. Chang, Q., et al., Nerve injury induces gap junctional coupling among axotomized adult motor neurons. *J Neurosci*, 2000. 20(2): 674–84.
169. Frantseva, M.V., et al., Specific gap junctions enhance the neuronal vulnerability to brain traumatic injury. *J Neurosci*, 2002. 22(3): 644–53.

170. de Pina-Benabou, M.H., et al., Blockade of gap junctions in vivo provides neuroprotection after perinatal global ischemia. *Stroke*, 2005. 36(10): 2232–7.
171. Damodaram, S., et al., Tonabersat inhibits trigeminal ganglion neuronal-satellite glial cell signaling. *Headache*, 2009. 49(1): 5–20.
172. Park, W.M., et al., Interplay of chemical neurotransmitters regulates developmental increase in electrical synapses. *J Neurosci*, 2011. 31(16): 5909–20.
173. Kostandy, B.B., The role of glutamate in neuronal ischemic injury: the role of spark in fire. *Neurol Sci*, 2012. 33(2): 223–37.
174. Decrock, E., et al., Connexin-related signaling in cell death: to live or let die? *Cell Death Differ*, 2009. 16(4): 524–36.
175. Cusato, K., et al., Gap junctions mediate bystander cell death in developing retina. *J Neurosci*, 2003. 23(16): 6413–22.
176. Chanson, M., et al., Interactions of connexins with other membrane channels and transporters. *Prog Biophys Mol Biol*, 2007. 94(1–2): 233–44.
177. Rash, J.E., Molecular disruptions of the panglial syncytium block potassium siphoning and axonal saltatory conduction: pertinence to neuromyelitis optica and other demyelinating diseases of the central nervous system. *Neuroscience*, 2010. 168(4): 982–1008.
178. Adriano, E., et al., A novel hypothesis about mechanisms affecting conduction velocity of central myelinated fibers. *Neurochem Res*, 2011. 36(10): 1732–9.
179. Eugenin, E.A., et al., The role of gap junction channels during physiologic and pathologic conditions of the human central nervous system. *J Neuroimmune Pharmacol*, 2012. 7(3): 499–518.
180. Nakase, T. and C.C. Naus, Gap junctions and neurological disorders of the central nervous system. *Biochim Biophys Acta*, 2004. 23: 1–2.
181. White, T.W. and D.L. Paul, Genetic diseases and gene knockouts reveal diverse connexin functions. *Annu Rev Physiol*, 1999. 61: 283–310.
182. Kelsell, D.P., J. Dunlop, and M.B. Hodgins, Human diseases: clues to cracking the connexin code? *Trends Cell Biol*, 2001. 11(1): 2–6.
183. Dere, E. and A. Zlomuzica, The role of gap junctions in the brain in health and disease. *Neurosci Biobehav Rev*, 2012. 36(1): 206–17.

5

ENZYME FILM ELECTROCHEMISTRY

JULEA N. BUTT,[1,2] ANDREW J. GATES,[2] SOPHIE J. MARRITT[1]
AND DAVID J. RICHARDSON[2]

[1]*School of Chemistry, University of East Anglia, Norwich, UK*
[2]*School of Biological Sciences, University of East Anglia, Norwich, UK*

5.1 INTRODUCTION

Redox enzymes are ubiquitous in nature. Their importance in respiratory processes is well recognized, and they make equally significant contributions to both the removal of cytotoxins and the elemental cycling of Fe, N, C, and S on the global scale. Consequently, there is much interest in providing quantitative descriptions of redox enzyme activity. In this regard, the ability of many enzymes to adsorb on an electrode surface in an orientation that allows for a direct and facile exchange of electrons is of note (Fig. 5.1). The flow of electrical current supported by the enzyme film can be quantified in solutions of infinitely variable chemical composition, for a range of temperature and across a broad window of electrochemical potential. In addition, modulations of activity arising from a change in any one of these variables can be resolved across the time domain. Consequently, the protein film electrochemistry (PFE) provides a powerful tool for resolving enzyme activity to inform descriptions of cellular function and their application in biotechnology.

Ideally, the enzyme adsorbs homogeneously on the electrode and as a monolayer, or less, of electroactive material. The oxidation state of the entire film is then rapidly defined and subsequently manipulated by variation of the electrode potential, and the corresponding flow of electrical current has a sign and magnitude describing the

Electrochemical Processes in Biological Systems, First Edition. Edited by Andrzej Lewenstam and Lo Gorton.

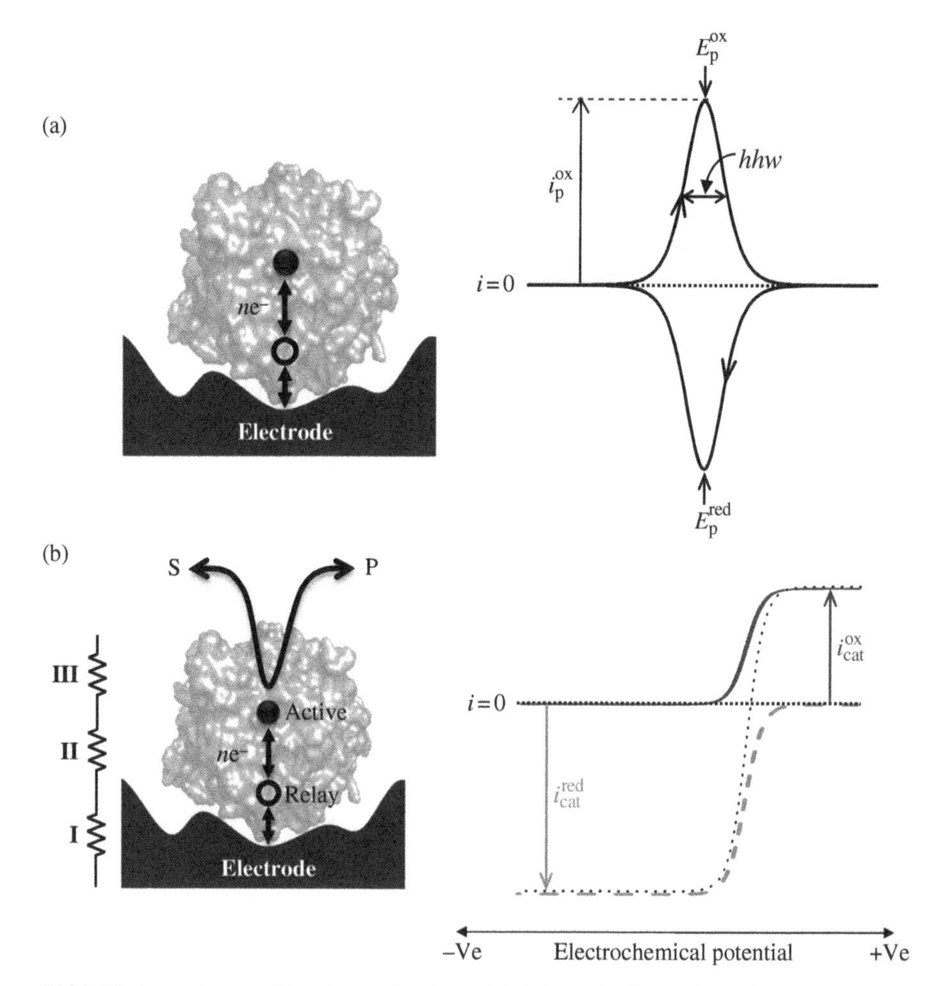

FIGURE 5.1 Enzyme film electrochemistry. (a) Schematic illustration of the nonturnover scenario for an enzyme with two redox cofactors, namely, the relay (open circle) and active site (closed circle) as indicated (left). The cyclic voltammetric response of current (i) versus potential (E) anticipated for a single redox cofactor undergoing reversible electron transfer where E_p indicates the peak potential for the oxidative (ox) and reductive (red) peaks and *hhw* indicates the half-height width of the oxidative peak (right). (b) Schematic illustration of the catalytic scenario for the enzyme in (a) (left). The catalytic current profiles that would be observed during cyclic voltammetry (right) for oxidative (continuous), reductive (dashed), and bidirectional (dotted) catalysis. The oxidative and reductive catalytic currents (i_{cat}) for each response are indicated. See text for details.

electron transfer capabilities of the enzyme. Significantly, this provides a powerful complement to long-standing biochemical analyses where the impact of electrochemical potential on catalysis was usually overlooked. PFE also has the added benefit of requiring only miniscule amounts, typically subpicomole, of the enzyme of interest.

The utility of the information afforded by PFE is readily demonstrated by its relatively rapid assimilation into the main stream of redox protein analysis. Suffice to say here that PFE has been applied successfully to proteins with diverse function and subunit and cofactor composition and from a variety of organisms as highlighted in several substantive reviews [1–3]. The possibilities for molecular resolution of redox enzyme activity through PFE with simultaneous *in situ* spectroscopy are excluded from this discussion [4, 5]. So too are the implications for orchestrated response to change of cellular redox poise and flux control contributions to cellular metabolism [6]. We also omit from our discussion the possibility for PFE to form a basis for the development of amperometric biosensors [7, 8]. Instead, we focus here on the wealth of biochemical information to be gained purely from the electrochemical experiment and illustrate this with examples drawn from our own research interests. However, first, we outline some of the practicalities to be considered when performing PFE of enzymes.

5.2 THE FILM ELECTROCHEMISTRY EXPERIMENT

PFE is typically performed using a cell configuration incorporating reference, counter, and working electrodes. One example of a glass electrochemical cell frequently used for PFE in our laboratories is illustrated in Figure 5.2. The reference electrode is housed in a side arm connected to the sample chamber by a Luggin capillary that minimizes physical mixing of the solutions in the sample and reference chambers while defining a sensing point for the reference electrode near the surface of the

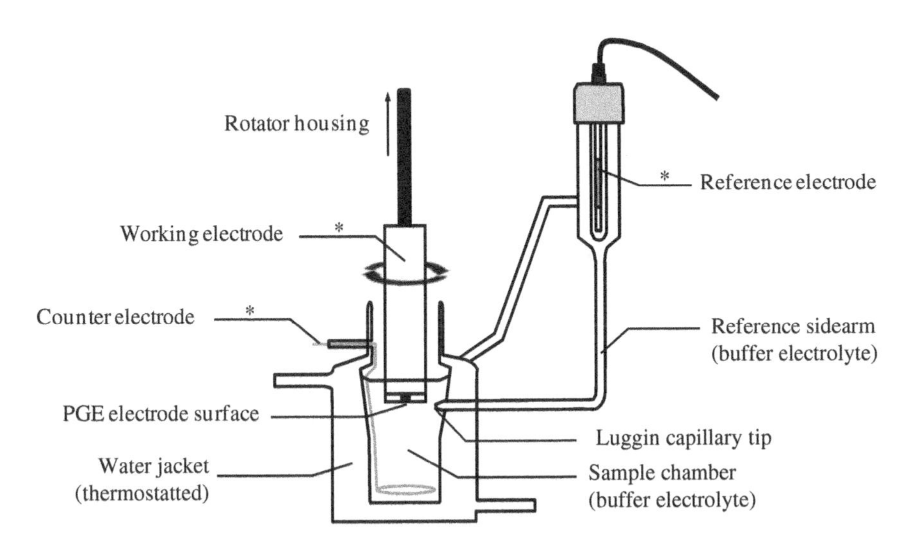

FIGURE 5.2 The three-electrode cell configuration typically used for protein film electrochemistry experiments with a PGE working electrode surface. The electrode–potentiostat interface is denoted by asterisks.

working electrode. The sample chamber contains the platinum wire counter electrode and is shaped to minimize solution turbulence during rapid working electrode rotation as is frequently required during studies of redox enzyme catalysis. The working electrode is prepared from a cylinder (3 mm diameter) of the desired electrode material mounted onto a brass rod with silver-loaded epoxy and subsequently sealed into a nylon sheath with epoxy resin. The sheath allows the working electrode to be mounted onto the shaft of a rotor that makes electrical contact with the working electrode via the brass rod. During experiments, the working electrode is positioned in the sample chamber at the level of the Luggin capillary and with its exposed face in contact with a 3.5 ml solution (Fig. 5.2). The entire electrochemical apparatus is placed inside a Faraday cage within a N_2-filled chamber to minimize electrical noise and the presence of oxygen, respectively. The temperature of the sample chamber is maintained by means of a jacket filled with circulating water under thermostat control.

Several electrode surfaces have been found to support nondestructive yet robust protein adsorption [1–3]. We favor either the edge or basal plane of pyrolytic graphite. These surfaces have been successful in adsorbing a variety of enzymes such that they retain similar redox and/or catalytic properties to those described in solution, which are the usual criteria for confirming the functional integrity of adsorbed protein. However, PFE is now sufficiently mature as a technique to recognize that cyclic voltammetry yielding well-behaved catalytic or nonturnover responses of the types we describe in the following should be taken as reflecting the behavior of a healthy enzyme even in the absence of independent corroboration by traditional biochemical methods. By this, we mean that exceptionally sharp and/or irregularly shaped features are unlikely to be present in the cyclic voltammetry of enzymes that have adsorbed without loss of structure or the presence of coadsorbing impurities.

To present a fresh graphite surface for protein adsorption, the electrode is lightly abraded immediately prior to experiments. Typically, this employs "polishing" with an aqueous 0.3 μm Al_2O_3 slurry that may be proceeded by abrasion with "wet and dry abrasive paper" of fine grade (English Abrasives & Chemicals Ltd, UK). This treatment exposes highly reactive edges of graphene sheets that become coated in a range of C–O functionalities, giving the surface an acidic, hydrophilic character that promotes a physical association with protein surfaces of a complementary character. In addition to its chemical heterogeneity, the graphite surface is microscopically rough, having an absolute surface area ca. 10^4-fold greater than the geometric surface area [9]. In addition, ca. 10–20% of the surface features that confer this extra surface area have widths greater than 10 nm, which is ample space to accommodate the molecular dimensions of many redox-active proteins and protein complexes. Following the mechanical abrasion of the graphite surface, it is rinsed and sonicated in high-purity water to remove traces of alumina, dried with a tissue, and covered with a few microliters of ice-cold, protein-containing solution from a chilled glass syringe. The protein solution is removed after a few seconds, and the electrode mounted on the rotor and then positioned in the electrochemical cell for experimentation.

The optimum composition of the solution for enzyme adsorption is usually found by exploring a range of ionic strengths, pH, buffer, and electrolyte identity in addition to enzyme concentration and where the inclusion of a coadsorbate, for example, positively charged neomycin, may help to secure an electroactive protein film. The time invested in exploring these conditions is rewarded when it comes to performing and analyzing experiments that probe enzyme function since both are greatly facilitated by the reproducible formation of stable films. Once the film is prepared, its catalytic performance can be readily assessed in a range of conditions using the cell illustrated in Figure 5.2. The design allows aliquots of reagents to be injected directly into the experimental solution from concentrated stock solutions, or the sample solution can be removed completely and replaced with a fresh solution of choice. Modification of the basic cell design allows for rapid equilibration of the sample chamber with gaseous agents of varying partial pressures [10]. In essence, the enzyme film can be subject to instantaneous dialysis by a variety of means and to great advantage as we illustrate later.

5.3 ENZYME FILM ELECTROCHEMISTRY: THE BASICS

When initiating studies of a new enzyme, we find cyclic voltammetry to be the most effective means to assess the electrochemical activity of a protein-exposed electrode. The voltammetric response of the protein is readily obtained from the experimental data by subtracting the response from an identical experiment in which the electrode was not exposed to protein. The presence of a film of a healthy, that is, functionally intact, electroactive protein will then be indicated by one of two general types of response. In one case, discrete peaks describing oxidation (positive current) and reduction (negative current) of a finite number of redox centers are observed as the molecules exchange electrons with the electrode, for example, Figure 5.1a. These are often referred to as nonturnover, that is, noncatalytic, signals. For sufficiently slow scan rates, typically of the order of 10 mV s^{-1}, these peaks will conform closely to the shape predicted for homogeneous arrays of adsorbed, isolated redox centers in Nernstian equilibration with the electrode potential [11]. There will be little separation of the oxidative $\left(E_{\mathrm{p}}^{\mathrm{ox}}\right)$ and reductive $\left(E_{\mathrm{p}}^{\mathrm{red}}\right)$ peak potentials whose average is the reduction potential of the redox couple engaged. The magnitude of the oxidative and reductive peaks will be equal at any given potential, and the half-height width (hhw in mV) of each peak will reflect the electron stoichiometry (n) of that couple where $n = 90.6 \div$ hhw at 25°C. For higher scan rates, an increased separation of the peak potentials and/or the narrowing of one peak relative to the other reflects a shift to measurements under nonequilibrium conditions such that the rate constants describing interfacial electron transfer (IET) and/or protein-induced gating of electron transfer can be quantified [12]. However, for all scan rates, the areas of the oxidative and reductive peaks will be equal since these reflect the moles of electrons, respectively, leaving and entering the enzyme film.

We will return briefly to the analysis of nonturnover signals at the end of this chapter, but for now, we turn our attention to the second general type of cyclic voltammogram displayed by enzyme films (Fig. 5.1b). In this case, the Faradaic current is nonzero at one or both extremes of the potential window, and its sign at any given potential is independent of the direction of the voltammetric sweep, whether to more negative or positive potentials. Here, the current describes a continuous flow of electrons through the enzyme film due to steady-state catalytic redox transformation of molecules in the sample chamber. Positive currents describe substrate oxidation and negative currents substrate reduction (Fig. 5.1b). Such currents may be observed by design following the addition of a defined concentration of substrate to the sample chamber. Alternatively, these catalytic currents may arise "by accident" due to the presence of previously unrecognized substrates, or impurities, in the buffer electrolyte. In either case, the stage is now set for detailed analysis of the catalytic performance of the enzyme.

It is usually of interest to define the kinetic parameters describing steady-state turnover, namely, k_{cat}, the maximum turnover frequency, and K_M, the Michaelis constant, for a given condition. Assuming a Michaelis–Menten description of enzyme-catalyzed conversion of S to P (Scheme 5.1),

$$\text{Enzyme} + \text{S} \leftrightarrow \text{enzyme} : \text{S} \rightarrow \text{enzyme} + \text{P}$$ **SCHEME 5.1**

the equation describing the rate of product formation from a consideration of steady-state enzyme kinetics can be equated to that describing the corresponding electron flow as catalytic current, i_{cat}, arising from the same process:

$$\frac{d[\text{P}]}{dt} = \frac{k_{cat}\Gamma_{total}[\text{S}]}{K_M + [\text{S}]} = \frac{i_{cat}}{nFA} \tag{5.1}$$

where Γ_{total} is the surface density of electroactive enzyme, F the Faraday constant, and A the electrode area. Rearranging Equation (5.1) yields an expression for the catalytic current as function of substrate concentration in terms of the kinetic parameters for steady-state catalysis:

$$i_{cat} = \frac{nFAk_{cat}(E)\Gamma_{total}[\text{S}]}{K_M(E) + [\text{S}]} = \frac{i_{max}(E)[\text{S}]}{K_M(E) + [\text{S}]} \tag{5.2}$$

where the dependence of certain parameters on the electrochemical potential (E) is introduced to recognize that the catalytic mechanism and associated rates are determined in part by the oxidation state of the enzyme [1, 2, 13, 14]. Experimentally, the quickest route to define i_{max} and K_M at a defined potential is through chronoamperometry. Here, the enzyme film is poised at a potential of interest, and i_{cat} is measured as defined aliquots of concentrated substrate solution are titrated into the sample chamber. The variation of i_{cat} with substrate concentration is then readily fitted to Equation (5.2) to yield i_{max} and K_M as illustrated for nitrate reduction by the enzyme NarGH from *Paracoccus pantotrophus* in Figure 5.3a.

The protocol described earlier is straightforward, but rigorous quantification of enzyme kinetics requires acknowledging the possible complexities intrinsic to

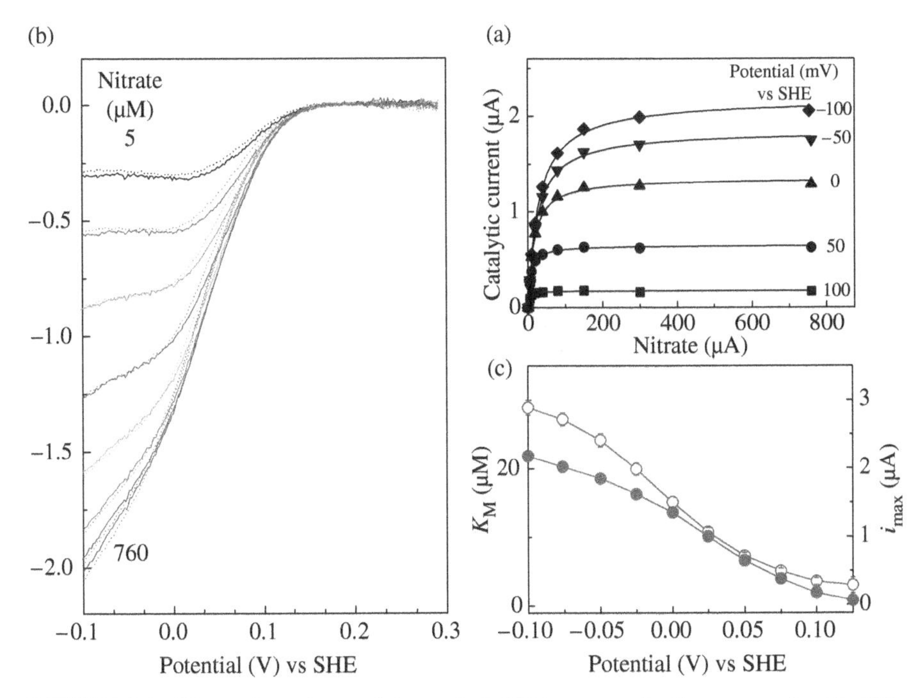

FIGURE 5.3 Nitrate dependence of the catalytic PFE response for *P. pantotrophus* NarGH. (a) Variation of catalytic current magnitude with nitrate concentration at the potentials as indicated. (b) Cyclic voltammograms recorded at 20 mV s^{-1} in the presence of nitrate for experiments performed in 25 mM Mes and 50 mM Na_2SO_4, pH 6, 20°C, with electrode rotation at 3000 rpm. Nitrate concentrations are 5, 10, 20, 40, 80, 150, 30, and 760 μM. (c) Dependence of K_M and i_{max} on electrochemical potential.

PFE. As a first step, any loss of catalytic current magnitude over time due to protein desorption or inactivation should be accounted for. Since such loss of magnitude is typically approximated by a first-order process, an appropriate correction to the current magnitude is readily performed. Conceptually, more challenging is the need to recognize the multiple steps that contribute to electrocatalysis and to identify those that are rate limiting. These steps are outlined in Figure 5.1b as (I) IET between the electrode and adsorbed protein, (II) processes intrinsic to turnover at the active site and internal electron transfer events between centers of the redox enzyme, and (III) substrate/product exchange between the protein film and bulk solution, termed mass transport [15]. Of these, it is (II) that is of interest to elucidate the biochemical properties of the enzyme, and it is fortunate that rate-limiting contributions from (I) and (III) can frequently be minimized through experimental design or mathematically accounted for as we outline in the following.

With regard to substrate/product mass transport, it is important to recognize that during catalysis by an enzyme film, the substrate is removed, and product generated, in the vicinity of the electrode. In a static solution, the substrate will diffuse to the vicinity of the electrode from bulk solution in an attempt to maintain the concentration of substrate around the enzyme. However, rates of substrate diffusion frequently fail to match those of catalysis such that the substrate concentration falls around the enzyme and so too the catalytic current as the experiment proceeds. To counter this effect, the electrode may be rotated rapidly, upward of 2000 rpm, such that forced convection maintains the defined bulk solution concentration around the enzyme. For very active enzymes and/or densely populated electrode surfaces, it may not be possible to rotate the electrode at a sufficiently high rate to alleviate all limitation by mass transport. In such cases, the Koutecky–Levich equation allows extrapolation of the rotation rate dependence of i_{cat} to yield the current at infinitely high rotation rate and so a value that is free from mass transport limitation (e.g., [15]). By measuring the catalytic current at three or four rotation rates for each substrate concentration, any limitation from mass transport is readily identified within the chronoamperometric experiment. The data can then be processed appropriately prior to plotting the variation of i_{cat} in the absence of mass transport limitation versus substrate concentration to define the values of K_M and i_{max} (e.g., [15]).

For PFE to provide insight into intrinsic features of enzyme catalysis, the catalytic current should also be free from limitation by the rate of IET. Here, the Butler–Volmer formulism provides the conceptual starting point since the reorganization energy relevant to describing a protein redox couple is typically 0.7 eV such that the more complex Marcus treatments may be unnecessary [16]. This formulism describes an exponential increase in IET rate constant with overpotential such that the catalytic current is most likely to be free from limitation due to IET kinetics at high overpotential. However, this condition will also ideally be met over a wide range of potential, including negligible overpotential, such that it is possible to take full advantage of PFE to quantify the intrinsic activity of the enzyme across a wide potential window. When nonturnover signals are resolved, the scan-rate dependence of the peak separation provides direct access to the rate constant describing IET at any given potential (*vide supra*). Nonturnover signals also define the population of electroactive enzyme such that absolute rates of catalysis can be abstracted from the catalytic currents. Thus, in such cases, it is possible to make a quantitative assessment of the extent to which rates of IET limit catalysis at any given electrode potential.

Unfortunately, it is more often the case that electrocatalytically active enzyme films fail to yield nonturnover signals. This is a situation that most likely arises from the low population of electroactive enzyme present in such experiments and that requires indirect assessment of whether the catalytic response is limited by rates of IET. This assessment is usually made by inspection of the cyclic voltammetric response of the enzyme [17]. If at moderate to high overpotential the catalytic currents take a constant value, as illustrated in Figure 5.1b, IET is unlikely to be rate limiting. By contrast, catalytic currents that fail to reach a limiting value at high overpotential are indicative of limiting IET. The result is a sloping relationship between i_{cat} and potential that may become more pronounced at higher substrate concentration and/or temperature as the increase in the intrinsic rate of catalysis is much greater than any increase in the rate of IET that may occur in parallel. When such sloping i_{cat} versus potential plots are observed, they can be described by considering that the response arises from a population of electroactive enzyme molecules displaying a distribution of interfacial rate constants due to a distribution of distances separating the electrode surface and enzyme redox cofactor. Fortunately, the slope of the catalytic response is proportional to the catalytic rate in these circumstances such that rigorous analysis can still be performed to yield the kinetic parameters as needed. However, it may also be possible to minimize this phenomenon experimentally by exploring a range of conditions for adsorption on graphite or employing more uniform electrode surfaces such as provided by atomically flat gold modified with alkane thiols.

5.4 MOLECULAR DETERMINANTS OF ENZYME ACTIVITY

Access to a variable, yet defined, electrode potential allows definition of K_M and i_{max} across a broad and continuous window of potential. A series of chronoamperometry experiments or more conveniently cyclic voltammetry, for example, Figure 5.3b, readily reveal how electrochemical potential modulates both kinetic parameters (Fig. 5.3c). Such modulations are relevant to integrating descriptions of enzyme activity into flux analyses of cellular kinetics since both parameters contribute to defining metabolic control over the total cellular flux through a particular enzyme [6]. However, the variation of activity with potential also provides a direct route to insights into the mechanism of catalysis. For NarGH, a peak of activity is resolved across the electrochemical potential domain in low nitrate concentrations that is lost when the nitrate concentration is increased (Fig. 5.3b). This observation prompted a reevaluation of the catalytic mechanism. The voltammetry may arise either from competition between an electron and nitrate for binding to the active site cofactor in its Mo(V) oxidation state or from the consequence of redox transformation of a site remote from the active site [6, 13]. It has not yet been possible to distinguish between these mechanisms. However, for *Escherichia coli* NrfA, a nitrite reductase, consideration of cofactor reduction potentials suggests that reduction of a heme cofactor remote from the active site is the most likely explanation for the attenuated activity observed at low potential (Fig. 5.4a) [6].

An additional means to exert metabolic control over enzyme activity is through interactions with molecules that inhibit or activate catalysis. PFE readily defines these interactions as the catalytic current magnitude from a defined substrate concentration will increase and decrease on addition of an activator and inhibitor, respectively. Returning the film to the initial solution conditions will establish whether the effect is reversible or results from irreversible covalent modification of the enzyme. For reversible modulators, their effects can be quantified and classified, for example, as competitive, mixed, etc., through Michaelis–Menten analyses defining K_M and i_{max} at several concentrations of the modulator [18–20]. As described in the previous section, such experiments are best performed with rapid electrode rotation to ensure effective dispersion of species in solution and to account for substrate mass transport limitations as necessary. During chronamperometry, a sharp, steplike response of the current versus time trace on addition of substrate indicates rapid equilibration of the enzyme, substrate, and modulator for each condition. Curvature of the current–time trace reports a slow relaxation to the steady-state condition that is most

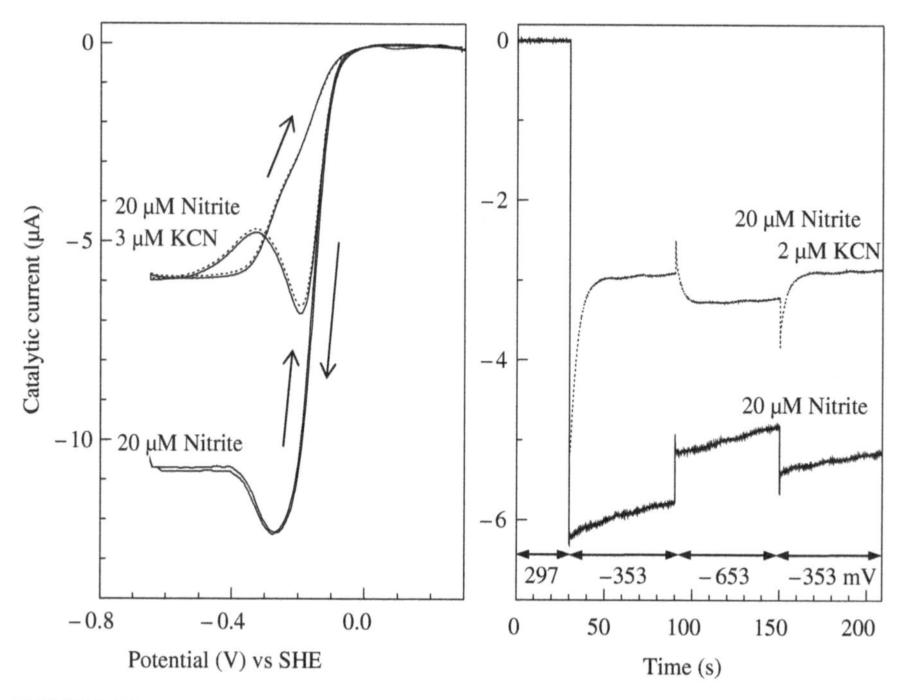

FIGURE 5.4 Inhibition of the nitrite reductase activity of *E. coli* NrfA by cyanide as revealed by PFE. (a) Cyclic voltammogram at 30 mV s^{-1} in the presence of 20 μM nitrite and consecutive voltammograms after the addition of 3 μM KCN as indicated. (b) Chronoamperometry in 20 μM nitrite with and without KCN as indicated. Both experiments performed in 50 mM Hepes and 2 mM $CaCl_2$, pH 7, 20°C, with electrode rotation at 3000 rpm.

likely due to relatively slow release of modulator from the enzyme when control experiments confirm that this is not due to reagent mixing [20]. In such cases, PFE can provide immediate access to rate constants describing enzyme–modulator interactions as we discuss in the following.

From the preceding discussion, it is a simple extrapolation to recognize that PFE can resolve thermodynamic and kinetic aspects of enzyme–modulator interactions across the electrochemical potential domain. Frequently, the results are complex and reflect a dependence of the affinity for the modulator and/or substrate on enzyme oxidation state [20, 21]. An example is found in the cyanide inhibition of nitrite reductase activity of *E. coli* NrfA (Fig. 5.4). Cyanide inhibition of NrfA nitrite reductase activity is immediately revealed by comparison of PFE in the absence and presence of cyanide. Hysteresis in the catalytic current potential profiles of the cyclic voltammetric sweeps demonstrates greater inhibition on returning to positive potentials than was noted on the excursion to negative potentials (Fig. 5.4a). Thus, cyanide binds relatively slowly and with higher affinity to reduced forms of NrfA than to the fully oxidized enzyme. This picture is reinforced by chronoamperometry employing a series of potential steps, or pulses, to explore the activity of NrfA in experiments with and without cyanide present (Fig. 5.4b). On stepping from +297 to −353 mV in the presence of cyanide, slower relaxation to steady-state catalysis is seen than in the absence of cyanide. Stepping to −653 mV shows increased in activity in the presence of cyanide, whereas in the absence of cyanide, the activity is decreased. Thus, the relative steady-state currents at −353 and −653 mV are reversed by the presence of cyanide. On returning to −353 mV, there is slow relaxation back to the level of inhibited steady-state current, indicating that cyanide was released from the enzyme poised at −653 mV. Thus, the potential pulses reveal the kinetics of modulator binding and release from an enzyme, and plots of current versus time can be analyzed by standard kinetic software to provide rate constants for the steps involved as elegantly illustrated by studies of hydrogenases (e.g., [22]) and fumarate reductase [23].

The interested reader is referred to a more detailed discussion of the PFE of NrfA to see how variation of voltammetric scan rate and cyanide/nitrite concentration in cyclic voltammetry and chronoamperometry resulted in a mechanistic interpretation of the interactions between these molecules [20]. Here, our purpose is simply to highlight the ease with which such intricacies are revealed by PFE. Hysteretic cyclic voltammetry has been seen in other situations where substrate mass transport is not limiting [23–25]. Here, the hysteresis is confined to the first voltammetric cycle, and it describes the need for redox cycling of the enzyme sample in order to activate it to display the steady-state response that is described in the subsequent cyclic voltammetry. The activation may be due to the release of an inhibitor and/or a conformational change triggered by accessing an oxidation state not present in the equilibrated sample with which the film was prepared. It may arise as a consequence of the enzyme purification protocol, an incomplete maturation of the enzyme, or a natural defense system put in place to protect the enzyme against exposure to damaging agents such as oxygen that are encountered during purification. In closing, we note that for enzymes with multiple substrates, the impact of competition between these substrates can provide a powerful route to catalytic insight. For NrfA, the much slower rate of sulfite compared to nitrite reduction allowed for the affinity of the former to be resolved for multiple oxidation states of the enzyme through PFE [21]. The results contributed to a more detailed description of catalysis than could be afforded by more traditional assays methods.

5.5 NONTURNOVER SIGNALS

Many enzymes support turnover at rates in excess of 100 s^{-1} such that the electrocatalytic population need only be on the order of 0.1 pmol cm^{-2} for informative experimentation. In the absence of substrate, this electroactive population is unlikely to give rise to detectable nonturnover responses on graphite electrodes. This is largely due to broad baseline features arising from electrochemistry intrinsic to the chemically heterogeneous graphite surface that masks small nonturnover peaks. The intrinsic electrochemical response of gold surfaces is much smaller than for graphite such that it may be easier to detect signals from a smaller population of electroactive enzyme adsorbed on these electrodes [26]. Nevertheless, a number of enzymes do adsorb on graphite at sufficiently high coverage, of the order of 10 pmol cm^{-2}, to resolve clear nonturnover peaks (e.g., [1–3, 27–29]). One example is the soluble domain of the quinol dehydrogenase, NapC, from *P. pantotrophus* (Fig. 5.5). This domain contains four *c*-type hemes, one of which serves as the site for oxidation of quinol with the extracted electrons being passed to the NapAB nitrate reductase during aerobic respiration [30, 31]. Cyclic voltammetry of a film of NapC reveals clear peaks for oxidation and reduction (Fig. 5.5). These peaks span a window of ca. 500 mV that is broader than that of ca. 200 mV expected from redox transformation of a single redox site exchanging one electron with the electrode. This, taken together with the shoulders seen on the flanks of the waves, supports the response arising from multiple centers within NapC, and given the proximity of neighboring hemes in this enzyme, it is likely that the voltammetry describes redox transformation of all four hemes as seen with other tetraheme cytochromes (e.g., [32]). At pH 7, the peaks for NapC redox activity are at more positive potentials than at pH 8, which indicates stabilization of the reduced enzyme by protonation. This can be quantified satisfactorily, given the inherent ambiguity of baseline subtraction, by assuming that each of the peak arises from four redox centers acting as independent sites for exchange

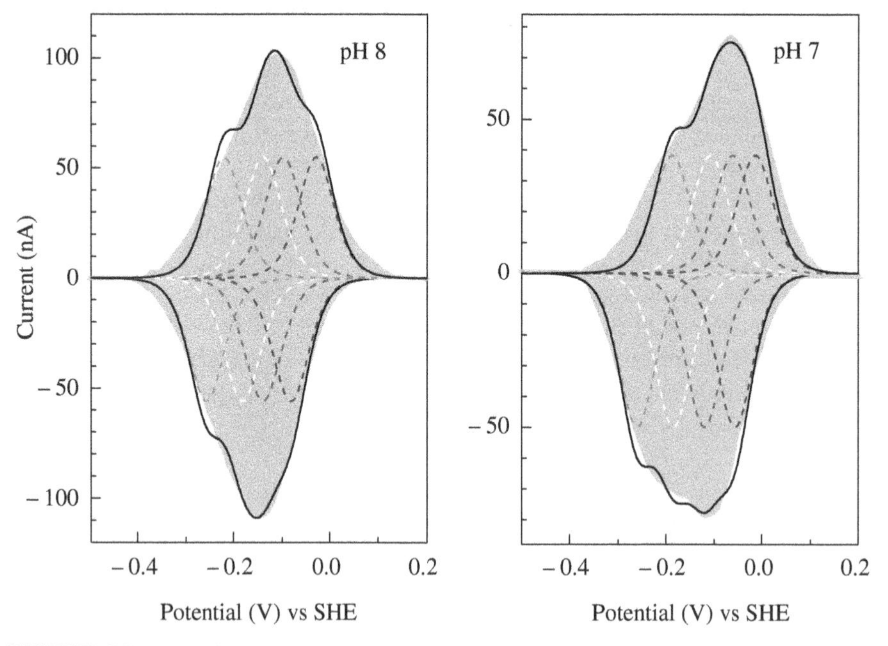

FIGURE 5.5 Nonturnover peaks displayed by the soluble domain of NapC during cyclic voltammetry of a protein film at the indicated pH values (gray area). Fit (continuous lines) of the peaks to the response from the sum of four independent centers acting as single-electron ($n = 1$) sites (broken lines). Reduction potentials of the four hemes taken as the average values from the oxidative and reductive fits are −35, −91, −146, and −222 (all ±10) mV at pH 7 and −56, −119, −160, and −239 (all ±10) mV at pH 8. Cyclic voltammetry performed at 50 mV s^{-1}, 20 °C.

of one electron (Fig. 5.5). Taking the average of the reductive and oxidative peak potentials to reflect midpoint potentials for the four hemes, these shift from −56, −119, −160 and −239 mV at pH 8 to −35, −91, −146, and −222 mV at pH 7 (all values ±10 mV). Further experiments will aim to resolve the role of each of these hemes in quinol oxidation.

5.6 CONCLUSION

The experimental setup for PFE is simple and versatile. Combined with a requirement for only miniscule amounts of enzyme, this allows for rapid interrogation of enzyme activity in numerous conditions. With a few examples from our research, we have illustrated how PFE may be used to resolve this activity in the concentration, electrochemical potential, and time domains. Most of all, we hope to have conveyed the versatility of PFE that is frequently able to resolve novel facets of redox enzyme activity.

ACKNOWLEDGMENTS

We would like to thank the present and former members of our research groups for their contributions to the work presented here and the UK Biotechnology and Biological Sciences Research Council for funding much of that work most recently through grants BB/G009228, 83/17233, 83/13842, and BBE0219991.

REFERENCES

1. Hirst, J. (2006) Elucidating the mechanisms of coupled electron transfer and catalytic reactions by protein film voltammetry, *Biochim.Biophys. Acta Bioenergetics* 1757, 225–239.
2. Léger, C., and Bertrand, P. (2008) Direct electrochemistry of redox enzymes as a tool for mechanistic studies, *Chem. Rev. 108*, 2379–2438.
3. Léger, C., Elliott, S. J., Hoke, K. R., Jeuken, L. J. C., Jones, A. K., and Armstrong, F. A. (2003) Enzyme electrokinetics: using protein film voltammetry to investigate redox enzymes and their mechanisms, *Biochemistry 42*, 8653–8662.
4. Marritt, S. J., Kemp, G. L., Xiaoe, L., Durrant, J. R., Cheesman, M. R., and Butt, J. N. (2008) Spectroelectrochemical characterization of a pentaheme cytochrome in solution and as electrocatalytically active films on nanocrystalline metal-oxide electrodes, *J. Am. Chem. Soc. 130*, 8588–8589.
5. Kemp, G. L., Marritt, S. J., Xiaoe, L., Durrant, J. R., Cheesman, M. R., and Butt, J. N. (2009) Opportunities for mesoporous nanocrystalline SnO_2 electrodes in kinetic and catalytic analyses of redox proteins, *Biochem. Soc. Trans. 37*, 368–372.
6. Gates, A. J., Kemp, G. L., To, C. Y., Mann, J., Marritt, S. J., Mayes, A. G., Richardson, D. J., and Butt, J. N. (2011) The relationship between redox enzyme activity and electrochemical potential-cellular and mechanistic implications from protein film electrochemistry, *PhysChem-ChemPhys 13*, 7720–7731.

7. Tasca, F., Zafar, M. N., Harreither, W., Noll, G., Ludwig, R., and Gorton, L. (2011) A third generation glucose biosensor based on cellobiose dehydrogenase from Corynascus thermophilus and single-walled carbon nanotubes, *Analyst 136*, 2033–2036.
8. Yakovleva, M., Buzas, O., Matsumura, H., Samejima, M., Igarashi, K., Larsson, P.-O., Gorton, L., and Danielsson, B. (2012) A novel combined thermometric and amperometric biosensor for lactose determination based on immobilised cellobiose dehydrogenase, *Biosens Bioelectron 31*, 251–256.
9. Blanford, C. F., and Armstrong, F. A. (2006) The pyrolytic graphite surface as an enzyme substrate: microscopic and spectroscopic studies, *J. Solid State Electrochem. 10*, 826–832.
10. Jones, A. K., Lamle, S. E., Pershad, H. R., Vincent, K. A., Albracht, S. P. J., and Armstrong, F. A. (2003) Enzyme electrokinetics: Electrochemical studies of the anaerobic interconversions between active and inactive states of *Allochromatium vinosum* [NiFe]-hydrogenase, *J. Am. Chem. Soc. 125*, 8505–8514.
11. Bard, A. J., and Faulkner, L. R. (2001) *Electrochemical methods: fundamentals and applications*, 2nd ed., John Wiley & Sons, Inc., New York.
12. Hirst, J., and Armstrong, F. A. (1998) Fast-scan cyclic voltammetry of protein films on pyrolytic graphite edge electrodes: Characteristics of electron exchange, *Anal. Chem. 70*, 5062–5071.
13. Anderson, L. J., Richardson, D. J., and Butt, J. N. (2001) Catalytic protein film voltammetry from a respiratory nitrate reductase provides evidence for complex electrochemical modulation of enzyme activity, *Biochemistry 40*, 11294–11307.
14. Heering, H. A., Hirst, J., and Armstrong, F. A. (1998) Interpreting the catalytic voltammetry of electroactive enzymes adsorbed on electrodes, *J. Phys.Chem. B 102*, 6889–6902.
15. Sucheta, A., Cammack, R., Weiner, J., and Armstrong, F. A. (1993) Reversible electrochemistry of fumarate reductase immobilized on an electrode surface. Direct voltammetric observations of redox centers and their participation in rapid catalytic electron transport, *Biochemistry 32*, 5455–5465.
16. Page, C. C., Moser, C. C., Chen, X. X., and Dutton, P. L. (1999) Natural engineering principles of electron tunnelling in biological oxidation-reduction, *Nature 402*, 47–52.
17. Léger, C., Jones, A. K., Albracht, S. P. J., and Armstrong, F. A. (2002) Effect of a dispersion of interfacial electron transfer rates on steady state catalytic electron transport in [NiFe]-hydrogenase and other enzymes, *J. Phys. Chem. 106*, 13058–13063.
18. Cornish-Bowden, A. (2004) *Fundamentals of enzyme kinetics*, Portland Press Ltd., London.
19. Gates, A. J., Richardson, D. J., and Butt, J. N. (2008) Voltammetric characterization of the aerobic energy-dissipating nitrate reductase of *Paracoccus pantotrophus*: exploring the activity of a redox-balancing enzyme as a function of electrochemical potential, *Biochem. J. 409*, 159–168.
20. Gwyer, J. D., Richardson, D. J., and Butt, J. N. (2004) Resolving complexity in the interactions of redox enzymes and their inhibitors: contrasting mechanisms for the inhibition of a cytochrome c nitrite reductase revealed by protein film voltammetry, *Biochemistry 43*, 15086–15094.
21. Kemp, G. L., Clarke, T. A., Marritt, S. J., Lockwood, C., Poock, S. R., Hemmings, A. M., Richardson, D. J., Cheesman, M. R., and Butt, J. N. (2010) Kinetic and thermodynamic resolution of the interactions between sulfite and the pentahaem cytochrome NrfA from *Escherichia coli*, *Biochem. J. 431*, 73–80.

22. Léger, C., Dementin, S., Bertrand, P., Rousset, M., and Guigliarelli, B. (2004) Inhibition and aerobic inactivation kinetics of *Desulfovibrio fructosovorans* NiFe hydrogenase studied by protein film voltammetry, *J. Am. Chem. Soc. 126*, 12162–12172.
23. Heering, H. A., Weiner, J. H., and Armstrong, F. A. (1997) Direct detection and measurement of electron relays in a multicentered enzyme: voltammetry of electrode-surface films of *E. coli* fumarate reductase, an iron-sulphur flavoprotein, *J. Am. Chem. Soc. 119*, 11628–11638.
24. Field, S. J., Thornton, N. P., Anderson, L. J., Gates, A. J., Reilly, A., Jepson, B. J. N., Richardson, D. J., George, S. J., Cheesman, M. R., and Butt, J. N. (2005) Reductive activation of nitrate reductases, *J. Chem. Soc. Dalton Trans, 3580–3586.*
25. Fourmond, V., Sabaty, M., Arnoux, P., Bertrand, P., Pignol, D., and Léger, C. (2010) Reassessing the strategies for trapping catalytic intermediates during nitrate reductase turnover, *J. Phys. Chem. 114*, 3341–3347.
26. Jeuken, L. J. C., and Armstrong, F. A. (2001) Electrochemical origin of hysteresis in the electron-transfer reactions of adsorbed proteins: Contrasting behavior of the "blue" copper protein, azurin, adsorbed on pyrolytic graphite and modified gold electrodes, *J. Phys. Chem. 105*, 5271–5282.
27. Clarke, T. A., Edwards, M. J., Gates, A. J., Hall, A., White, G. F., Bradley, J., Reardon, C. L., Shi, L., Beliaev, A. S., Marshall, M. J., Wang, Z., Watmough, N. J., Fredrickson, J. K., Zachara, J. M., Butt, J. N., and Richardson, D. J. (2011) Structure of a bacterial cell surface decaheme electron conduit, *Proc. Nat. Acad. Sci. 108*, 9384–9389.
28. Hartshorne, R., Jepson, B., Clarke, T., Field, S., Fredrickson, J., Zachara, J., Shi, L., Butt, J., and Richardson, D. (2007) Characterization of *Shewanella oneidensis* MtrC: a cell-surface decaheme cytochrome involved in respiratory electron transport to extracellular electron acceptors, *JBIC 12*, 1083–1094.
29. Hartshorne, R. S., Reardon, C. L., Ross, D., Nuester, J., Clarke, T. A., Gates, A. J., Mills, P. C., Fredrickson, J. K., Zachara, J. M., Shi, L., Beliaev, A. S., Marshall, M. J., Tien, M., Brantley, S., Butt, J. N., and Richardson, D. J. (2009) Characterization of an electron conduit between bacteria and the extracellular environment, *Proc. Nat. Acad. Sci. 106*, 22169–22174.
30. Roldán, M. D., Sears, H. J., Cheesman, M. R., Ferguson, S. J., Thomson, A. J., Berks, B. C., and Richardson, D. J. (1998) Spectroscopic characterization of a novel multiheme c-type cytochrome widely implicated in bacterial electron transport, *J. Biol. Chem. 273*, 28785–28790.
31. Cartron, M. L., Roldán, M. D., Ferguson, S. J., Berks, B. C., and Richardson, D. J. (2002) Identification of two domains and distal histidine ligands to the four haems in the bacterial c-type cytochrome NapC; the prototype connector between quinol/quinone and periplasmic oxido-reductases, *Biochemical J. 368*, 425–432.
32. Pulcu, G. S., Elmore, B. L., Arciero, D. M., Hooper, A. B., and Elliott, S. J. (2007) Direct electrochemistry of tetraheme cytochrome c_{554} from *Nitrosomonas europaea*: Redox cooperativity and gating, *J. Am. Chem. Soc. 129*, 1838–1839.

6

PLANT PHOTOSYSTEM II AS AN EXAMPLE OF A NATURAL PHOTOVOLTAIC DEVICE

WIESŁAW I. GRUSZECKI

Department of Biophysics, Institute of Physics, Maria Curie-Skłodowska University, Lublin, Poland

6.1 INTRODUCTORY REMARKS ON PHOTOSYNTHESIS

Life on our planet is driven by the energy of the sun, and photosynthesis is practically the exclusive process able to convert the energy of electromagnetic solar radiation to the forms that can be directly utilized by living organisms to drive their metabolic reactions [1]. Plant photosynthesis is also a source of molecular oxygen that we and most living organisms require for respiration. Since the pioneering experiments of Joseph Priestley carried out in England in 1770s, researches had to pass a long way to understand the relationship between illumination of a plant, oxygen evolution, and CO_2 fixation. Even today, several questions regarding realization, and in particular regulation of the photosynthetic reactions, at the molecular and submolecular level remain open and are a challenge for contemporary researchers. There are several interesting phenomena associated with the photosynthesis reactions, which stimulated research activity of physicists and biophysicists. Among such phenomena are certainly the nonradiative excitation energy transport and intermolecular electric charge transport. Some aspects of those two physical processes will be addressed in the following text and discussed, taking as an example the plant photosystem II (PSII), responsible for capturing light quanta, for the initial electric charge separation and

Electrochemical Processes in Biological Systems, First Edition. Edited by Andrzej Lewenstam and Lo Gorton.

for the light-driven electron and proton transport. Moreover, all those processes in PSII are associated with the photosynthetic water splitting, which results in the molecular oxygen evolution.

6.2 PHOTOSYNTHETIC EXCITATION ENERGY TRANSFER

The photovoltaic effect in a solid semiconductor, as we know from our solar cells, consists in the light-induced electric charge separation right in the place in where the incident light quantum has been absorbed. A very different strategy has been developed by photosynthesizing organisms during millions of years of biological evolution. The photosynthetic charge separation takes place only in the specialized reaction centers, which are the pigment–protein complexes containing very few pigment molecules [1]. Obviously, taking into account the typical daylight photon flux density, which is in the order of magnitude of 10^{-4} mol photons $m^{-2}s^{-1}$, one may not expect fluent and efficient capture of sunlight in the photosynthetic apparatus, owing to the fact that the photosynthetic reaction center proteins occupy relatively small fraction of the surface of the chloroplast membranes. In order to maximize light energy absorbed by the photosynthetic apparatus, plants synthesize and accumulate several pigment–protein complexes that are specialized in capturing of light quanta and transferring the electronic excitation energy toward the photosynthetic reaction centers. Owing to such an activity, hundreds of accessory pigments are involved in powering of a single photoreaction center. The plant light-harvesting pigment–protein complex of PSII, called LHCII, is the most abundant photosynthetic antenna and, at the same time, the most abundant membrane protein in the biosphere [2, 3] (see Fig. 6.1). It comprises approximately half of chlorophyll molecules on Earth. Single molecule of LHCII binds 8 molecules of chlorophyll *a*, 6 molecules of chlorophyll *b*, and 4 molecules of carotenoid pigments: 2 luteins, 1 violaxanthin, and 1 neoxanthin. The packing

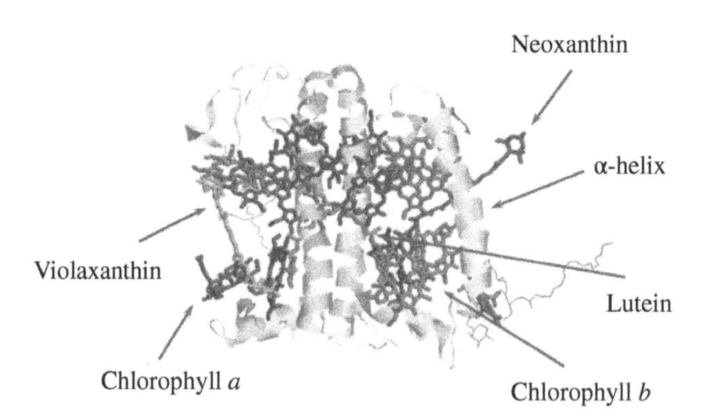

FIGURE 6.1 Model of the structure of the light-harvesting pigment–protein complex of photosystem II based on crystallographic data deposited in PDB (ID: 1RWT). The basic constituents of the complex are marked. See the text for more explanations.

density of the pigments is extremely high. The chlorophyll concentration, calculated per volume of the complex, corresponds to 0.3 M [4]. A chlorophyll solution in organic solvents, prepared at such a high concentration, is characterized by relatively high excitation quenching, manifested as dumping of a chlorophyll fluorescence, referred to as a "concentration quenching." Owing to the precise binding and localization of the pigment molecules within the protein bed, in terms of the interchromophore distance and orientation, the light energy is very efficiently absorbed by the pigments and without losses transferred out of the complex toward the reaction centers. The excitation energy transfer itself is also an amazing physical phenomenon. It used to be simplified in the several academic textbooks as an energy emission from the donor molecule associated with the act of the energy absorption by the acceptor molecule. In fact, the electron excitation energy transfer is a nonradiative process, despite the fact that several requirements, typical for radiative energy transfer, have to be satisfied. In fact, there are two essentially different nonradiative excitation energy transfer processes. One of them operates at relatively long distances, referred to as the Förster mechanism [5], and the other one, called the Dexter mechanism [6], takes place only when the donor and acceptor molecules remain practically in the van der Waals contact. This latter mechanism is, in fact, both the combined excitation and electron transfer: the transfer of the electron from the excited state of the donor molecule to the acceptor is associated with the transfer of the ground-state electron in the opposite direction—from the acceptor molecule to the donor. This mechanism operates within each separate photosynthetic antenna complex to exchange the excitation energy between the pigments. On the other hand, the long-distance nonradiative excitation energy exchange operates between the individual antenna complexes (e.g., between the monomers in the trimeric structure of LHCII) and is particularly important in the energy transfer between the antenna complexes and the reaction centers. This type of excitation energy transfer is referred to as a long-distance mechanism, which may also cause a kind of confusion. In fact, this process is highly distance dependent since the rate of the Förster-type energy transfer depends on R^{-6}, where R is a distance between the centers of the transition dipoles of the donor and acceptor. The Förster-type energy transfer has also one very interesting and important feature: despite the fact that it does not rely on a fluorescence emission, its rate depends formally on the fluorescence quantum yield. According to this, carotenoid pigments, which are characterized by very low fluorescence quantum yield, can act as accessory pigments exclusively owing to the dense packing with chlorophylls within the photosynthetic antenna complexes. The presence of the orange and yellow carotenoid pigments in the photosynthetic apparatus has an important advantage: they absorb light from the part of the spectrum that is not absorbed by chlorophylls. Owing to better coverage of the sunlight spectrum by combination of various photosynthetic pigments, the conversion of incident light can be more effective. There are at least two other important aspects of the presence of carotenoids in the photosynthetic apparatus. One of them is associated with a structural stabilization of the pigment–protein complexes [7] and the lipid membranes [8–10], and the other one is associated with protection against photooxidative damage. One of the most important physical mechanisms, via which carotenoid pigments can realize their photoprotective

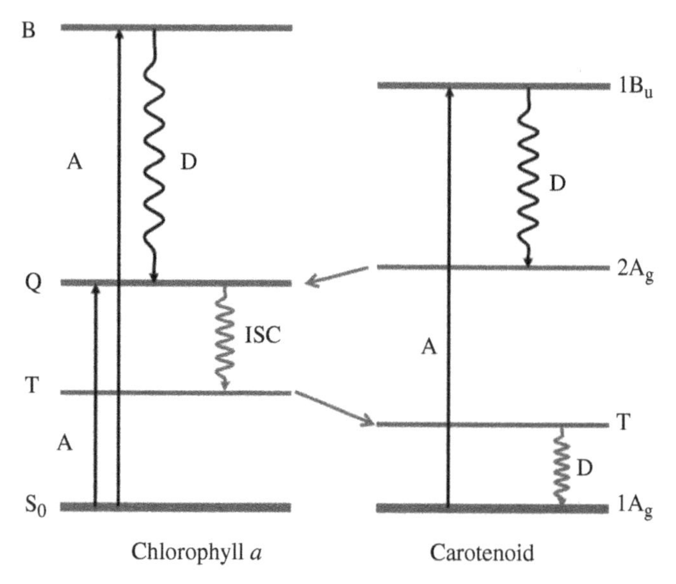

FIGURE 6.2 A comparative Jablonski diagram of the energy levels of chlorophyll *a* and of a carotenoid with indicated basic photophysical processes, including the singlet–singlet excitation energy transfer from carotenoid to chlorophyll (considered as antenna function) and the triplet–triplet excitation energy transfer from chlorophyll to carotenoid (considered as photoprotection). A stands for light absorption, D thermal energy dissipation, and ISC intersystem crossing. For simplicity, the bands B and Q of chlorophyll are represented by single energy levels. See the text for more explanations.

function, is quenching of the excited triplet states of chlorophylls [11]. This quenching is realized via the triplet–triplet nonradiative excitation energy transfer (see Fig. 6.2). The chlorophyll triplets are one of the most efficient photosensitizers, mediating generation of singlet oxygen, which is one of the most toxic species that can be produced in the photosynthetic apparatus. Relevant from the standpoint of photosynthesis is the excitation energy transfer between the singlet-excited energy levels. Such a transfer is faster (and therefore competitive) than the radiative deactivation of the lowest singlet-excited energy states ($\sim 10^{-9}$ s). Excitations that arrive to the photosynthetic reaction centers from the entire antenna systems give rise to the electric charge separation, the central mechanism of photosynthesis both from the photophysical and photochemical points of view.

6.3 PHOTOSYNTHETIC ELECTRON AND PROTON TRANSPORT

The photosensitive center of the PSII is a pigment called historically P680, due to the fact that it has its red most light absorption band with the maximum at 680 nm. In fact, this pigment is a chlorophyll *a* dimer, called also the special pair [1, 12, 13]. Excitation of P680 to P680* results in charge separation and electron transfer to the nearest

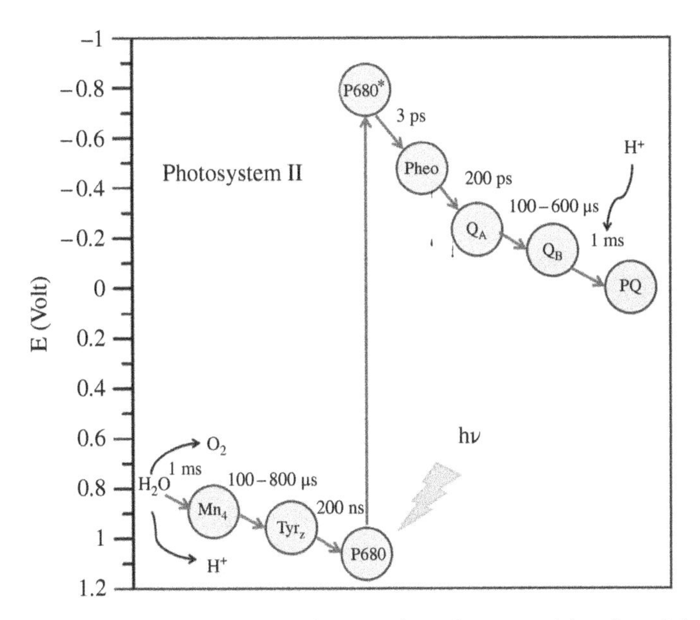

FIGURE 6.3 The schematic representation, on the redox potential scale, of the electron transport in photosystem II. Typical time-scale parameters of individual electron transfer steps are also presented. See the text for further explanations.

electron acceptor, called the primary electron acceptor, which is a molecule of pheophytin (Phe; see Fig. 6.3). This process is extremely fast and takes only 3×10^{-12} s [1, 12, 13]. The subsequent electron transfer from Phe^- to Q_A (protein-bound plastoquinone (PQ)) is also relatively fast and proceeds within 200×10^{-12} s [1, 12, 13]. As we can see from the diagram presented in Figure 6.3, the electron traveling along the redox potential cascade, from P680 to Phe and Q_A, loses its energy gained from the electronic excitation. On the other hand, in a relatively short time period, it is separated spatially from the initial donor ($P680^+$), which prevents the back reaction and the electric charge recombination. From Q_A^-, the electron is transferred to another PQ electron acceptor, Q_B, and this step of the electron transfer chain is much slower since it takes as long as $100–600 \times 10^{-6}$ s [1, 12, 13]. In contrast to Phe and Q_A, which are characterized by a one-electron reduction, PQ Q_B can accept two electrons during subsequent charge separation in P680. Double-reduced Q_B is an anion biradical, and therefore, it can bind two protons at the outer thylakoid membrane–water interface. The molecule Q_BH_2 is lipid soluble and electrically neutral, and therefore, it may be detached from its docking niche in the PSII reaction center protein and freely diffuse within the membrane as a plastohydroquinone pool (PQ) [1, 12, 13]. From this point, we may speak not only about the electron transfer chain but also about the proton transfer chain (see Fig. 6.4). PQ meets on its diffusion the cytochrome b_6f protein complex (Cyt b_6f), which can accept electrons from PQ at the inner membrane side, liberating protons into the thylakoid lumen. The process of electron exchange between

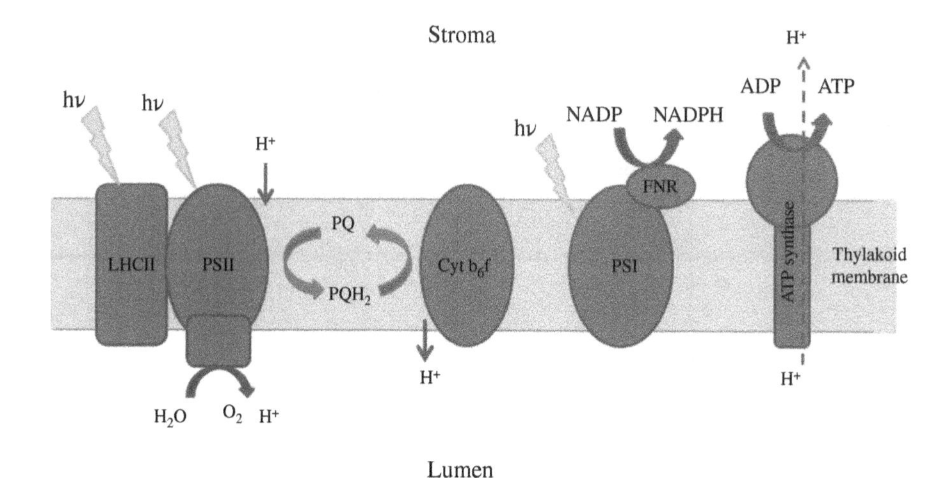

FIGURE 6.4 Schematic model of the photosynthetic electric charge transport within the thylakoid membrane. See the text for explanations.

the PQ and Cyt b_6f pair is in fact more complex and proceeds via the so called two-electron gate [1, 12, 13]. The reoxidized Q_B may again bind to its docking site in the PSII reaction center, and the electrons from Cyt b_6f complex can be transferred to the reaction center of the photosystem I (PSI) via the mobile small electron transfer protein plastocyanin (PC). Excitations of PSI are necessary to transfer electrons along the electron transfer chain elements within PSI to the enzyme ferredoxin–NADPH oxidoreductase (FNR), which catalyzes synthesis of NADPH. NADPH is called by some researchers a "reducing power" and is required for the photosynthetic CO_2 fixation and for many other metabolic reactions as a proton donor. Let us come back to the oxidized PSII special pair. The $P680^+$ is a very strong oxidizing agent, and therefore, very efficiently it pulls out an electron from the nearby donor, the amino acid residue tyrosine, called Tyr_Z. The PSII reaction center protein is associated at its donor side with the protein containing a cluster composed of four manganese atoms (Mn_4) and playing a role of the water-splitting enzyme called also the oxygen-evolving center [14]. Tyr_Z^+ can easily accept electrons from Mn, oxidizing step by step the cluster [1, 12, 13]. When all four Mn atoms of the cluster are positively charged, this provides a strong enough "oxidative power" to split simultaneously two water molecules, bound to the active enzymatic center. This reaction results in a concerted release of one molecule O_2 and four H^+ ions. The origin of such an idea is based on a beautiful experiment of Pierre Joliot, which he performed in the early 1960s of the last century [15]. Joliot noticed that photosynthetic oxygen evolution induced by a series of saturating light flashes, given 300 ms apart, displays a certain pattern of yield, which is characterized by a period of four (see Fig. 6.5). Such a phenomenon has been interpreted in 1970 by Bessel Kok with coworkers [17] in terms of the model based on four-step charge accumulation:

$$S_0 \rightarrow S_1 \rightarrow S_2 \rightarrow S_3 \rightarrow S_4 \rightarrow S_0 + O_2 \text{ evolution}$$

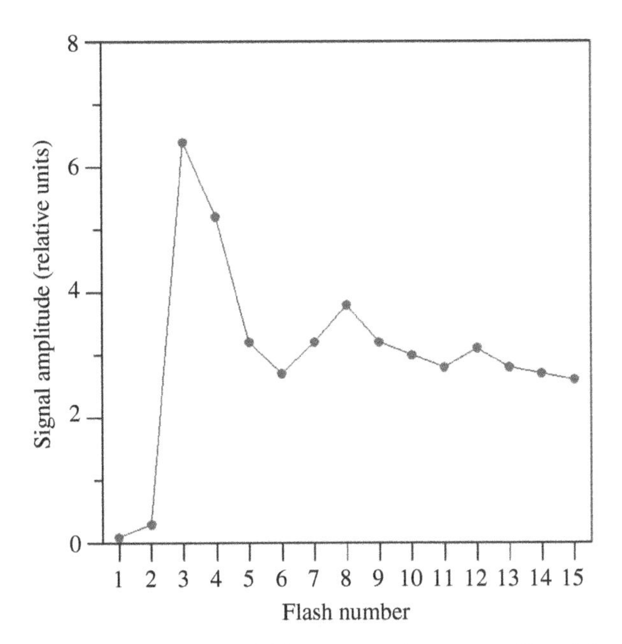

FIGURE 6.5 Flash-induced oxygen yield pattern in photosystem II particles isolated from tobacco, expressed as amplitudes of the polarographic signal. See Ref. [16].

In the model, S represents a redox state of oxygen-evolving center, and the subscript corresponds to a positive charge of the complex.

The water-splitting reaction is a source of proton gradient across the thylakoid membrane. As we remember, protons were also transferred from the outer into the inner thylakoid space by the PQ pool. According to the chemiosmotic mechanism, such a transmembrane proton gradient can drive the reaction of ATP synthesis by the ATP-synthase enzymatic complex located in the thylakoid membrane. In the outer thylakoid space (stroma), the pH remains at the level close to 8, but as a result of the photophysical and photochemical photosynthetic reactions, the pH of the thylakoid lumen drops to the level of 5. This means that the proton concentration inside the thylakoid space is higher by a factor of 1000! This number looks impressive, but, in fact, owing to a relatively small volume of the lumen, such a dramatic change in the H^+ concentration can be reached by a transfer of just few protons into the luminal space. On the other hand, due to the buffering capacity of the lumenal surface of the transmembrane pigment–protein antenna complexes, a high number of protons, which appear due to the water-splitting reaction and operation of the PQ shuttle, can be accommodated within the thylakoid membrane interior and drive photosynthetic ATP production. In this sense, the thylakoid membrane can be compared to a capacitor charged by operation of the light-driven electric charge separation (see Fig. 6.6).

A total of 8 light photons (4 in PSII and 4 in PSI) are required to transfer 4 electrons from 2 molecules of water to 2 molecules of NADPH. At the same time, the proton gradient generated can give rise to synthesis of at least 2 molecules of ATP.

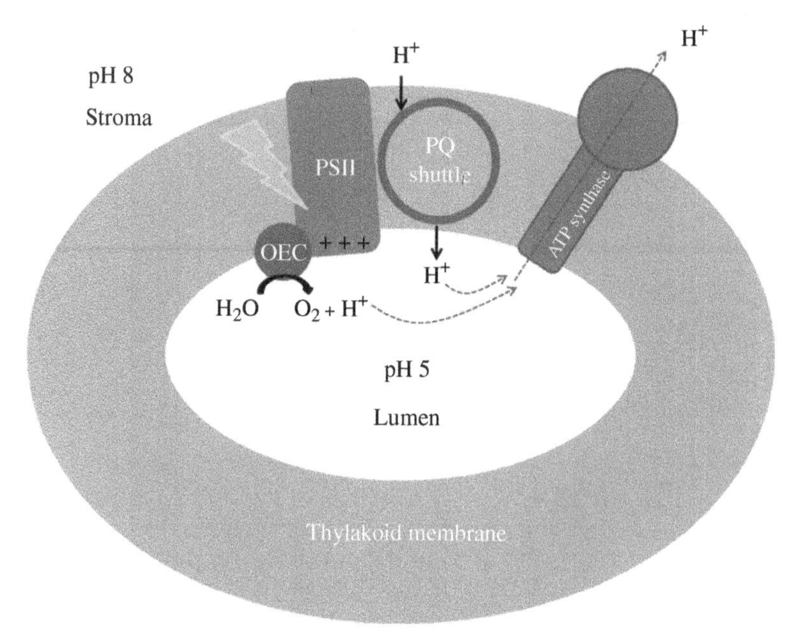

FIGURE 6.6 Simplified scheme of the thylakoid membrane with indicated proton flows. PSII, photosystem II; OEC, oxygen-evolving complex; PQ, plastoquinone pool.

Similarly as NADPH, ATP can be utilized not only to power the photosynthetic CO_2 fixation but also to provide energy for numerous other biochemical reactions. The effective energetic yield of photosynthesis is not considered to be much higher than 30%, also owing to the absorption of more energetic shorter wavelengths than those directly utilized to drive photosynthetic charge separation. On the other hand, the quantum yield of the photophysical and photochemical processes in the PSII reaction center is practically 100%. This efficiency makes the PSII reaction center a unique and perfect photovoltaic device, optimized by nature during millions of years of biological evolution.

6.4 PERSPECTIVES OF BIOMIMETIC APPLICATIONS

The results of the research on structure of the photosynthetic apparatus inspired the researches over the past decades to mimic photosynthesis and to construct artificial devices able to convert light energy to other forms, including electric current. One of the approaches was directly based on chlorophyll molecules being efficient light absorbers (the extinction coefficient ~10^5 Mol^{-1} cm^{-1} at the absorption maximum) and covering broad part of the sunlight spectrum. Chlorophyll molecules were placed into the artificial lipid membranes [18] and other model systems such as monomolecular layers [19–21]. Unfortunately, it appeared that despite the reasonable efficiencies, the chlorophyll-based solar cells were not stable enough owing to the fact that pigments were subjected to relatively quick photooxidation. The device stability has not been also much improved by admixing carotenoids, known to play a role of photoprotectors *in vivo* [22]. This has shown again that not only composition but also a precise construction (at the molecular level) is crucial for efficient, fluent and durable light-induced electric charge separation in biomimetic systems. Advances in the crystallographic techniques, in recent years, enabled to gain precise information on a structure of the photosynthetic apparatus of many organisms at atomic resolution. This, in principle, could facilitate construction of biomimetic solar cells, but it rather has shown the enormous complexity and precision one is going to face in order to approach the efficiency and other "technical" parameters of the natural photosynthetic apparatus while engineering photosynthesis-like photovoltaic devices. Another approach, which seems to be quite popular in recent years, is to adopt light-harvesting antenna or other intact proteins isolated from the photosynthetic apparatus, as elements of constructed photovoltaic devices. For example, Paulsen with coworkers has demonstrated immobilization of LHCII complexes on functionalized surfaces of gold, which resulted in efficient excitation energy exchange between the light-harvesting complex and the surface plasmon excitations [23]. The authors have concluded that this could be a prerequisite for using photosynthetic pigment–protein complexes as light harvesters in photovoltaic devices. Interaction of intact photosynthetic light-harvesting complexes with plasmonic platforms and metal nanoparticles is also a subject of interest in the aspect of "improving" the design and functioning of artificial light-harvesting systems [24]. One of the most simple photovoltaic device, based on pigment–protein complexes isolated from the photosynthetic apparatus, would be a Grätzel-type cell [25] in which the intact photosynthetic reaction centers or antenna complexes adsorbed on TiO_2 nanoparticles play a role of photosensitizing dyes (see Fig. 6.7).

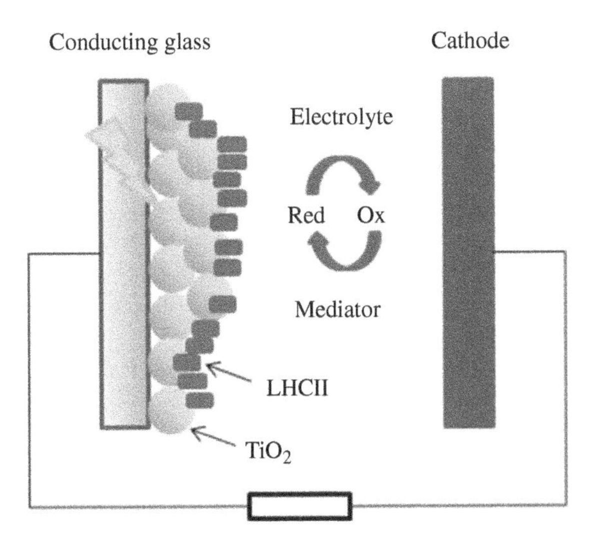

FIGURE 6.7 Simplified scheme of the dye-sensitized solar cell based on the photosynthetic antenna complexes.

Most probably, the "clean" energy shortage we face nowadays will stimulate research on removable energy sources based on photovoltaic devices that mimic or use as constituents elements of the photosynthetic apparatus, the nature-optimized light energy conversion device.

ACKNOWLEDGMENTS

The research on regulation of the photosynthetic energy transfer in the author laboratory is carried out within the project "Molecular Spectroscopy for BioMedical Studies" financed by the Foundation for Polish Science within the TEAM program.

REFERENCES

1. Ke, B. 2001. *Photosynthesis. Photobiochemistry and Photobiophysics.* Edited by Govindjee. Vol. 10, *Advances in Photosynthesis.* Dordrecht/Boston/London: Kluwer Academic Publishers.
2. Kühlbrandt, W., D. N. Wang, and Y. Fujiyoshi. 1994. Atomic model of plant light-harvesting complex by electron crystallography. *Nature* 367:614–621.
3. Liu, Z., H. Yan, K. Wang, T. Kuang, J. Zhang, L. Gui, X. An, and W. Chang. 2004. Crystal structure of spinach major light-harvesting complex at 2.72 A resolution. *Nature* 428 (6980):287–292.
4. Barros, T., and W. Kuhlbrandt. 2009. Crystallisation, structure and function of plant light-harvesting Complex II. *Biochimica Et Biophysica Acta-Bioenergetics* 1787 (6):753–772.
5. Förster, T. 1959. Transfer mechanisms of electronic excitation. *Faraday Discussions of the Chemical Society* 27:7–17.
6. Dexter, L. D. 1953. A theory of sensitized luminescence in solids. *The Journal of Chemical Physics* 21:836–850.
7. Britton, G. 2008. Functions of intact carotenoids. In *Carotenoids. Natural Functions*, edited by G. Britton, S. Liaaen-Jensen and H. Pfander. Basel, Boston, Berlin: Birkhauser.
8. Gruszecki, W. I., and K. Strzalka. 2005. Carotenoids as modulators of lipid membrane physical properties. *Biochimica et Biophysica Acta* 1740 (2):108–115.
9. Gruszecki, W. I. 2010. Carotenoids in lipid membranes. In *Carotenoids. Physical, Chemical and Biological Functions and Properties*, edited by J. T. Landrum. London, New York, Boca Raton FL: CRC Press.
10. Gruszecki, W I. 1999. Carotenoids in membranes. In *The Photochemistry of Carotenoids*, edited by H. A. Frank, A. J. Young, G. Britton and R. J. Cogdell. Dordrecht: Kluwer Academic Publishers.
11. Krinsky, N I. 1979. Carotenoid protection against oxidation. *Pure and Applied Chemistry* 51:649–660.
12. van Rensen, J. J. S., C. H. Xu, and Govindjee. 1999. Role of bicarbonate in photosystem II, the water-plastoquinone oxido-reductase of plant photosynthesis. *Physiologia Plantarum* 105 (3):585–592.

13. Wydrzynski, T.J., and K. Satoh, eds. 2006. *Photosystem II: The light-driven water: Plastoquinone oxidoreductase*. Vol. 22, *Advances in Photosynthesis and Respiration*. Dordrecht: Springer.
14. Ferreira, K. N., T. M. Iverson, K. Maghlaoui, J. Barber, and S. Iwata. 2004. Architecture of the photosynthetic oxygen-evolving center. *Science* 303 (5665):1831–1838.
15. Joliot, P. 2005. Period-four oscillations of the flash-induced oxygen formation in photosynthesis. In *Discoveries in Photosynthesis*, edited by Govindjee, J. T. Beatty, H. Gest and J. F. Allen. Dordrecht: Springer.
16. Gruszecki, W. I., K. Strzalka, A. Radunz, and G. H. Schmid. 1997. Cyclic electron flow around photosystem II as examined by photosynthetic oxygen evolution induced by short light flashes. *Zeitschrift Fur Naturforschung C-a Journal of Biosciences* 52 (3–4):175–179.
17. Kok, B., B. Forbush, and M. McGloin. 1970. Cooperation of charges in photosynthetic oxygen evolution. A linear four step mechanism. *Photochemistry and Photobiology* 11:457–475.
18. Tien, H. T. 1980. Pigmented bilayer lipid-membranes as a biomimetic system—model for understanding the quantum conversion in photosynthesis. *Journal of the Electrochemical Society* 127 (3):C119–C120.
19. Desormeaux, A., and R. M. Leblanc. 1990. Effect of light-intensity and temperature on the photovoltaic parameters of chlorophyll-B Langmuir-Blodgett-films. *Lower-Dimensional Systems and Molecular Electronics* 248:557–562.
20. Desormeaux, A., J. J. Max, S. Hotchandani, M. Ringuet, and R. M. Leblanc. 1987. Photovoltaic properties of sandwich cells of chlorophyll-a, chlorophyll-b and zinc porphyrin complexes. *Biophysical Journal* 51 (2):A127–A127.
21. Desormeaux, A., J. J. Max, and R. M. Leblanc. 1993. Photovoltaic and electrical-properties of Al/Langmuir-Blodgett films/Ag sandwich cells incorporating either chlorophyll-a, chlorophyll-b, or zinc porphyrin derivative. *Journal of Physical Chemistry* 97 (25):6670–6678.
22. Diarra, A., S. Hotchandani, J. J. Max, and R. M. Leblanc. 1986. Photovoltaic properties of mixed monolayers of chlorophyll a and carotenoid canthaxanthin. *Journal of the Chemical Society-Faraday Transactions Ii* 82:2217–2231.
23. Lauterbach, R., J. Liu, W. Knoll, and H. Paulsen. 2010. Energy Transfer between Surface-immobilized Light-Harvesting Chlorophyll a/b Complex (LHCH) Studied by Surface Plasmon Field-Enhanced Fluorescence Spectroscopy (SPFS). *Langmuir* 26 (22):17315–17321.
24. Mackowski, S., S. Wormke, A. J. Maier, T. H. Brotosudarmo, H. Harutyunyan, A. Hartschuh, A. O. Govorov, H. Scheer, and C. Brauchle. 2008. Metal-enhanced fluorescence of chlorophylls in single light-harvesting complexes. *Nano Letters* 8 (2):558–64.
25. O'Regan, B., and M. Grätzel. 1991. A low-cost, high-efficiency solar cell based on dye-sensitized colloidal TiO_2 films. *Nature* 353:737–740.

7

ELECTROCHEMICAL ACTIVATION OF CYTOCHROME P450

Andrew K. Udit,[1] Michael G. Hill[1] and Harry B. Gray[2]
[1]*Department of Chemistry, Occidental College, Los Angeles, CA, USA*
[2]*Beckman Institute, California Institute of Technology, Pasadena, CA, USA*

7.1 INTRODUCTION

The cytochromes P450 (P450s) are heme-containing monooxygenases that perform regio- and stereospecific reactions under physiological conditions. First discovered in 1960, to date, more than 500 different P450s have been identified, cloned, and sequenced, resulting in a plethora of information on P450 structure, function, and biochemistry [1–3]. P450s play critical roles in mammalian cellular pathways, as they are responsible for the biosynthesis of a variety of signaling molecules [4–6]. Human P450s act on more than 90% of all drugs currently marketed, making their activity crucial for determining the fate of many xenobiotics. By extension, P450 activity and expression are central considerations in the emerging medical field of pharmacogenetics: indeed, as of January 2011, the US Food and Drug Administration labels nearly 30 drugs with P450 biomarker information (primarily the 2D6 and 2C9 isoforms) for genotype-specific dosing and/or clinical pharmacology precautions [7].

The hallmark reaction of this enzyme class involves hydrocarbon oxidation utilizing dioxygen and NAD(P)H:

$$R-H+NAD(P)H+H^{+}+O_{2}\xrightarrow{P450}R-OH+NAD(P)^{+}+H_{2}O \qquad (7.1)$$

Electrochemical Processes in Biological Systems, First Edition. Edited by Andrzej Lewenstam and Lo Gorton.

A sample of the vast repertoire of P450-catalyzed chemical reactions is shown in Figure 7.1, highlighting the potential utility of P450 activity for *in vitro* applications. For example, P450s are capable of catalyzing many industrially relevant oxidations, converting inert hydrocarbons into alcohols and ketones that can be used as precursors for more complex molecules. These enzymes offer significant advantages over typical oxidation catalysts, which usually require the use of potent reagents that lead to uncontrolled oxidation, poor regio- and stereoselectivity, low efficiency, and various side reactions [8, 9]. The harsh conditions required to drive these chemical conversions can also lead to degradation of the desired products—a particular concern for medicinal compounds that possess multiple functionalities.

Despite the many advantages of P450-catalyzed chemistry, routine application of P450s for commercial use has not been realized. This is due in large part to the complex electron transfer (ET) machinery that characterizes the native system, which includes several intermediate proteins and cofactors that are required for ET and catalysis. Moreover, *in vitro* biocatalysis that relies on stoichiometric consumption of NAD(P)H would be outrageously expensive (at commercial prices, more than $100,000 mol^{-1}). Herein, we review efforts to develop catalytically competent P450 systems in which a simple electrode replaces NAD(P)H, and in some instances native reductase proteins, in the catalytic cycle.

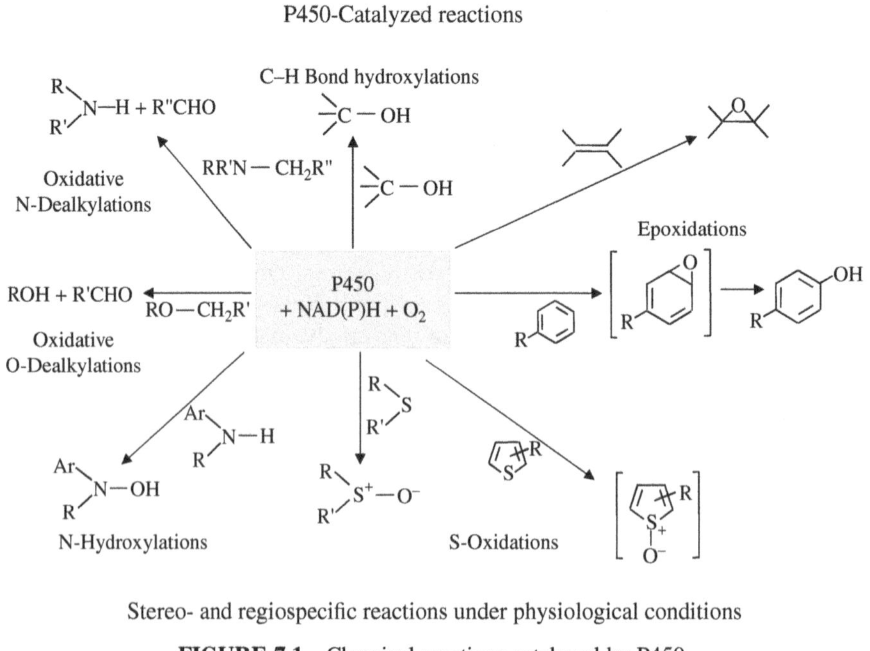

FIGURE 7.1 Chemical reactions catalyzed by P450.

7.1.1 Mechanism

Thiol ligation to the heme iron is the critical factor that separates P450-type reactivity from other heme proteins by modulating the iron redox properties to activate dioxygen for substrate oxidation [10]. The scheme shown in Figure 7.2 represents the generally accepted catalytic mechanism for P450 [2]. The resting state of the enzyme features a six-coordinate low-spin Fe^{III} ($E°$ ~ −350 mV vs. NHE), with the porphyrin occupying the four planar positions and the axial positions occupied by a cysteine (proximal)

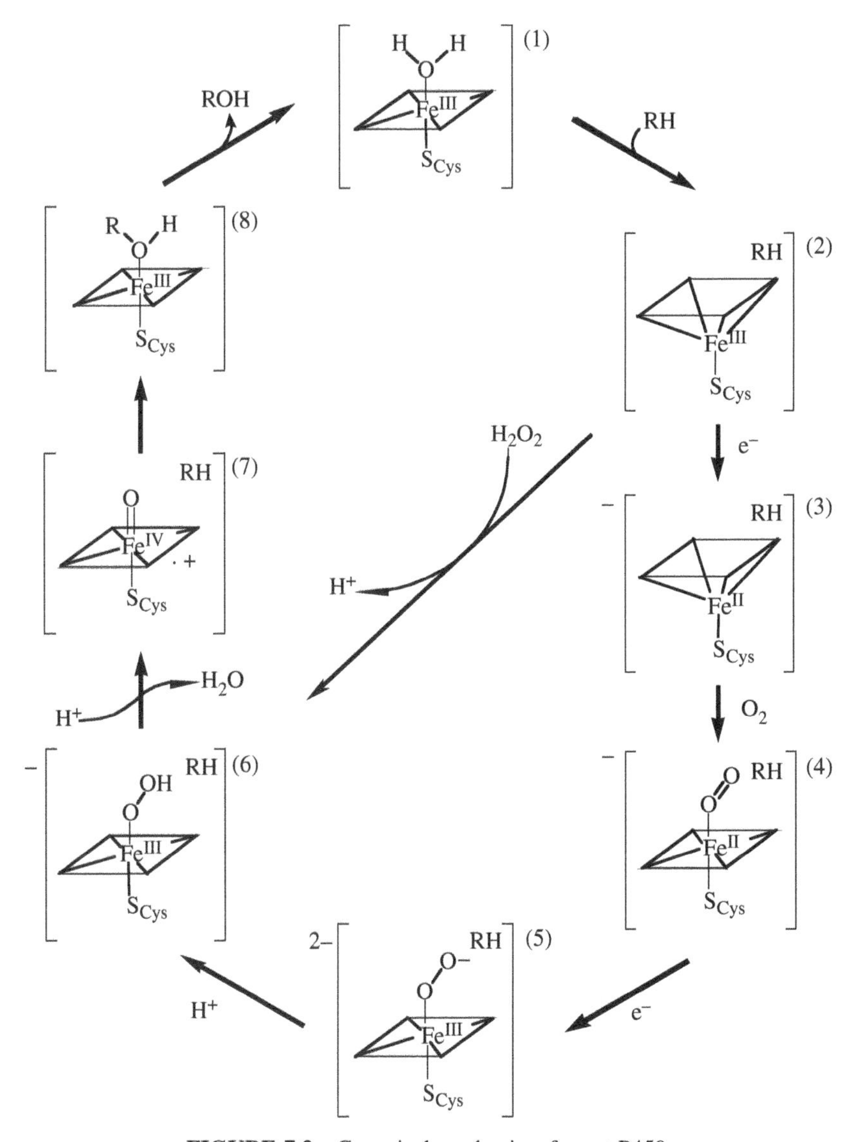

FIGURE 7.2 Canonical mechanism for cyt P450.

and water (distal). Substrate binding in the protein active site displaces the axial water, causing the six-coordinate low-spin Fe^{III} to convert to five-coordinate high-spin Fe^{III} and a change in heme potential (>+100 mV), which initiates ET from native reductase proteins [11]. Reduction to Fe^{II} leads to dioxygen binding, followed by a second proton-coupled reduction to yield the peroxy complex, $Fe^{III}{-}OOH^-$ (compound 0). Although compound 0 may directly oxidize some very active substrates (e.g., thioethers), for most applications—and certainly for C—H bond activation—compound 0 is further protonated to give water and a high-valent iron–oxo complex known as compound I. The nature of this species was recently shown to be $Fe^{IV}{=}O$ with a radical cation delocalized over the porphyrin and axial thiolate ligands [12]. Oxidation then occurs via the "rebound" reaction: compound I abstracts hydrogen from substrate to form Fe(IV)=OH, which subsequently hydroxylates the substrate radical by OH· transfer. Notably, an alternative pathway termed the peroxide shunt also results in substrate oxidation (stage (8) in Fig. 7.2). In this mechanism, reduced dioxygen is provided in the form of hydrogen peroxide, which binds to the Fe^{III} heme. The ferric-peroxy species then leads to formation of compound I through the native mechanism.

Native P450 systems *in vivo* generally exhibit highly coupled mechanisms, whereby product formation and ET are stoichiometric. This is a consequence of the catalytic cycle described earlier: substrate binding results in heme dehydration and a spin shift yielding redox potentials that now favor heme reduction from native reductase proteins. Such substrate-gated catalysis is a key feature of P450 that prevents futile redox cycling and wasting of reducing equivalents; in an aerobic environment, this gating mechanism also serves to prevent the formation of reactive oxygen species that may cause damage *in vivo*. As discussed in the following text, a major challenge of *in vitro* catalysis is to achieve coupling efficiencies comparable to the native cycle.[1]

7.1.2 P450 Electrochemistry

Given the native catalytic mechanism, seemingly the most straightforward way of activating P450 *in vitro* is by direct heme reduction at an electrode. In general, electrochemistry of large proteins is problematic as the intervening peptide medium greatly diminishes the electronic coupling between redox cofactors and the electrode surface, rendering heterogeneous ET too slow to measure [14–16]. This is true for both P450 redox cofactors and the heme domain itself, where the iron center is buried deep within the polypeptide. Thus, several strategies for promoting direct electrochemistry of P450s have been explored.

Hill and coworkers reported the first electrochemistry of P450 from *Pseudomonas putida* (CAM) in solution using an edge-plane graphite electrode [17], establishing the possibility of electrochemically induced P450 activity. Consistent with spectroscopic redox titrations, the electrochemically measured reduction potential of Fe^{III} shifts upward by approximately 100 mV upon addition of camphor to the solution.

[1] A detailed discussion of the major uncoupling pathways based on quantum-mechanical molecular modeling has been published (see Ref. [13]).

In the presence of dioxygen, the electrochemical response is dominated by dioxygen reduction.

Since that initial study, numerous reports have been published that describe electrochemical reduction of P450s by utilizing electrodes modified with clays [18], polymer films [19], self-assembled monolayers (SAMs) [20], and flavin cofactors [21], among other systems. However, few of the investigated systems are catalytically competent, and even fewer show potential for practical application (e.g., substrate turnover is too slow, the electrode/protein combination is unstable, or the system exhibits nonnative activity).

Notably, in evaluating the successes of P450 electrocatalytic methods, there is an important distinction to be made between mammalian and bacterial systems. Recent advances (described in the following text) have led to native-like chemistry for several mammalian P450s driven by electrochemical activation, whereas analogous turnover by bacterial systems remains elusive. In general, mammalian P450s are difficult to handle *in vitro* and exhibit significantly slower catalytic rates ($1–10\ s^{-1}$) and correspondingly lower turnover numbers. By contrast, bacterial systems tend to be more robust, exhibit rapid rates of turnover ($100–1000\ s^{-1}$), appear more tolerant of mutation, and can be overexpressed easily for scale-up. As a result, applications for mammalian P450s tend toward drug screening, whereas large-scale commercial applications have focused on bacterial systems [22].

In the following text, we examine the collective groups of techniques investigators have used to explore P450 bioelectrocatalysis and highlight some of the lessons learned. While some mention of notable mammalian P450 electrocatalytic systems is made, we focus primarily on bacterial systems and specifically on the NADPH-dependent flavocytochrome P450 from *Bacillus megaterium* (BM3). BM3 is comprised of separate heme and diflavin reductase domains fused together on a single polypeptide chain [23]. This arrangement facilitates rapid ET between the reductase and heme domains, resulting in high catalytic turnover rates (hundreds to thousands per minute, depending on substrate). BM3 has many advantages as a candidate for commercial applications, as the enzyme (i) is soluble; (ii) can be expressed in large quantities in *E. coli*; (iii) has relatively broad substrate specificity, while (iv) mutants of BM3 are even more active and have broader substrate specificity [5, 24, 25]; (v) has a partial crystal structure available [26]; and (vi) can be easily assayed for activity [27]. Furthermore, BM3's homology to comparable mammalian systems—which are typically membrane bound and less robust—makes it suitable for potential pharmaceutical applications in addition to selective hydrocarbon oxidation reactions [28].

7.2 HOMOGENEOUS SYSTEMS: SMALL-MOLECULE ELECTROCHEMICAL MEDIATORS

Harnessing P450 activity for *in vitro* applications may be most simply accomplished with electrochemical systems utilizing soluble mediators. Estabrook and coworkers discovered a promising system that utilized a platinum electrode and the water-soluble cobalt(III) sepulchrate (Co(sep)) cage complex ($E° = -550$ mV vs. AgCl/Ag) as the electron shuttle (Table 7.1) [29]. Co(sep)-mediated catalysis was demonstrated with a variety of P450s (mammalian and bacterial), with rates approaching that of NAD(P)H-driven systems: as an example, reactions with BM3 and lauric acid yielded turnover rates of 110 min^{-1} (as compared with 900 min^{-1} for reactions with NADPH) [29]. While encouraging, practical limitations of the Co(sep) system include production of reactive oxygen species [30], difficulty in synthetically manipulating Co(sep) to tune the mediator to different reaction conditions (e.g., solvent, pH), and aggregation/precipitation at functional Co(sep) concentrations (typically, 1–10 mM) [31].

TABLE 7.1 Electrochemical mediators for HTP catalysis.

Mediator	Structure	Best rate
Cobalt(III) sepulchrate	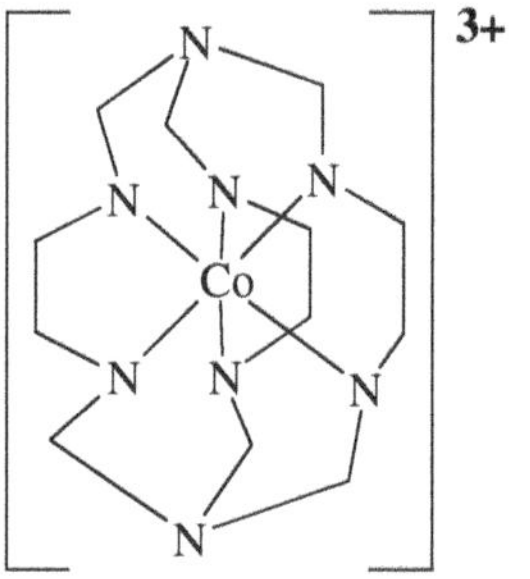 [Co(sep)]$^{3+}$ (cage: Co coordinated by six N)	110 min^{-1} with P450 BM3 and lauric acid
1,1′-Dicarboxycobaltocenium	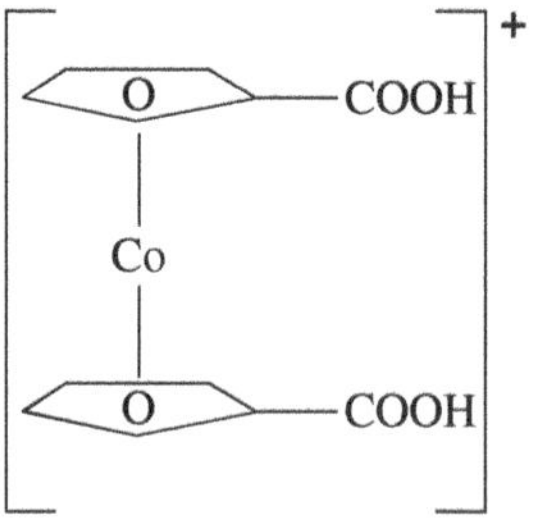 [Co(C$_5$H$_4$COOH)$_2$]$^{+}$ (COOH, Co, COOH)	16.5 min^{-1} with P450 BM3 and lauric acid
Putidaredoxin	2Fe-2S protein, 11.6 kDa	36 min^{-1} for P450 CAM with camphor
Cp∗Rh(bpy)(H_2O)Cl_2	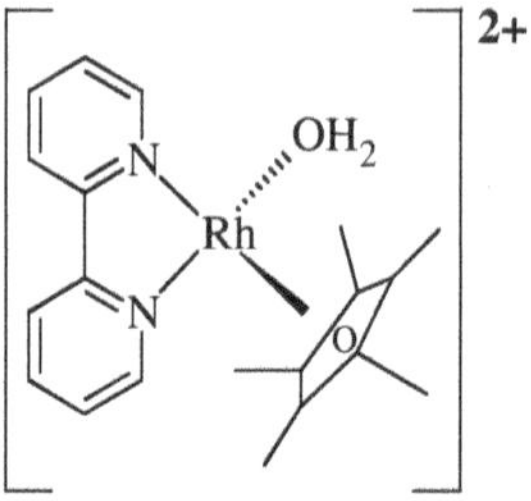 [Cp∗Rh(bpy)(OH$_2$)]$^{2+}$ (N, N, Rh, OH$_2$)	6.4 mM h^{-1} for StyA with styrene; HTP activity not determine

TABLE 7.2 Rates and total turnovers for the electrochemical biocatalytic reactions.

Enzyme	Mediator	Rate (nmol product/nmol enzyme/min)	Total turnover (nmol product/nmol enzyme)
BM3	Cobaltocene cation	16.4 ± 0.6	224 ± 7
BM3	Cobalt(III) sepulchrate	37.8 ± 0.3	835 ± 7
hBM3	Cobaltocene cation	1.8 ± 0.5	58 ± 7
hBM3	Cobalt(III) sepulchrate	2.2 ± 0.1	76 ± 7

Inspired by Estabrook's work and by previous work with glucose oxidase utilizing a ferrocene derivative to mediate enzyme oxidation [32], we decided to try the analogous reductant cobaltocene as a scaffold to construct a suitable mediator. Anticipating that the dicarboxy derivative could improve water solubility and disfavor aggregation, we synthesized 1,1′-dicarboxycobaltocenium hexafluorophosphate (M_{ox}) (Table 7.1) and evaluated its ability to mediate bioelectrochemical catalysis with P450 BM3 (Table 7.2) [33]. The surrogate mediator was observed to reduce both the FAD and FMN in the P450 BM3 reductase domain, as well as the iron in the heme domain. Electrolysis reactions with the holoprotein resulted in lauric acid hydroxylation (16.5 nmol product/nmol enzyme/min). Similar ET and catalysis were observed in reactions with just the wild-type BM3 heme domain (hBM3) and the metallocene (1.8 nmol product/nmol enzyme/min). However, as with the Co(sep) system, our M_{ox} system could not overcome the problem of direct dioxygen reduction by the reduced mediator. This problem was most evident from the observed coupling efficiency, which revealed that only 2% of the total current passed resulted in product formation.

Vilker described an electrochemical system with CAM using putidaredoxin (Pdx), the native ET partner, as the electrochemical mediator (Table 7.1) [34]. Direct reduction of Pdx with an antimony-doped tin oxide electrode initiated the ET cascade that ultimately resulted in dioxygen activation and camphor hydroxylation. An average turnover rate of 36 min^{-1} and 2600 total turnovers were reported. Notably, reactor performance was optimal when the concentration of Pdx was approximately two orders of magnitude higher than CAM. Once again, the cosubstrate dioxygen proved problematic: the working electrode had to be screened with a platinum mesh to disproportionate peroxide formed from direct dioxygen reduction, while the necessary dioxygen for catalysis was generated *in situ* through water oxidation at another electrode.

Shumyantseva *et al.* reported a technique that covalently attached a synthetic flavin cofactor to the surface of two mammalian P450 heme domains, 2B4 and 1A2 [21]. The flavin functioned as an electronic relay, mimicking the contact that the native reductase would make with the heme domains. Reduction of the flavin at a rhodium-graphite electrode resulted in ET through the protein sheath and into the heme. Although ET rates were not reported, they did observe electrode-driven catalysis at rates comparable to NADPH-driven systems (e.g., ~0.5 turnovers per minute for aminopyrine N-demethylation by 2B4).

Clearly, the key obstacle in the mediated systems is dioxygen reduction, resulting in futile redox cycling and production of reactive oxygen species. A potential way to overcome this is described in studies by Hollmann *et al.* where they have demonstrated the efficacy of $Cp^{*}Rh(bpy)(H_2O)Cl_2$ (Rh) (Table 7.1) as a recyclable hydride transfer mediator, capable of regenerating nicotinamide-(NADH, NADPH) and isoalloxazine-based (FAD, FMN) cofactors in solution [35–37]. The key advantage is the relatively low rate of uncoupling from direct reduction of dioxygen; this was demonstrated in reactions using Rh to reduce the flavins in the catalytic cycle of styrene monooxygenase, which had a coupling efficiency of 60% while displaying substrate turnover rates comparable to the native system [37]. While activity of Rh with P450 has yet to be demonstrated, engineered enzymes and/or similar mediators may yield better results.

7.3 HETEROGENEOUS SYSTEMS: SURFACE-CONFINED P450 FILMS

In addition to the problem of nonenzymatic dioxygen reduction, systems featuring homogeneous protein solutions are inherently limited by slow protein diffusion as well as electrode passivation and fouling. To overcome these obstacles, some researchers have investigated heterogeneous systems where the protein is immobilized on an electrode surface. Given a sufficiently robust P450 immobilized in a manner that preserves protein integrity, such systems have the potential to be used/reused continuously. An added advantage is facile product purification as the catalytic unit remains surface bound. Achieving a general system that is sufficiently robust and active for high-throughput catalysis remains challenging. Indeed, key obstacles must be overcome before functional, electrode-driven P450 biocatalysis can be realized:

1. *Protein integrity.* For surface-confined systems, questions arise regarding protein fold, stability, and substrate discriminatory properties. Maintaining native-like structure is crucial for achieving native-like activity.
2. *Protein dynamics.* Attenuated polypeptide flexibility resulting from adsorption and/or protein aggregation on the electrode surface may hinder substrate access and restrict conformational changes that are necessary for catalysis.
3. *Dioxygen activation.* Catalysis in the native system relies on tightly regulated electron and proton transfers that are coupled to substrate binding and protein conformational changes. All of these events must occur in a specific sequence in order to achieve proper activation of dioxygen at the precise moment for oxygen atom transfer. In contrast, electrode-driven catalytic dioxygen reduction by P450 often produces reduced oxygen species (one-, two-, or four-electron reduction reactions), leading to uncoupling of ET and subsequent oxidative damage to the enzyme. Mediated systems are further complicated by the requirement for powerful reducing agents in the presence of dioxygen, leading to direct dioxygen reduction before P450 activation occurs.

7.3.1 Surfactant Films

Rusling pioneered the use of surfactant films on carbon surfaces for direct electrochemistry of heme proteins [38–40]. The surfactant is deposited onto the electrode, resulting in the formation of bilayers and micelles into which the protein is incorporated. The end result is a system that supports rapid and reversible ET between the electrode and the enzyme; typically, standard rate constants between 50 and 300 s^{-1} are observed. Comprehensive reviews on thin-film electrochemistry are available [41, 42].

Protein–surfactant film voltammetry has been widely used for studying the redox chemistry of P450s. Flavocytochromes [43], engineered mutants [44], as well as bacterial [45] and mammalian [46] variants have been investigated, and the methodology continues to be exploited [47, 48]. Among the first instances of catalysis, Rusling and coworkers reported epoxidation of styrene (although inefficiently, ~1–10 turnovers/h) by the peroxide shunt, with peroxide generated from electrochemically mediated P450-catalyzed dioxygen reduction [49]. Notably, Farmer's group has conducted extensive investigations of the thermophilic P450 from *Sulfolobus solfataricus* [50, 51]. In addition to the $Fe^{III/II}$ redox couple, they also have accessed the $Fe^{II/I}$ couple (at −1040 mV vs. AgCl/Ag), demonstrating that the Fe^{I} species is capable of catalytic dehalogenation of certain substrates. In work at elevated temperatures (>50°C), they were able to achieve efficient dechlorination of carbon tetrachloride, producing methane.

7.3.2 Dioxygen Reduction

We have used surfactant films on carbon electrodes to explore the reactivity of the truncated heme domain from BM3, hBM3 [38, 43, 49, 52]. hBM3, which can readily be incorporated into surfactant films, exhibits rapid ET between the ferric heme and electrode surface. A representative voltammogram revealing the $Fe^{III/II}$ redox couple ($E^{\circ\prime} = -195$ mV vs. AgCl/Ag, pH 7) for hBM3 in didodecyldimethylammonium polystyrenesulfonate (DDAPSS) on basal plane graphite (BPG) is shown in Figure 7.3; addition

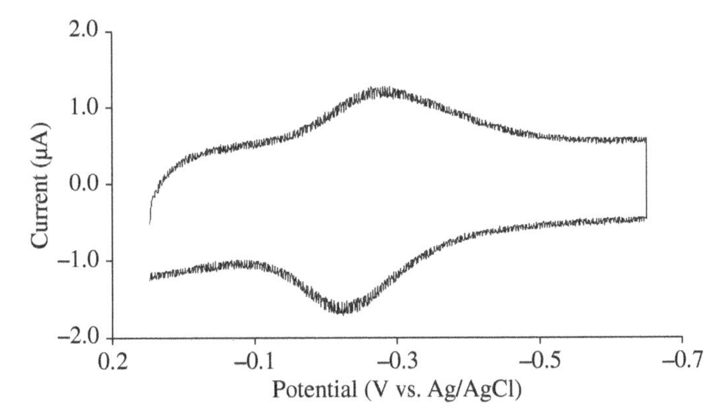

FIGURE 7.3 Cyclic voltammogram of hBM3–DDAPSS film at a basal plane graphite (BPG) electrode. Recorded in 0.05 M KPi/20 mM KCl, pH 7, at a scan rate of 0.5 mV s^{-1}.

of O_2 results in large catalytic currents at the onset of Fe^{III} reduction, demonstrating the ability of ferrous P450 to catalyze dioxygen reduction.

The stability of DDAPSS films to mechanical stress [53] permitted us to use a rotating disk electrode (RDE) to investigate the mechanism of dioxygen reduction by hBM3. At an RDE, the diffusion–convection-limited current (i_L) depends on the angular rotational velocity (ω) and the bulk concentration of substrate (O_2 in this case) according to the Levich equation [54]:

$$i_L = 0.62\ nFAD^{2/3}[O_2]\nu^{-1/6}\omega^{1/2} \tag{7.2}$$

where n is the number of electrons transferred, F is Faraday's constant, A is the area of the electrode, ν is the kinematic viscosity of the solution, and D is the diffusion constant of the substrate. RDE experiments were carried out by performing electrolysis at –0.5 V (vs. AgCl/Ag) with hBM3–DDAPSS films and measuring the limiting current for each rotation rate. The data are plotted in Figure 7.4a (points), along with theoretical lines for two- and four-electron reduction of dioxygen (solid lines).

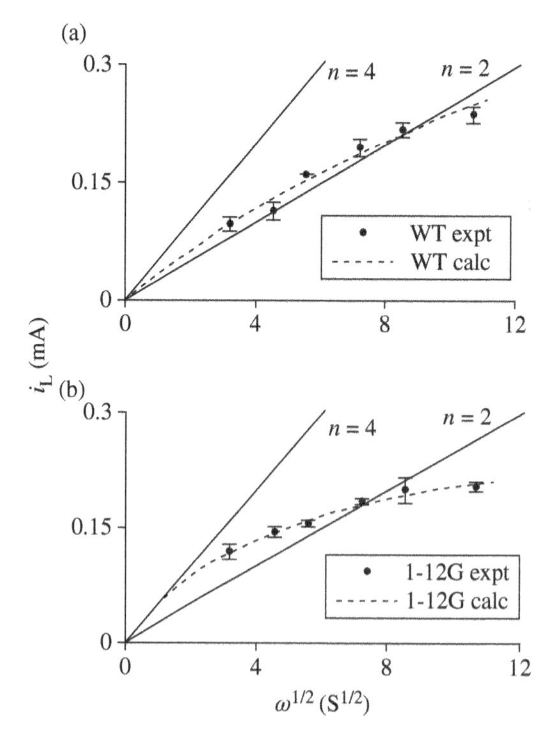

FIGURE 7.4 Levich plots for wt (a) and 1-12G (b) hBM3 in DDAPSS on BPG. Straight lines represent the theoretical dependence of i_L on angular velocity assuming a two- and four-electron process. Dashed lines represent best fits to Equation (7.4), assuming respective values of n and k_{obs} of 2.7 and $1.4 \times 10^6\ M^{-1}\ s^{-1}$ (wt) and 4.7 and $1 \times 10^5\ M^{-1}\ s^{-1}$ (1-12G).

Note that as the flux of O_2 to the electrode is increased (higher rotation rates), the overall reaction cannot keep up with mass transport and the limiting currents level off. The current density that describes the reaction between the surface-bound mediator (P450 in this case) and dioxygen, i_k, is then given by

$$i_k = nFA[O_2]\Gamma_{P450}k_{obs} \tag{7.3}$$

where Γ_{P450} is the surface coverage of electroactive enzyme at the electrode (experimentally accessible via coulometry) and k_{obs} is the pseudo-second-order rate constant. In the case of a simple, one-step irreversible reaction, the overall RDE current is determined by i_L and i_k according to Equation (7.4) [55, 56]. Thus, a plot of $i_{lim}{}^{-1}$ versus $\omega^{-1/2}$ (Koutecky–Levich plot) should yield a straight line whose slope gives the number of electrons transferred and whose intercept yields k_{obs} directly:

$$\frac{1}{i_{lim}} = \frac{1}{i_L} + \frac{1}{i_k} \tag{7.4}$$

Indeed, the resulting Koutecky–Levich plot is linear with slope $n = 2.7$ (Fig. 7.5a) and intercept giving k_{obs} of $1.4 \times 10^6\ M^{-1}\ s^{-1}$ for hBM3.[2] Values for n and k_{obs} were then used to generate the calculated Levich plot for hBM3 (Fig. 7.4a, broken line), taking into account the rate of the catalytic reaction.

In principle, dioxygen can be reduced by one, two, or four electrons to form superoxide, peroxide, or water. Our calculated value for n is probably overestimated: within the hydrophobic film, dioxygen is likely more concentrated than in solution. With this in mind, it appears that hBM3 reduces dioxygen primarily by two electrons, consistent with other electrochemical studies of hBM3 and point mutants [43, 57]. Thus, dioxygen reduction to peroxide appears to be the major uncoupling pathway for hBM3 within DDAPSS films on electrode surfaces.

7.3.3 Mutant P450s, ET, and Dioxygen Reduction

For comparison, we also used the DDAPSS films to investigate the redox chemistry of hBM3 mutant 1-12G. Fifteen mutations distinguish 1-12G from hBM3 and are collectively responsible for the unique catalytic properties of this derivative, including its ability to regio- and stereoselectively hydroxylate linear hydrocarbons and small gaseous alkanes [58]. We suggest that these differences in catalysis are due in part to differences in redox properties conferred by the mutations. Indeed, although the two proteins have virtually identical $Fe^{III/II}$ redox potentials, they differ markedly in other ET properties: analyzing variable scan-rate voltammograms according to Laviron's model [59] gives respective $k°$ values of 250 and $30\ s^{-1}$ for hBM3 and 1-12G.

[2] Importantly, the intercept is linear with surface concentration of hBM3 P450, validating the use of Equation (7.4).

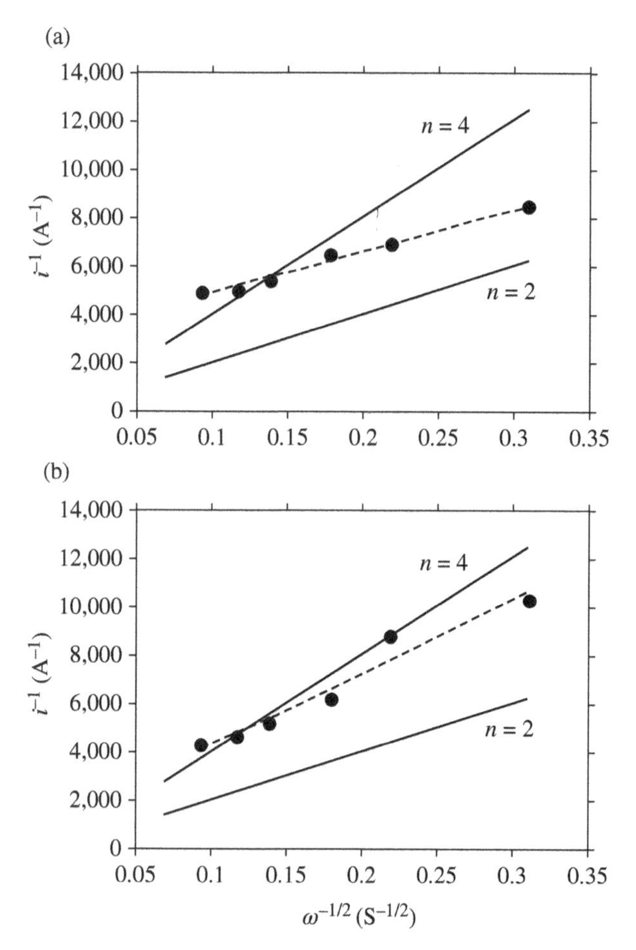

FIGURE 7.5 Koutecky–Levich plots for wt (a) and 1-12G (b) hBM3 in DDAPSS on BPG. Solid lines correspond to response calculated for $n = 2$ and $n = 4$.

As for hBM3, RDE experiments were carried out for 1-12G. The resulting Levich and Koutecky–Levich plots are shown in Figures 7.4b and 7.5b. Values for n and k_{obs} are 4.7[3] and $1 \times 10^5\ M^{-1}\ s^{-1}$, respectively. In stark contrast to hBM3, 1-12G reduces dioxygen by four electrons. Note that P450-catalyzed four-electron reduction of dioxygen has been interpreted [5, 60] within the canonical mechanism through reduction of compound I by two electrons ("oxidase" activity) [61]. Despite this activity, dioxygen reduction by 1-12G is much slower than for hBM3 (14-fold drop in k_{obs}).

Results from our electroanalytical treatment can be correlated with the biochemistry of the proteins. First, there is the 10-fold difference in heterogeneous ET rate

[3] As for hBM3, this value is likely to be overestimated.

constants (k°). Indeed, preliminary stopped-flow experiments using holoproteins and NADPH as reductant reveal that heme reduction in hBM3 occurs approximately 5 times faster (Unpublished results). In addition, substrate turnover rates for hBM3 [62] are generally higher (up to 10-fold) than for 1-12G [58]. Since ET is the rate-limiting step, the relative k° and k_{obs} values are consistent with the catalytic oxidation rates. Second, there is the striking observation that under similar conditions 1-12G is capable of reducing dioxygen by four electrons. This difference can be reconciled if the iron-peroxy species is longer-lived in 1-12G. For hBM3, peroxide apparently dissociates before subsequent conversion to compound I. However, the slower rate of ET in 1-12G, coupled with the observation of catalytic currents corresponding to four-electron transfer to dioxygen, argues for an iron-peroxy complex that is longer-lived in 1-12G relative to that of hBM3. This finding could potentially explain the high degree of stereo- and regioselectivity that 1-12G displays: a longer-lived iron-peroxy species allows substrates more time to adopt their lowest-energy conformation within the active site and orient a specific C—H bond for subsequent hydroxylation. This argument is in accord with that previously postulated for 1-12G [58], which reasoned that the mutations (especially A328V and A82L in the active site) keep the substrate in a more fixed orientation, allowing specific products to be generated.

For hBM3, dissociation of peroxide apparently occurs before the ferric-peroxy complex can form compound I. The results with 1-12G, however, are more intriguing: four-electron reduction of dioxygen to water suggests oxidase activity, implying the transient formation of compound I. As the reaction between compound I and bound substrate would surely occur many of orders of magnitude faster than ET (which is milliseconds at best), the system ought to be catalytically competent. However, turnover experiments with an extensive series of substrates (including styrene, octane, lauric acid, dimethyl ether, thioanisole, hexylmethylether, trichloroacetic acid, 1-octanol, and hexane) yielded no products.

An alternative hypothesis to oxidase activity is that 1-12G catalyzes the heterogeneous four-electron reduction of dioxygen to water, without ever reaching compound I. In fact, our results are reminiscent of analogous work with cobalt porphyrins, where the product of dioxygen reduction (two-electron or four-electron pathways) can be selected by controlling the ET rate and the accessibility of the metal–oxo complex to solvent [63, 64]. Thus, taken together, the hBM3 and 1-12G experiments further underscore the need for regulating ET in P450–electrode systems: the ferric-peroxy complex must remain intact long enough for decay into compound I, while ET to this same complex must be prevented to avoid reduction to water. (This is true for both the ferric-peroxy complex and compound I.)

7.3.4 Protein Integrity

Another striking aspect of P450 electrochemistry in surfactant films is the dramatic shift of the $Fe^{III/II}$ couple to positive potentials (~ +200 to +300 mV cf. solution). Indeed, such upshifts are not uncommon for surface-confined P450 electrodes (although the surfactant systems display some of the most dramatic effects). The close relationship between structure and function, which is a recurring theme in

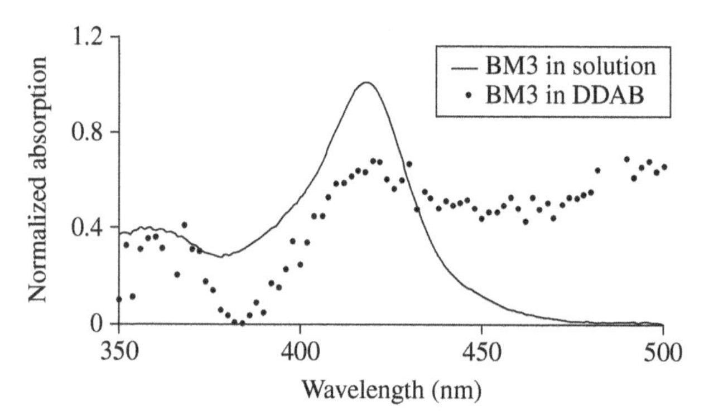

FIGURE 7.6 Absorption spectra of hBM3 in solution and in DDAB on glass.

biochemistry, is well illustrated by P450 (and other cysteine-ligated heme proteins, e.g., iNOS [65]). Activity depends critically on structure, so the significant positive shift in $Fe^{III/II}$ potentials of P450–electrode systems (cf. in solution) suggests that the heme is not in a native environment. Questions then arise regarding whether these shifts are attributable to protein unfolding (perhaps even heme dissociation) or other more subtle structural perturbations.

We undertook studies to examine the factors that lead to the high-potential species in order to shed light on the poor activity displayed by electrode-bound P450 [66]. To probe the origin of the potential shifts, we performed voltammetry of hBM3 within a didodecyldimethylammonium bromide (DDAB) surfactant film on graphite electrodes. As summarized in the following text, changes in enthalpy and entropy accompanying reduction of ferric hBM3 are consistent with a conformationally stabilized ferrous state of the enzyme, giving rise to a significant positive shift in the heme reduction potential.

DDAB films are readily prepared on BPG by depositing aqueous solutions of DDAB (10 mM) onto the electrode surface, followed by slow drying in air. Protein can be incorporated into the film by soaking the coated electrode in an approximately 10 μM protein solution (in 50 mM potassium phosphate buffer, pH 7) at 8°C for 30 min. Cyclic voltammetry of the resulting film yields a well-defined, chemically reversibly process for the $Fe^{III/II}$ couple at −260 mV (vs. SCE)[4] [65]; the analogous value in solution is approximately 300 mV more negative for the six-coordinate, water-ligated low-spin heme.

To probe the heme microenvironment within the film, we prepared hBM3–DDAB films on fused silica slides and recorded electronic absorption spectra. Reproducible spectra were obtained on these slides: in a typical spectrum (Fig. 7.6), a distinct peak is observed at 418 nm, identical to the position of the Soret band in the spectrum of low-spin Fe^{III} hBM3 in solution. This spectrum clearly indicates a thiol-ligated

[4] Notably, at faster scan rates, a complex voltammogram with two cathodic peaks is observed. This behavior is analogous to that found for nitric oxide synthase in DDAB, indicative of water-free and water-bound hemes (see Ref. [65] for a detailed discussion).

protein-bound heme, suggesting that the protein retains its native conformation within the film.

Additional information concerning the nature of immobilized hBM3 was obtained by evaluating the entropy change that accompanies ET at the heme reaction center (ΔS^{o}_{rc}, Eq. 7.5). This parameter is accessible from the temperature dependence of the redox potential—which, in turn, is conveniently measured using a nonisothermal electrochemical cell configuration [67]. Voltammograms were recorded at slow sweep rates, approximately 25 mV s^{-1}, between 18.5 and 40°C. As DDAB gel to liquid-crystal film transitions occur between 9 and 17°C [39], we performed our experiments above this range in order to eliminate phase-transition effects. The data reported below represent the results of at least five independent experiments for each set of conditions:

$$\Delta S^{\circ}_{rc} = S^{\circ}_{Fe(II)} - S^{\circ}_{Fe(III)} = F\left(\frac{dE^{\circ}}{dT}\right) \tag{7.5}$$

A plot of $E^{\circ\prime}$ versus temperature revealed that the $Fe^{III/II}$ redox potential shifts approximately −1.02 mV C^{-1} (Fig. 7.7). The change in entropy accompanying heme reduction is equal to the slope of this line multiplied by the Faraday constant, which yields ΔS°_{rc} = −98 J mol^{-1} K^{-1} [68]. Taking $S^{\circ}(H_2)$ = 130.4 J·mol K^{-1} and assigning $S^{\circ}(H_2)$ = 0 [68], we calculated an overall ΔS^{o} for the complete cell reaction (adjusted to the NHE scale) of −163 J·mol K^{-1}, fixing the value of ΔH° at −47 kJ mol^{-1} from the measured $E^{\circ\prime}$ at 25°C. Table 7.3 lists thermodynamic parameters for hBM3 along with values reported for other heme proteins [68].

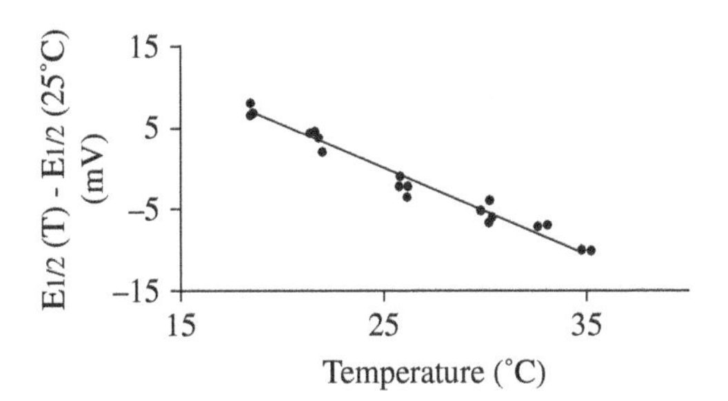

FIGURE 7.7 Temperature dependence of the FeIII/II redox couple of hBM3 in DDAB films on BPG. Voltammograms were recorded in 50 mM KP_i/50 mM KCl, pH 7 buffer.

TABLE 7.3 Thermodynamic parameters for reduction of selected heme proteins.

	ΔS^{o} (J mol^{-1} K^{-1})	ΔH^{o} (kJ mol^{-1})
BM3[a]	−163	−47
BM3 + imidazole[a]	−73	−21
Myoglobin[b] (Ref. [69])	−148	−38
Horse heart cytochrome *c*[b] (Ref. [68])	−127	−64

[a] In DDAB film on BPG.

[b] In solution.

ET to ferric hBM3 within DDAB is accompanied by substantial decreases in both enthalpy and entropy. Negative entropy changes following protein reduction normally are attributed to tightening of the protein structure, resulting in a more rigid conformation [68, 70, 71]. In our case, reduction of six-coordinate water-ligated Fe^{III} yields five-coordinate Fe^{II} and expulsion of the axial water ligand [72], leading to collapse of the protein around the active site into a more compact structure [69, 71, 73]. Further, the additional negative charge in the polypeptide matrix following heme reduction produces an overall tightening of the H-bonding network, thereby increasing the rigidity of the protein [68]. The corresponding loss in enthalpy indicates that the protein is stabilized in the reduced form, again consistent with polypeptide collapse to a more rigid conformation.

We performed similar electrochemical experiments with hBM3–DDAB films in the presence of 500 mM imidazole (Table 7.3). Imidazole replaces water as the heme axial ligand and, unlike water, remains bound to the heme in both Fe^{III} and Fe^{II} oxidation states [74]. Under these conditions, the $Fe^{III/II}$ $E^{o'}$ is virtually temperature independent. Calculated values for $\Delta S°$ and $\Delta H°$ are $-73\ \mathrm{J\ mol^{-1}\ K^{-1}}$ and $-21\ \mathrm{kJ\ mol^{-1}}$, respectively. Strikingly, there is very little change in entropy upon heme reduction with imidazole bound to the iron, in contrast to our findings for the water-ligated species. This dramatic difference is consistent with the likely structural rearrangements (or lack thereof) that accompany these two redox reactions: whereas reduction of a six-coordinate axially aquated heme triggers water dissociation and hydrophobic collapse, ET to an imidazole-bound heme produces no change in coordination, so there is minimal nuclear reorganization. The smaller negative change in enthalpy can be attributed to the preference of imidazole, a better electron donor than water, to bind and stabilize iron in the +3 oxidation state.

Our electrochemical analysis sheds new light on the origin of the large positive potential shifts that accompany encapsulation of heme–thiolate proteins in surfactant films. First, there is the hydrophobic effect. Dehydration of the protein likely occurs within the bilayer environment, resulting in hydrophobic collapse and stabilization of the Fe^{II} (five-coordinate) heme. Examination of crystal structures provides some insight into the extent of dehydration that may occur; for example, P450 CAM has six water molecules within the active site and substrate access channel [75]. Since the extent of hydration is known to affect heme potentials [76], it is likely that expulsion of several water molecules from the P450 active site would result in large heme potential shifts.

Second, tightening of the H-bonding network within the polypeptide upon reduction yields a more compact structure, which would be favored within the sterically hindered bilayer environment. Notably, a "perturbed" ferrous-oxy intermediate in P450 CAM observed in the presence of reduced Pdx was attributed to an intermediate with enhanced H bonding to the axial cysteine thiolate in the P450-reductase complex [77]. It is interesting that the surfactant on the electrode surface may play a similar role to Pdx in solution.

To further probe the electronic structure of this high-potential species, we prepared hBM3 Fe^{II}—CO ($\lambda_{max} = 448$ nm) and monitored its absorption in DDAB over time. Notably, after several minutes, a shift in λ_{max} to 420 nm occurs

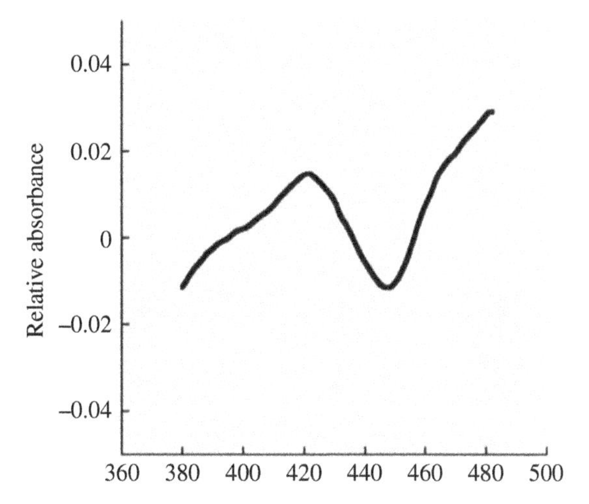

FIGURE 7.8 Difference spectrum of the ferrous hBM3—CO complex in DDAB two minutes after incorporating hBM3—CO into the film.

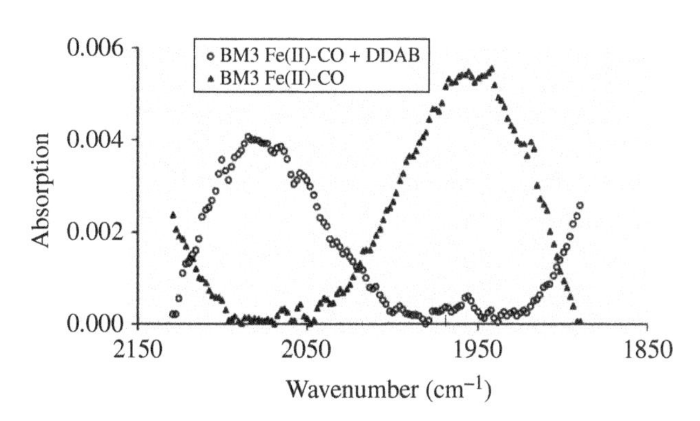

FIGURE 7.9 IR spectra of the ferrous hBM3 complex in the absence and presence of DDAB (2.5 mM in 50 mM KP_i, pH 7). Reproduced from Ref. [66] by permission of the American Chemical Society. © American Chemical Society.

(Fig. 7.8): a 420 nm Soret band for P450 ferrous carbonyl has been ascribed to a folded but biologically inactive form of the enzyme (P420) [78–80].

We next measured the IR spectrum of hBM3 Fe^{II}—CO in the absence and presence of DDAB. Prior work with P450 Fe^{II}—CO has established a range for the CO stretch between 1900 and 1950 cm^{-1} [81, 82]; our value of 1950 cm^{-1} for hBM3 in solution (Fig. 7.9) is within this range. Addition of DDAB caused this stretch to shift >100 cm^{-1} to higher frequency, implying greatly diminished π-backbonding into the carbonyl ligand (Fig. 7.10). We suggest that this shift is due to enhanced hydrogen bonding to the axial cysteine ligand within the hydrophobic DDAB film: a weakened iron–thiolate bond would decrease electron density at the metal center,

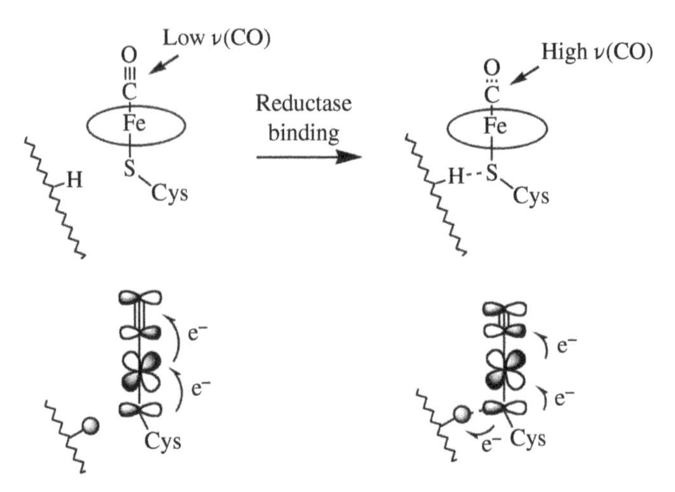

FIGURE 7.10 Hydrogen bonding to the proximal cysteine sulfur reduces the electron density around the iron center. As the CO stretching frequency in ferrous hBM3—CO depends on π-backbonding from the metal center, it is a convenient reporter for changes in the relative push effect of the thiolate ligand. Reproduced from Ref. [66] by permission of the American Chemical Society. © American Chemical Society.

shifting the redox potential to higher values [83] (diminished "push" effect [84]), thereby affecting the catalytic activity of the enzyme [85].

Our results are consistent with results from parallel surface-enhanced resonance Raman investigations by Todorovic *et al.*, which suggest that immobilization of P450 on modified silver surfaces perturbs the heme potential by dehydration, ultimately producing P420 [86]. Thus, formation of P420 on electrode surfaces is a likely explanation for the characteristically poor catalytic activity displayed by P450–electrode systems [87].

7.3.5 Direct Oxidation of P450

Problems that appear inherent to cathodic activation of P450 include producing catalytically active ferryl species and direct dioxygen reduction by the electrode to form reactive oxygen species. One approach to overcome these issues would be to run the catalytic cycle *backward*, generating reactive ferryl species directly from the ferric-aquo resting state (Fig. 7.11), thus circumventing the problems with reductive activation of dioxygen. Indeed, precedent for this can be found in work with chloroperoxidase [88] and horseradish peroxidase [89]: chemically oxidizing the heme by one and two electrons results in rapid conversion of $Fe^{III}—OH_2$ to the ferryl species $Fe^{IV}=O—H$ (compound II) and $Fe^{V}=O$ (compound I).

To investigate the feasibility of direct heme oxidation, we turned to the P450–DDAB surfactant system. Preliminary studies using CAM in this system revealed a quasireversible $1e^-$ oxidation of Fe^{III} near the solvent limit (Fig. 7.12) [90]. We assign this oxidation to proton-coupled oxidation of P450 to $Fe^{IV}=OH$ based on its

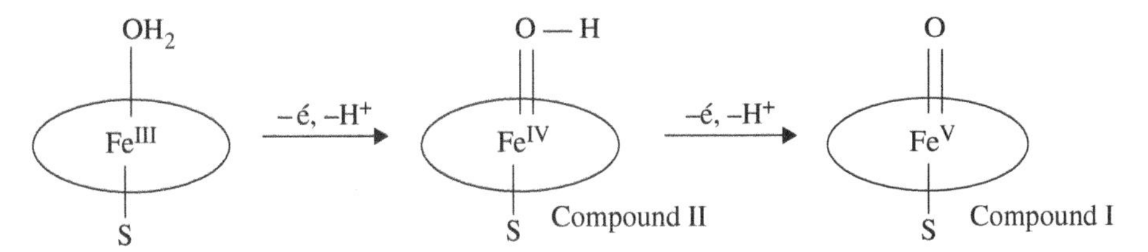

FIGURE 7.11 Direct electrochemical oxidation pathways and the resulting heme species.

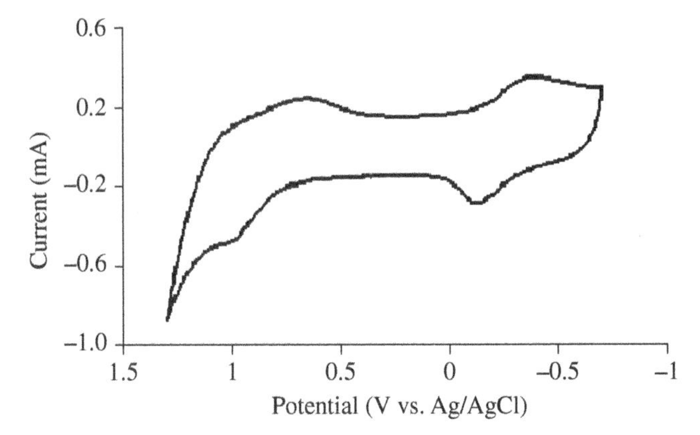

FIGURE 7.12 Cyclic voltammogram of P450 CAM in DDAB at a BPG electrode in 50 mM KP_i, pH 8. Reproduced from Ref. [90] by permission of Elsevier. © Elsevier.

pH dependence (−47 mV pH^{-1}, close to the ideal value of −59 mV pH^{-1}) and its inhibition by imidazole (which displaces the distal aquo ligand). These two experiments rule against a porphyrin-centered redox reaction, which should be relatively insensitive to changes in solution conditions [50]. It is noteworthy that these electrochemical data are fully consistent with EXAFS measurements carried out on chloroperoxidase compound II [10]: a relatively long Fe—O bond distance (1.82 Å) coupled to a contracted Cys—S—Fe^{IV} bond (2.39 Å) strongly suggests that the Fe^{IV}=O unit in compound II is protonated at neutral pH, with a pKa greater than 8.

To test whether the putative Fe^{IV} species would be catalytically active, we performed electrolysis reactions with CAM—DDAB films on graphite electrodes. Of the substrates investigated (including C_3–C_{12} alkanes, fatty acids, and styrene), only thioanisole was oxidized by the system. Clearly, the Fe^{IV}=OH unit is not powerful enough to perform more challenging alkane or alkene oxidations (in DDAB). This is hardly unexpected, as the active heme center in the native system (compound I) is one full oxidation state higher than our electrochemically accessible compound II.

Mayer has proposed [9] that the key determinant for C—H bond activation by metal–oxo species is the strength of the resulting $M^{(n-1)+}$—O—H bond following hydrogen atom abstraction: if the O—H bond (D(O—H)) in the formally one-electron-reduced

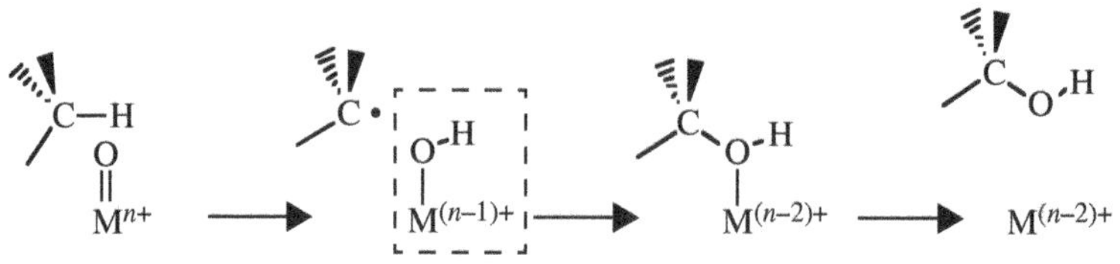

FIGURE 7.13 Rebound mechanism for metal–oxo C—H bond activation. The key O—H bond is highlighted.

$$\mathrm{Fe^{IV}{=}O{-}H + e^- \rightleftharpoons Fe^{III}{-}O{-}H}$$

$$\mathrm{1/2H_2 \rightleftharpoons H^+ + e^-}$$

$$\mathrm{Fe^{III}{-}O{-}H + H^+ \rightleftharpoons Fe^{III}{-}OH_2}$$

$$\mathrm{H\bullet \rightleftharpoons 1/2H_2}$$

$$\mathrm{Fe^{IV}{=}O{-}H + H\bullet \rightleftharpoons Fe^{III}{-}OH_2}$$

FIGURE 7.14 Thermodynamic cycle for calculating D(O—H) for Fe(III)—OH_2. Reproduced from Ref. [90] by permission of Elsevier. © Elsevier.

"rebound" complex is stronger than the C—H bond (D(C—H)) being broken, the reaction can proceed (Fig. 7.13). D(O—H) is equal to $-\Delta H°$ for the reaction: $M^{n+} = O + H\bullet \rightleftharpoons M^{(n-1)+}{-}OH$. This value can be estimated using a simple thermodynamic cycle (Fig. 7.14, where $M^{n+}{=}O$ is $Fe^{IV}{=}OH$). Values for the individual reactions in this cycle are available either experimentally (e.g., the potential of the metal redox couple ($Fe^{IV/III}$) and pK_a for $Fe^{III}{-}OH_2$) or have been calculated (e.g., the bond energy of H_2 in aqueous solution). Bordwell [91] has proposed an expression for D(O—H) based on this cycle:

$$\mathrm{D(O{-}H)(kcal/mol)} = 23.06 \times \left[E°\left(\mathrm{Fe^{IV/III}}\right)\right] + 1.37 \times \mathrm{p}K_\mathrm{a}\left(\mathrm{Fe^{III}{-}OH_2}\right) + 57 \tag{7.6}$$

Green et al. used this approach to estimate D(O—H) for compound II in chloroperoxidase at 98 kcal mol^{-1}, making the corresponding compound I potent enough to oxidize most alkanes (D(C—H) ~ 95–99 kcal mol^{-1}) [10]. In contrast, D(O—H) for $Fe^{III}{-}OH_2$ in CAM–DDAB films (calculated at pH 6) is only approximately 94 kcal mol^{-1} (based on $E°$ for $Fe^{IV/III}$ of 1.14 V vs. NHE and an electrochemically measured value for pK_a ($Fe^{III}{-}OH_2$) of 9). This provides an explanation for the inability of the system to perform reactions more demanding than thioanisole oxidation. Indeed, were it not for the substantial upshift in potential that occurs upon immobilizing the enzyme on electrode surfaces, the electrochemically generated compound II would be even less active: assuming that the $Fe^{IV/III}$ couple is shifted by an amount comparable to that for $Fe^{III/II}$ (~300 mV vs. solution), surface effects account for a boost of nearly 7 kcal/mol in D(O—H) for Fe^{III}—OH_2. As such, by exploiting surface effects, it may be possible to turn the electrochemically accessible compound II into a more potent oxidant or even access compound I in the absence of dioxygen.

7.3.6 Other P450–Electrode Films

In addition to surfactant film systems, several other techniques have been described for achieving P450 electrochemistry with surface-confined methodologies. Polyelectrolyte films have been used with various P450s for electrochemical studies [49, 52]. The polyions are coadsorbed with the protein in distinct layers: alternate layer-by-layer adsorption of the molecules is achieved by multiple rounds of soaking the electrode in a polyion solution, washing, and then soaking in an enzyme solution. Good electronic coupling can be achieved with these systems with standard ET rate constants (k°, $\Delta G^\circ = 0$) typically between 100 and 300 s^{-1}. The immobilized P450s catalytically epoxidize styrene to styrene oxide (1–6 mol substrate/mol enzyme/h); the reaction operates through the peroxide shunt, with peroxide being generated from electrochemically mediated P450-catalyzed dioxygen reduction. Polystyrenesulfonate layers have been used to study electrochemistry with mammalian P450 1A2 in experiments that yielded catalytic turnover of styrene [92]. Styrene oxidation also was studied in films with CAM and DNA as the polyanionic film [93]; in this case, styrene oxide-mediated DNA damage was found. Generally, styrene oxidation rates in these systems are on the order of 10 turnovers/h.

Lei et al. studied electrochemistry with CAM using sodium montmorillonite (carbon-based polyanionic clay) to coat the surface of an electrode [18]. The anionic clay was proposed to interact electrostatically with a region on the surface of CAM that is positively charged; this region also is presumed to be the binding site of the native reductase. Rapid, reversible ET ($k^\circ \sim 150\ s^{-1}$) as well as catalytic reduction of dioxygen was observed. In other work, Shumyantseva et al. cofilmed P450 2B4 with montmorillonite and the nonionic detergent Tween 80 [94]. Tween 80 keeps 2B4 monomeric: the slow ET in the absence of Tween 80 was attributed to protein aggregation that blocked charge transfer. Differential catalytic currents for 2B4 substrates aminopyrine and benzphetamine paralleled enzymatic specificity for these substrates, demonstrating biosensor potential for the system.

Fantuzzi et al. utilized self-assembled mercaptan films on gold to achieve ET with mammalian P450 2E1 [20]. Indeed, alkanethiol monolayers on gold electrodes are commonly used to provide surfaces that interact favorably with various proteins to achieve ET [95]. For 2E1, the gold-mercaptan surface was further derivatized using either maleimide or poly(diallyldimethylammonium chloride). While electronic coupling was not as strong as in the aforementioned systems (k° between 2 and 10 s^{-1}), catalytic substrate oxidation was observed. The group has since followed up this work with analogous films and single-surface cysteine mutants to show that the orientation of 2E1 on the electrode is important for activity: orienting the enzyme with the substrate access channel toward the solution and the putative ET transfer site adjacent to the electrode yields good substrate oxidation [96]. Although this type of system is not practical for large-scale substrate turnover [97],[5] it could potentially be used for biosensing.

7.3.7 Electrochemical Wiring

Direct electrochemical wiring of P450 can be accomplished through covalent modification of the protein. Attaching a molecular electronic relay to the protein surface can result in a system well coupled for ET from the electrode to the heme. Site-specifically tagging the protein with a molecular anchor can define the orientation of the enzyme on the electrode. Further, varying the point of attachment among adjacent surface amino acids should permit varying the electronic coupling to the heme while having a minimal effect on surface orientation. However, once surface confined, aggregation and surface adsorption (leading to denaturation) would hinder native polypeptide dynamics, thereby affecting catalytic function. Achieving effective wiring includes (i) controlling the spacing of the protein on the surface to disfavor aggregation, (ii) defining the nature of the surface to prevent protein adsorption, (iii) controlling protein orientation on the surface to allow the substrate access channel to bind substrates, and (iv) choosing the precise position for protein modification as the wire must be attached at a site that is well coupled for ET.

Gilardi et al. utilized flavodoxin as an electrochemical wire for reduction of hBM3 [98]. A chimeric gene consisting of hBM3 and flavodoxin was created by fusing the 3′ end of the hBM3 gene to the 5′ end of the flavodoxin gene through a short nucleotide that encoded the native linker between hBM3 and its reductase. The linker allowed the two proteins some separation so they could fold independently and retain native solution dynamics. Modeling studies revealed that the closest approach between the two cofactors is 12 Å, putting the molecules in proximity for efficient ET. Photochemical experiments on the resulting fusion protein using deazariboflavin to reduce flavodoxin showed that the flavin to heme ET rate is 370 s^{-1}, verifying that the two proteins are electronically well coupled. Electrochemical experiments at a glassy carbon electrode resulted in enhanced ET to hBM3.

In an effort to achieve direct wiring of hBM3 to an electrode (Fig. 7.15), we followed Sevrioukova et al. who observed rapid hBM3 photoreduction (2.5×10^6 s^{-1}) by covalently tethering a ruthenium diimine to an engineered cysteine (N387C) on the protein [99]. The position of the ruthenium complex was selected to mimic the interaction between hBM3 and its reductase. It occurred to us that wiring the N387C hBM3 mutant to an electrode through the engineered cysteine also could yield high electron tunneling rates. Previously, Katz had utilized *N*-(1-pyrene)iodoacetamide (Py) (thiol specific) to anchor and electronically connect a photosynthetic reaction center to a BPG electrode [100]. Thus, we made the hBM3 N387C mutant, attached Py to the cysteine, and successfully achieved ET employing a Py-modified BPG electrode

[5] There are several difficulties with scaling up this system, most of which are related to properly preparing the gold surface (see Ref. [97] for a discussion on this problem).

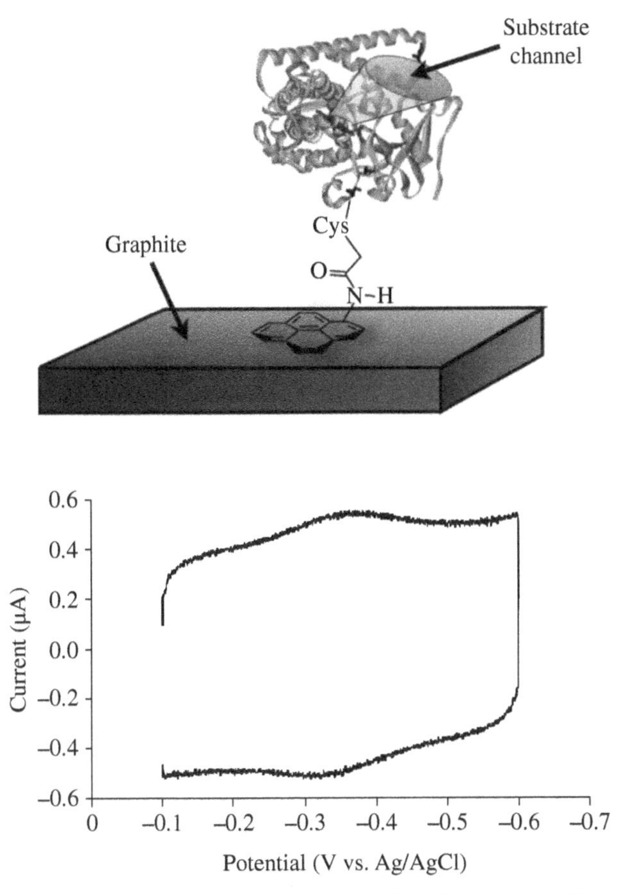

FIGURE 7.15 Schematic representation (top) and cyclic voltammogram (bottom) of pyrene-wired hBM3 to BPG electrode. Integration of the CV trace yields a surface coverage of 1.2 pmol cm^{-2} or approximately 40% monolayer coverage based on the heme-protein dimensions.

[101]. The voltammogram in Figure 7.15 shows that the $Fe^{III/II}$ couple is at –340 mV (vs. AgCl/Ag). AFM images revealed that only pyrene-wired enzyme molecules are adsorbed to the electrode, resulting in protein clusters on the surface 2–5 nm high and 30–40 nm wide. The enzyme–electrode system undergoes rapid and reversible ET: $k°$ was estimated to be 650 s^{-1}, underscoring the relatively strong coupling we achieved through the Cys^{387}-wired enzyme. Voltammetry in the presence of dioxygen resulted in large catalytic reduction currents. Independent analytical assays using fluorescence and RDE techniques demonstrated that the wired enzyme catalyzes four-electron reduction of dioxygen to water. However, catalytic substrate oxidation was not observed. Instead of dioxygen activation, we believe our system rapidly reduces dioxygen to water without forming catalytically active ferryl species. Notably, while our initial goal was to achieve rapid ET, it is now clear that control, rather than speed,

is the key for proper O_2 activation. Cys^{387} evidently taps into an ET pathway that is too fast: before the two-electron-reduced ferric-peroxy species can produce compound I, reduction by an additional two electrons likely occurs, leading to formation of water. ET must be properly coordinated to a series of chemical events that ultimately leads to formation of compound I at the right moment for oxygen atom transfer.

Pyrene wiring not only can be exploited as a general method for P450 wiring [102], but it also can be used as a tool to systematically probe P450 surfaces for ET "hot-spots." The location of the protein labeling site should have a profound effect on electronic coupling between the electrode and the heme [14]. Our pyrene-wiring technique therefore provides a method to fine-tune the ET rate by systematically interrogating potential labeling sites on the enzyme surface. In an effort to create a topological map of potential ET pathways for use in the subsequent design of a catalytic electrode, we investigated different single-surface cysteine hBM3 mutants. Our goal was to find a system in which the ET rate is *slower* than formation of compound I from the $2e^-$-reduced ferric-hydroperoxy adduct (to prevent further dioxygen reduction to water) yet fast enough to ensure both generation of the ferric-hydroperoxy complex in the first place and a practical rate of catalytic turnover. Of no less importance was to find surface labeling sites such that the substrate binding channel and proton-delivery conduits were not perturbed following surface immobilization.

In exploratory investigations [103], we prepared hBM3 surface cysteine mutants at positions 62, 97, 383, and 397. Of these mutants, only Cys^{62} and Cys^{383} (Fig. 7.16) yielded electrochemical responses following labeling with Py. The respective side chain distances from Cys^{62} and Cys^{383} to the heme-iron center are 16 and 25 Å (cf. Cys^{387} spacing of 19 Å). Notably, while Cys^{62} is spatially closest to the heme, it does not provide a through-bond pathway for reduction; the crystal structure shows

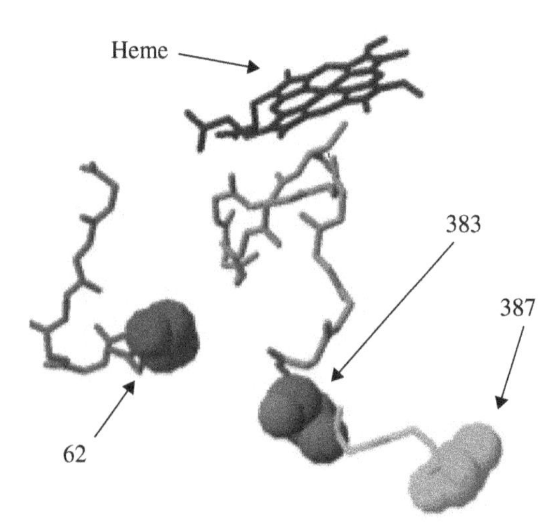

FIGURE 7.16 Crystal structure of the cytochrome P450 BM3 heme domain. The heme and the position of the amino acid (aa) substitutions to surface cysteines are shown.

a 5.5 Å gap between the heme edge and Lys^{69}—the residue closest to the heme along the peptide fold in proximity to Cys^{62}.

Each of these Py-labeled P450 mutants, which bind strongly to graphite (~50% surface coverage as determined by coulometry, similar to the Cys^{387} mutant), exhibits a reversible $Fe^{III/II}$ redox couple under anaerobic conditions at approximately −340 mV versus AgCl/Ag. Based on variable scan-rate cyclic voltammetric measurements [104], we calculate standard ($\Delta G^\circ = 0$) heterogeneous ET rate constants, k_s°, of 0.8(1) s^{-1} and 50(5) s^{-1} for Py-Cys^{383} and Py-Cys^{62}. Experimentally validated timetables [14] predict that the rate of ET through a protein matrix should drop exponentially with distance (r) according to (7.7), with a distance-decay constant, β, bracketed between 1 and 1.3 $Å^{-1}$ (depending on secondary structure elements within the tunneling pathway):

$$k_r = k_o \exp(-\beta r) \quad (7.7)$$

Indeed, the ET rate for Py-Cys^{383} is virtually identical with that predicted by (7.7), assuming a β of 1 $Å^{-1}$ and an increased tunneling distance of 6 Å relative to Py-Cys^{387} (Fig. 7.17). In contrast, Py-Cys^{62} exhibits a k_s° value approximately 10 times less than that for Py-Cys^{387}, even though the cysteine/heme spacing is smaller. As observed in previous ET-pathway studies [14], the through-space "jump" in the direct pathway between Cys^{62} and the heme center apparently drastically decreases the electronic coupling, thereby attenuating the ET rate.

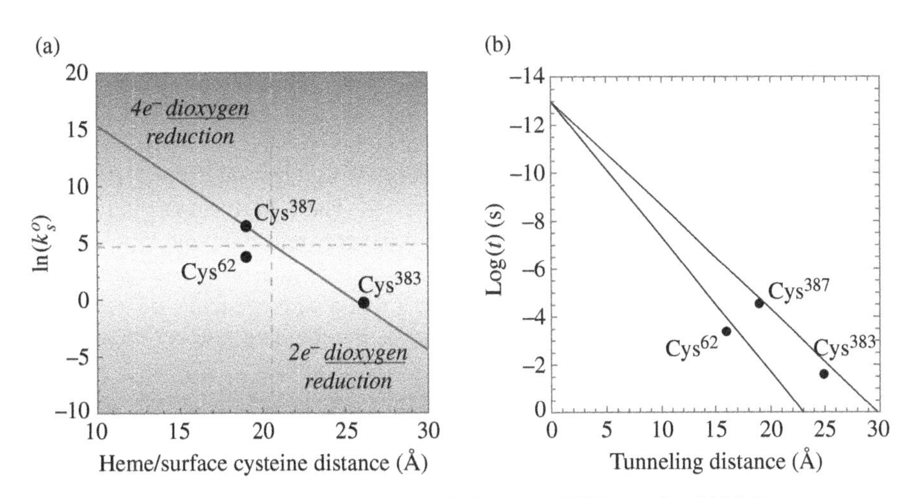

FIGURE 7.17 (a) Distance dependence of k° for the FeIII/II couple of hBM3 mutants as a function of surface cysteine/heme spacing. The data suggest that systems in which k° is greater than ~10 s^{-1} exhibit 4e⁻ dioxygen reduction, while those with slower rates result primarily in 2e⁻ reduction to peroxide. (b) Calculated activationless ($-\Delta G^\circ = \lambda$) ET tunneling times for hBM3 mutants based on a heme reorganizational energy, λ, of 0.8 eV. The straight lines correspond to theoretical activationless tunneling times assuming $\beta = 1$ $Å^{-1}$ (top) and 1.3 $Å^{-1}$ (bottom). Reproduced from Ref. [103] by permission of Elsevier. © Elsevier.

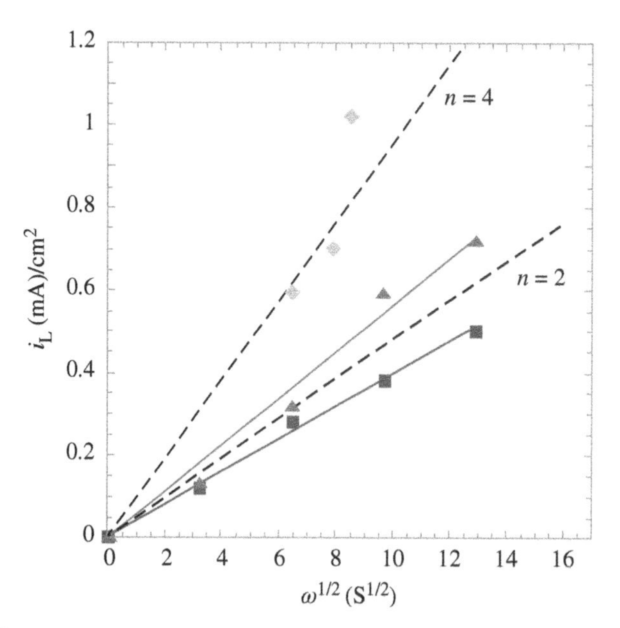

FIGURE 7.18 Levich plots for hBM3 387 (diamonds, Ref. [14]), 62 (triangles), and 383 (squares). Dashed lines corresponding to current densities predicted for $n = 2$ and $n = 4$ are shown for comparison. Reproduced from Ref. [103] by permission of Elsevier. © Elsevier.

Both Py-Cys^{383} and Py-Cys^{62} exhibit catalytic dioxygen reduction at applied potentials sufficiently negative to produce the ferrous heme centers, confirming their ability to activate dioxygen. We examined the electrochemistry of Py-Cys^{383} and Py-Cys^{62} films using graphite RDE to determine the products of catalytic dioxygen reduction (water and/or peroxide). Levich plots for Py-Cys^{62} and Py-Cys^{383}, along with data for Py-Cys^{387}, are shown in Figure 7.18. These RDE studies indicate that in contrast to the findings for Py-Cys^{387}, dioxygen reduction by Py-Cys^{62} and Py-Cys^{383} occurs primarily via 2e$^-$ reduction to peroxide ($n = 2.5(4)$ and 1.8(2), respectively). By extension, these data suggest that a heterogeneous ET rate on the order of 10 s^{-1} for ferric heme reduction may be the dividing line between 2e$^-$ and 4e$^-$ dioxygen reactivities for this system.

Clearly correlating k_s° with the products of dioxygen reduction has important consequences in designing electrode-bound P450s that can be activated via electrochemical methods to generate compound I. The rate of ET into ferric P450 must be tuned such that the hydroperoxy "compound 0" species has sufficient time to rearrange into compound I *before* delivery of additional electrons that ultimately will reduce dioxygen to water. In addition to uncoupling pathways exhibited by the native enzyme system, rapid and unregulated electrochemical reduction of the heme center constitutes a 4e$^-$-uncoupling mechanism at electrode surfaces. Yet as illustrated earlier, engineering surface cysteine mutants suitable for pyrene labeling offers a straightforward approach to modulate electronic coupling into the heme and hence tune the rate of

ET into the iron center. These studies therefore provide a preliminary roadmap for designing hBM3 mutants that undergo electrochemical reduction at rates suitable for native-like activity.

7.4 THOUGHTS ABOUT THE FUTURE OF P450 ELECTROCHEMISTRY

Enormous benefits could come from exploiting P450 activity *in vitro*, and work along these lines remains a central theme in the field. Several reviews have outlined the various uses of P450s in biotechnology, as well as the challenges and potential solutions of capturing this oxygenase activity *in vitro* [22, 105–110]. Perhaps, the greatest obstacle in the way of practical application is an intricate redox machinery that includes lengthy ET chains that utilize several intermediate proteins and cofactors, each of which makes a specific interaction in order to achieve catalysis. This problem has motivated studies aimed at replacing the ET machinery required for P450 catalysis, some of which we have discussed.

Electrochemical methods represent the simplest way of providing the necessary reducing equivalents for P450 catalysis. The first reports of P450 electrochemistry were published approximately 30 years ago; determination of the redox potential of rabbit liver P450 with electrochemical mediators [111] was followed shortly by direct electrochemistry with a mercury electrode [112]. Although great advances in P450 electrochemistry have been made in the last few years, a practical system has yet to be found. We can step back at this point and ask: why not? First, direct electrochemical wiring is not practical, owing to the difficult step of protein modification, which is both inefficient and harsh for a marginally stable P450. Regarding thin-film systems, these are best suited for biosensors, perhaps something that can be used on a chip for screening purposes. However, confining P450s to surfaces often calls into question the integrity of the protein: is the protein properly folded, does it retain its native catalytic and substrate discriminatory properties, and what is its long-term stability? Finally, mediated electrochemical P450 systems are perhaps the best bets for large-scale biocatalysis. Unfortunately, the problem of uncoupling through dioxygen reduction by mediators has not been satisfactorily addressed.

Key questions regarding the feasibility of a functional P450–electrode system are as follows:

- *Can substrates access the heme of electrode-bound P450s?* If a compact protein conformation is indeed favored on the electrode surface, as suggested from our experiments with BM3, then it would be difficult for substrates to access the P450 active site. This problem is compounded if the active site is blocked through interactions with the electrode surface, neighboring proteins, or if the protein is positioned incorrectly (i.e., substrate access channel not oriented directly opposite to the electrode surface). To address this problem, it is imperative that we control the orientation of the protein on the electrode surface.

- *Do we need to regulate P450 ET?* An electrode poised at a negative potential is a continuous source of electrons for P450, resulting in unregulated ET into the heme. Regulating ET may be the key to achieving substrate oxidation: recall that the native system utilizes a series of cofactors (e.g., NADPH, FAD, FMN) and ET proteins (e.g., reductases, ferredoxins), as well as several internal thermodynamic controls (e.g., heme spin shift that accompanies substrate binding) to accomplish this feat. Thus, simply "pumping" electrons into the P450 heme may never work. Instead, it may be necessary to identify and utilize discrete ET pathways that result in heme reduction at physiologically relevant rates. Employing such strategy could minimize uncoupling due to ET that is either too slow (formation of superoxide) or too rapid (formation of water), thereby allowing generation of the ferric-peroxy species with a natural lifetime sufficiently long to permit subsequent conversion to compound I.
- *Can we solve problems associated with dioxygen?* Direct dioxygen reduction at the electrode, uncoupling of ET resulting from improper dioxygen activation by the enzyme, and supplying dioxygen in sufficient quantities for continuous turnover are matters that must be considered.

At this point, we highlight important recent reports. Expanding on prior thin-film efforts, Rusling coadsorbed a large stoichiometric excess of purified mammalian P450s (separately, 1A2 and 2E1) with microsome-bound P450 reductase in an alternate layer-by-layer system using several different polyions [113]. The films were layered to promote vectorial ET from the electrode to the reductase and finally the heme domains, with the idea of maintaining native ET pathways and utilizing electrochemically accessible reductase cofactors for the electrode-mediated reaction. While the authors noted considerable layer intermixing (hence negating layer-by-layer deposition), their data suggest native-like ET. Significant findings include reductase-mediated heme reduction, essentially no uncoupling, and a 1.5-fold larger turnover rate for the electrochemical versus NADPH reaction. Notably, this methodology combines elements of the thin-film and mediated systems to yield a highly coupled, native-like catalytic P450 electrode. While successful, as mentioned earlier, we should point out that slow mammalian turnover rates combined with inherent instabilities (further underscored by the need for reductase microsomes) will disfavor practical application. Nonetheless, this work represents a step forward.

Gilardi's group has established a foundation for genetic manipulation of P450 ("molecular Lego") and modification of gold electrodes to achieve P450 electrochemistry. Building on this work, the group has taken a fundamental step toward creating a P450 catalytic device by constructing a microfluidic cell [22]. The gold chip working electrode was modified with a mercaptan SAM followed by covalent immobilization of a 3A4–flavodoxin construct generated by molecular Lego. Although details are pending, this preliminary description points toward a new direction for P450 electrochemistry that could lead to the ultimate goal of realizing a practical catalytic system.

ACKNOWLEDGMENTS

Work at Occidental College is supported by the donors of the American Chemical Society Petroleum Research Fund (AKU and MGH), the Dreyfus Foundation (AKU), and the Howard Hughes Medical Institute, and work at Caltech by NIH DK019038.

REFERENCES

1. Lewis, D. F. V. (2001) *Guide to Cytochromes P450: Structure and Function*, Taylor and Francis, New York.
2. Ortiz de Montellano, P. R. (Ed.) (2004) *Cytochrome P450: Structure, Mechanism, and Biochemistry*, 3rd ed., Springer, New York.
3. Phillips, I. R. and Shephard, E. A. (Eds.) (1998) *Cytochrome P450 Protocols*, Vol. 107, Humana Press, Totowa.
4. Guengerich, F. P. (2003) Cytochromes P450, drugs, and diseases, *Mol. Interv. 3*, 194–204.
5. Glieder, A., Farinas, E. T., and Arnold, F. H. (2002) Laboratory evolution of a soluble, self-sufficient, highly active alkane hydroxylase, *Nat. Biotechnol. 20*, 1135–1139.
6. Joseph, S., Rusling, J. F., Lvov, Y. M., Friedberg, T., and Fuhr, U. (2003) An amperometric biosensor with human 3A4 as a novel drug screening tool, *Biochem. Pharmacol. 65*, 1817–1826.
7. U.S. Food and Drug Administration. Table of Pharmacogenomic Biomarkers in Drug Labeling. http://www.fda.gov/Drugs/ScienceResearch/ResearchAreas/Pharmacogenetics/ucm083378.htm (accessed on September 3, 2014).
8. Labinger, J. A. and Bercaw, J. E. (2002) Understanding and exploiting C—H bond activation, *Nature 417*, 507–514.
9. Mayer, J. M. (1998) Hydrogen atom abstraction by metal-oxo complexes: understanding the analogy with organic radical reactions, *Acc. Chem. Res. 31*, 441–450.
10. Green, M. T., Dawson, J. H., and Gray, H. B. (2004) Oxoiron(IV) in chloroperoxidase compound II is basic: implications for P450 chemistry, *Science 304*, 1653–1656.
11. Daff, S., Chapman, S., Turner, K., Holt, R., Govindaraj, S., Poulos, T., and Munro, A. (1997) Redox control of the catalytic cycle of flavocytochrome P-450 BM3, *Biochemistry 36*, 13816–13823.
12. Rittle, J. and Green, M. T. (2010) Cytochrome P450 compound I: capture, characterization, and C—H bond activation kinetics, *Science 330*, 933–937.
13. Altarsha, M., Benighaus, T., Kumar, D., and Thiel, W. (2010) Coupling and uncoupling mechanisms in the methoxythreonine mutant of cytochrome P450cam: a quantum mechanical/molecular mechanical study, *J. Biol. Inorg. Chem. 15*, 361–372.
14. Gray, H. B. and Winkler, J. R. (2003) Electron tunneling through proteins, *Quart. Rev. Biophys. 36*, 341–372.
15. Gray, H. B. and Winkler, J. R. (1996) Electron transfer in proteins, *Annu. Rev. Biochem. 65*, 537–561.
16. Hawkridge, F. M. and Taniguchi, I. (2000) Electrochemistry of heme proteins. In *The Porphyrin Handbook* (Kadish, K. M., Smith, K. M., and Guilard, R., Eds.), Academic Press, San Diego, CA.

17. Kazlauskaite, J., Westlake, A. C. G., Wong, L.-L., and Hill, H. A. O. (1996) Direct electrochemistry of cytochrome P450CAM, *Chem. Commun.*, 2189–2190.

18. Lei, C., Wollenberger, U., Jung, C., and Scheller, F. W. (2000) Clay-bridged electron transfer between cytochrome P450CAM and electrode, *Biochem. Biophys. Res. Commun. 268*, 740–744.

19. Holtmann, D., Mangold, K.-M., and Schrader, J. (2009) Entrapment of cytochrome P450 BM-3 in polypyrrole for electrochemically-driven biocatalysis, *Biotechnol. Lett. 31*, 765–770.

20. Fantuzzi, A., Fairhead, M., and Gilardi, G. (2004) Direct electrochemistry of immobilized human cytochrome P450 2E1, *J. Am. Chem. Soc. 126*, 5040–5041.

21. Shumyantseva, V., Bulko, T., Bachmann, T., Bilitewski, U., Schmid, R., and Archakov, A. (2000) Electrochemical reduction of flavocytochromes 2B4 and 1A2 and their catalytic activity, *Arch. Biochem. Biophys. 377*, 43–48.

22. Sadeghi, S., Fantuzzi, A., and Gilardi, G. (2011) Breakthrough in P450 bioelectrochemistry and future perspectives, *Biochim. Biophys. Acta 1814*, 237–248.

23. Munro, A. W., Leys, D. G., McLean, K. J., Marshall, K. R., Ost, T. W. B., Daff, S., Miles, C. S., Chapman, S. K., Lysek, D. A., Moser, C. C., Page, C. C., and Dutton, P. L. (2002) P450 BM3: the very model of a modern flavocytochrome. *Trends Biochem. Sci. 27*, 250–257.

24. Dietrich, J. A., Yoshikuni, Y., Fisher, K. J., Woolard, F. X., Ockey, D., McPhee, D. J., Renninger, N. S., Chang, M. C. Y., Baker, D., and Keasling, J. D. (2009) A novel semi-biosynthetic route for artemisinin production using engineered substrate-promiscuous P450BM3. *ACS Chem. Biol. 4*, 261–267.

25. Lewis, J. C. and Arnold, F. H. (2009) Catalysts on demand: selective oxidations by laboratory-evolved cytochrome P450 BM3, *Chimia 63*, 309–312.

26. Ravichandran, K. G., Boddupalli, S. S., Hasemann, C. A., Peterson, J. A., and Deisenhofer, J. (1993) Crystal structure of hemoprotein domain of P450-BM3, a prototype for microsomal P450s, *Science 261*, 731–736.

27. Schwaneberg, U., Otey, C., Cirino, P. C., Farinas, E., and Arnold, F. H. (2001) Cost-effective whole-cell assay for laboratory evolution of hydroxylases in *Escherichia coli*, *J. Biomol. Screen. 6*, 111.

28. Schewe, H., Holtmann, D., and Schrader, J. (2009) P450BM-3-catalyzed whole-cell biotransformation of a-pinene with recombinant *Escherichia coli* in an aqueous-organic two-phase system, *Appl. Microbiol. Biotechnol. 83*, 849–857.

29. Estabrook, R., Faulkner, K., Shet, M., and Fisher, C. (1996) Application of electrochemistry for P450-catalyzed reactions, *Methods Enzymol. 272*, 44–51.

30. Faulkner, K., Shet, M., Fisher, C., and Estabrook, R. (1995) Electrocatalytically driven ω-hydroxylation of fatty acids using cytochrome P450 4A1, *Proc. Natl. Acad. Sci. USA 92*, 7705–7709.

31. Schwaneberg, U., Appel, D., Schmitt, J., and Schmid, R. (2000) P450 in biotechnology: zinc driven ω-hydroxylation of p-nitrophenoxydodecanoic acid using P450 BM-3 F87A as a catalyst, *J. Biotechnol. 84*, 249–257.

32. Heller, A. and Degani, Y. (1988) Direct electrical communication between chemically modified enzymes and metal electrodes. 2. Methods for bonding electron-transfer relays to glucose oxidase and d-amino-acid oxidase, *J. Am. Chem. Soc. 110*, 2615–2620.

33. Udit, A. K., Arnold, F. H., and Gray, H. B. (2004) Cobaltocene-mediated catalytic monooxygenation using holo and heme domain cytochrome P450 BM3, *J. Inorg. Biochem. 98*, 1547–1550.

34. Reipa, V., Mayhew, M. P., and Vilker, V. L. (1997) A direct electrode-driven P450 cycle for biocatalysis, *Proc. Natl. Acad. Sci. USA 94*, 13554–13558.

35. Hollmann, F., Schmid, A., and Steckhan, E. (2001) The first synthetic application of a monooxygenase employing indirect electrochemical NADH regeneration, *Angew. Chem. Int. Ed. 40*, 169–171.

36. Hollmann, F., Witholt, B., and Schmid, A. (2002) $[Cp^*Rh(bpy)(H_2O)]^{2+}$: a versatile tool for efficient and non-enzymatic regeneration of nicotinamide and flavin coenzymes *J. Mol. Catal. B Enzym. 19–20*, 167–176.

37. Hollmann, F., Lin, P.-C., Witholt, B., and Schmid, A. (2003) Stereospecific biocatalytic epoxidation: the first example of direct regeneration of a FAD-dependent monooxygenase for catalysis, *J. Am. Chem. Soc. 125*, 8209–8217.

38. Zhang, Z., Nassar, A.-E., Lu, Z., Schenkman, J. B., and Rusling, J. F. (1997) Direct electron injection from electrodes to cytochrome P450CAM in biomembrane-like films. *J. Chem. Soc., Faraday Trans. 93*, 1769–1774.

39. Rusling, J. F. and Nassar, A.-E. F. (1993) Enhanced electron transfer for myoglobin in surfactant films on electrodes, *J. Am. Chem. Soc. 115*, 11891–11897.

40. Zu, X., Lu, Z., Zhang, Z., Schenkman, J. B., and Rusling, J. F. (1999) Electroenzyme-catalyzed oxidation of styrene and cis-β-methylstyrene using thin films of cytochrome P450cam and myoglobin, *Langmuir 15*, 7372–7377.

41. Farmer, P., Lin, R., and Bayachou, M. (1998) Electrochemistry and catalysis by myoglobin in surfactant films. *Comments Inorg. Chem. 20*, 101–120.

42. Rusling, J. F. (1998) Enzyme bioelectrochemistry in cast biomembrane-like films, *Acc. Chem. Res. 31*, 363–369.

43. Fleming, B. D., Tian, Y., Bell, S. G., Wong, L., Urlacher, V., and Hill, H. A. O. (2003) Redox properties of cytochrome P450 BM3 measured by direct methods, *Eur. J. Biochem. 270*, 4082–4088.

44. Udit, A. K., Hindoyan, N., Hill, M. G., Arnold, F. H., and Gray, H. B. (2005) Protein-surfactant film voltammetry of wild type and mutant cytochrome P450 BM3, *Inorg. Chem. 44*, 4109–4111.

45. Aguey-Zinsou, K., Bernhardt, P. V., Voss, J. J. D., and Slessor, K. E. (2003) Electrochemistry of P450cin: new insights into P450 electron transfer, *Chem. Commun.*, 418–419.

46. Shukla, A., Gillam, E. M., Mitchell, D. J., and Bernhardt, P. V. (2005) Direct electrochemistry of enzymes from the cytochrome P450 2C family, *Electrochem. Commun. 7*, 437–442.

47. Zhang, L., Liu, X., Wang, C., Liu, X., Cheng, G., and Wu, Y. (2010) Expression, purification and direct electrochemistry of cytochrome P450 6A1 from the house fly, *Musca domestica*, *Protein Expr. Purif. 71*, 74–78.

48. Rhieu, S., Ludwig, D., Siu, V., and Palmore, G. (2009) Direct electrochemistry of cytochrome P450 27B1 in surfactant films, *Electrochem. Commun. 11*, 1857–1860.

49. Munge, B., Estavillo, C., Schenkman, J. B., and Rusling, J. F. (2003) Optimization of electrochemical and peroxide-driven oxidation of styrene with ultrathin polyion films containing cytochrome P450CAM and myoglobin, *ChemBioChem 4*, 82–89.

50. Immoos, C. E., Di Bilio, A. J., Cohen, M. S., Van der Veer, W., Gray, H. B., and Farmer, P. J. (2004) Electron-transfer chemistry of Ru-linker-(heme)-modified myoglobin: rapid intraprotein reduction of a photogenerated porphyrin cation radical. *Inorg. Chem. 43*, 3593–3596.
51. Blair, E., Greaves, J., and Farmer, P. J. (2004) High-temperature electrocatalysis using thermophilic P450 CYP119: dehalogenation of CCl_4 to CH_4, *J. Am. Chem. Soc. 126*, 8632–8633.
52. Lvov, Y. M., Lu, Z., Schenkman, J. B., Zu, X., and Rusling, J. F. (1998) Direct electrochemistry of myoglobin and cytochrome P450CAM in alternate layer-by-layer films with DNA and other polyanions, *J. Am. Chem. Soc. 120*, 4073–4080.
53. Ma, H. and Hu, N. (2001) Electrochemistry and electrocatalysis with myoglobin in 2C12N +PSS-multibilayer composite films, *Anal. Lett. 34*, 339–361.
54. Levich, V. (Ed.) (1962) *Physicochemical Hydrodynamics*, Prentice Hall, Englewood Cliffs, NJ.
55. Koutecky, J., and Levich, V. (1956) The application of the rotating disc electrode to studies of kinetic and catalytic processes, *Zh. Fiz. Khim. 32*, 1565–1575.
56. Andrieux, C., and Saveant, J.-M. (1992) *Molecular Design of Electrode Surfaces*, John Wiley & Sons, Inc, New York.
57. Noble, M., Miles, C., Chapman, S., Lysek, D., Mackay, A., Reid, G., Hanzlik, R., and Munro, A. (1999) Roles of key active-site residues in flavocytochrome P450 BM3, *Biochem. J. 339*, 371–379.
58. Peters, M. W., Meinhold, P., Glieder, A., and Arnold, F. H. (2003) Regio- and enantioselective alkane hydroxylation with engineered cytochromes P450 BM-3, *J. Am. Chem. Soc. 125*, 13442–13450.
59. Laviron, E. (1979) General expression of the linear potential sweep voltammogram in the case of diffusionless electrochemical systems, *J. Electroanal. Chem. 101*, 19–28.
60. Kadkhodayan, S., Coulter, E. D., Maryniak, D. M., Bryson, T. A., and Dawson, J. H. (1995) Uncoupling oxygen transfer and electron transfer in the oxygenation of camphor analogues by cytochrome P450-CAM, *J. Biol. Chem. 270*, 28042–28048.
61. Wong, L. L., Westlake, C. G., and Nickerson, D. P. (1997) Metal sites in proteins and models, In *Structure and Bonding*, pp. 175–207, Springer-Verlag, Berlin.
62. Boddupalli, S., Estabrook, R., and Peterson, J. (1990) Fatty acid monooxygenation by cytochrome P-450 BM-3, *J. Biol. Chem. 265*, 4233–4239.
63. Anson, F. C., Shi, C., and Steiger, B. (1997) Novel multinuclear catalysts for the electroreduction of dioxygen directly to water, *Acc. Chem. Res. 30*, 437–444.
64. Chang, C. J., Loh, Z.-H., Shi, C., Anson, F. C., and Nocera, D. G. (2004) Targeted proton delivery in the catalyzed reduction of oxygen to water by bimetallic pacman porphyrins, *J. Am. Chem. Soc. 126*, 10013–10020.
65. Udit, A. K., Belliston-Bittner, W., Glazer, E. C., Nguyen, Y. H. L., Gillan, J. M., Hill, M. G., Marletta, M. A., Goodin, D. B., and Gray, H. B. (2005) Redox couples of inducible nitric oxide synthase, *J. Am. Chem. Soc. 127*, 11212–11213.
66. Udit, A. K., Hagen, K. D., Gillan, J. M., Goldman, P. J., Star, A., Gray, H. B., and Hill, M. G. (2006) Spectroscopy and electrochemistry of cytochrome P450 BM3-surfactant films assemblies, *J. Am. Chem. Soc. 128*, 10320–10325.
67. Yee, E. L., Cave, R. J., Guyer, K. L., Tyma, P. D., and Weaver, M. J. (1979) A survey of ligand effects upon the reaction entropies of some transition metal redox couples, *J. Am. Chem. Soc. 101*, 1131–1137.

68. Taniguchi, V. T., Sailasuta-Scott, N., Anson, F. C., and Gray, H. B. (1980) Thermodynamics of metalloprotein electron transfer reactions. *Pure Appl. Chem.* *52*, 2275–2281.
69. Liu, X., Huang, Y., Zhang, W., Fan, G., Fan, C., and Li, G. (2005) Electrochemical investigation of redox thermodynamics of immobilized myoglobin: ionic and ligation effects, *Langmuir 21*, 375–378.
70. Liu, H., Wang, L., and Hu, N. (2002) Direct electrochemistry of hemoglobin in biomembrane-like DHP-PDDA polyion-surfactant composite films, *Electrochim. Acta 47*, 2515–2523.
71. Grealis, C. and Magner, E. (2003) Comparison of the redox properties of cytochrome c in aqueous and glycerol media, *Langmuir 19*, 1282–1286.
72. Wilker, J. J., Dmochowski, I. J., Dawson, J. H., Winkler, J. R., and Gray, H. B. (1999) Substrates for rapid delivery of electrons and holes to buried active sites in proteins. *Angew. Chem. Int. Ed. 38*, 90–92.
73. Battistuzzi, G., Borsari, M., Cowan, J. A., Ranieri, A., and Sola, M. (2002) Control of cytochrome c redox potential: axial ligation and protein environment effects, *J. Am. Chem. Soc. 124*, 5315–5324.
74. Dawson, J. H., Andersson, L. A., and Sono, M. (1983) The diverse spectroscopic properties of ferrous cytochrome P450-CAM-ligand complexes, *J. Biol. Chem. 258*, 13637–13645.
75. Poulos, T. L., Finzel, B. C., and Howard, A. J. (1986) Crystal structure of substrate-free *Pseudomonas putida* cytochrome P-450, *Biochemistry 25*, 5314–5322.
76. Tezcan, F. A., Winkler, J. R., and Gray, H. B. (1998) Effects of ligation and folding on reduction potentials of heme proteins, *J. Am. Chem. Soc. 120*, 13383–13388.
77. Glascock, M. C., Ballou, D. P., and Dawson, J. H. (2005) Direct observation of a novel perturbed oxyferrous catalytic intermediate during reduced putidaredoxin-initiated turnover of cytochrome P-450-CAM, *J. Biol. Chem. 280*, 42134–42141.
78. Yu, C.-A., and Gunsalus, I. (1974) Cytochrome P450CAM II: interconversion with P420, *J. Biol. Chem. 249*, 102–106.
79. Martinis, S. A., Blanke, S. R., Hager, L. P., Sligar, S. G., Hoa, G. H. B., Rux, J. J., and Dawson, J. H. (1996) Probing the heme iron coordination structure of pressure-induced cytochrome P450cam, *Biochemistry 35*, 14530–14536.
80. Perera, R., Sono, M., Sigman, J. A., Pfister, T. D., Lu, Y., and Dawson, J. H. (2003) Neutral thiol as a proximal ligand to ferrous heme iron: implications for heme proteins that lose cysteine thiolate ligation on reduction, *Proc. Natl. Acad. Sci. USA 100*, 3641–3646.
81. O'Keefe, D. H., Ebel, R. E., Peterson, J. A., Maxwell, J. C., and Caughey, W. S. (1978) An infrared spectroscopic study of carbon monoxide bonding to ferrous cytochrome P-450, *Biochemistry 17*, 5845–5852.
82. Nagano, S., Shimada, H., Tarumi, A., Hishiki, T., Kimata-Ariga, Y., Egawa, T., Suematsu, M., Park, S.-Y., Adachi, S., Shiro, Y., and Ishimura, Y. (2003) Infrared spectroscopic and mutational studies on putidaredoxin-induced conformational changes in ferrous CO-P450cam, *Biochemistry 42*, 14507–14514.
83. Low, D. W. and Hill, M. G. (2000) Backbone-engineered high-potential iron proteins: effects of active-site hydrogen bonding on reduction potential, *J. Am. Chem. Soc. 122*, 11039–11040.
84. Yoshioka, S., Takahashi, S., Ishimori, K., and Morishima, I. (2000) Roles of the axial push effect in cytochrome P450cam studies with the site-directed mutagenesis at the heme proximal site, *J. Inorg. Biochem. 81*, 141–151.

85. Ost, T. W. B., Clark, J., Mowat, C. G., Miles, C. S., Walkinshaw, M. D., Reid, G. A., Chapman, S. K., and Daff, S. (2003) Oxygen activation and electron transfer in flavocytochrome P450 BM3, *J. Am. Chem. Soc. 125*, 15010–15020.

86. Todorovic, S., Jung, C., Hildebrandt, P., and Murgida, D. H. (2006) Conformational transitions and redox potential shifts of cytochrome P450 induced by immobilization, *J. Biol. Inorg. Chem. 11*, 119–127.

87. Panicco, P., Astuti, Y., Fantuzzi, A., Durrant, J. R., and Gilardi, G. (2008) P450 versus P420: correlation between cyclic voltammetry and visible absorption spectroscopy of the immobilized heme domain of cytochrome P450 BM3. *J. Phys. Chem. B 44*, 14063–14068.

88. Berglund, J., Pascher, T., Winkler, J. R., and Gray, H. B. (1997) Photoinduced oxidation of horseradish peroxidase, *J. Am. Chem. Soc. 119*, 2464–2469.

89. Egawa, T., Proshlyakov, D. A., Miki, H., Makino, R., Ogura, T., Kitagawa, T., and Ishimura, Y. (2001) Effects of a thiolate axial ligand on the pi to pi* electronic states of oxoferryl porphyrins: a study of the optical and resonance Raman spectra of compounds I and II of chloroperoxidase. *J. Biol. Inorg. Chem. 6*, 46–54.

90. Udit, A. K., Hill, M. G., and Gray, H. B. (2006) Electrochemical generation of a high-valent state of cytochrome P450, *J. Inorg. Biochem. 100*, 519–523.

91. Bordwell, F. G., Cheng, J., Ji, G., Satish, A. V., and Zhang, X. (1991) Bond dissociation energies in DMSO related to the gas phase, *J. Am. Chem. Soc. 113*, 9790–9795.

92. Estavillo, C., Lu, Z., Jansson, I., Schenkman, J. B., and Rusling, J. F. (2003) Epoxidation of styrene by human cyt P450 1A2 by thin film electrolysis and peroxide activation compared to solution reactions, *Biophys. Chem. 104*, 291–296.

93. Zhou, L., Yang, J., Estavillo, C., Stuart, J. D., Schenkman, J. B., and Rusling, J. F. (2003) Toxicity screening by electrochemical detection of DNA damage by metabolites generated in situ in ultrathin DNA-enzyme films, *J. Am. Chem. Soc. 125*, 1431–1436.

94. Shumyantseva, V. V., Ivanov, Y. D., Bistolas, N., Scheller, F. W., Archakov, A. I., and Wollenberger, U. (2004) Direct electron transfer of cytochrome P450 2B4 at electrodes modified with nonionic detergent and colloidal nanoparticles, *Anal. Chem. 76*, 6046–6052.

95. Niki, K. (2002) Interprotein electron transfer: an electrochemical approach, *Electrochemistry 70*, 82–90.

96. Mak, L., Sadeghi, S., Fantuzzi, A., and Gilardi, G. (2010) Control of human cytochrome P450 2E1 electrocatalytic response as a result of unique orientation on gold electrodes, *Anal. Chem. 82*, 5357–5362.

97. Tanimura, R., Hill, M. G., Margoliash, E., Niki, K., Ohno, H., and Gray, H. B. (2002) Active carboxylic acid-terminated alkanethiol self-assembled monolayers on gold bead electrodes for immobilization of cytochromes c, *Electrochem. Solid State Lett. 5*, E67–E70.

98. Sadeghi, S. J., Tsotsou, G. E., Fairhead, M., Meharenna, Y. T., and Gilardi, G. (2001) Rational design of P450 enzymes for biotechnology. In *Focus in Biotechnology* (Bulte, J. and De Cuyper, M., Eds.), Kluwer Academic Publisher, Dordrecht.

99. Sevrioukova, I. F., Immoos, C. E., Poulos, T. L., and Farmer, P. J. (2000) Electron transfer in the ruthenated heme domain of cytochrome P450BM-3, *Isr. J. Chem. 40*, 47–53.

100. Katz, E. (1994) Application of bifunctional reagents for immobilization of proteins on a carbon electrode surface: oriented immobilization of photosynthetic reaction centers, *J. Electroanal. Chem. 365*, 157–164.
101. Udit, A. K., Hill, M. G., Bittner, V. G., Arnold, F. H., and Gray, H. B. (2004) Reduction of dioxygen catalyzed by pyrene-wired heme domain cytochrome P450 BM3 electrodes, *J. Am. Chem. Soc. 126*, 10218–10219.
102. Mhaske, S., Ray, M., and Mazumdar, S. (2010) Covalent linkage of CYP101 with the electrode enhances the electrocatalytic activity of the enzyme: vectorial electron transport from the electrode, *Inorg. Chim. Acta 363*, 2804–2811.
103. van der Felt, C., Hindoyan, K., Choi, K., Javdan, N., Goldman, P., Bustos, R., Hunter, B., Hill, M., Nersissian, A., and Udit, A. (2011) Electron-transfer rates govern product distribution in electrochemically-driven P450-catalyzed dioxygen reduction, *J. Inorg. Biochem. 105*, 1350–1353.
104. Weber, K. and Creager, S. E. (1994) Voltammetry of redox-active groups irreversibly adsorbed onto electrodes – treatment using the Marcus relation between rate and overpotential, *Anal. Chem. 66*, 3164–3172.
105. Vilker, V., Reipa, V., Mayhew, M., and Holden, M. (1999) Challenges in capturing oxygenase activity in vitro, *J. Am. Oil Chem. Soc. 76*, 1283–1289.
106. van Beilen, J. B., Duetz, W. A., Schmid, A., and Witholt, B. (2003) Practical issues in the application of oxygenases, *Trends Biotechnol. 21*, 170–177.
107. Guengerich, F. P. (2002) Cytochrome P450 enzymes in the generation of commercial products, *Nat. Rev. Drug Discov. 1*, 359–366.
108. Urlacher, V. and Schmid, R. D. (2002) Biotransformations using prokaryotic P450 monooxygenases, *Curr. Opin. Biotechnol. 13*, 557–564.
109. Urlacher, V. B., Lutz-Wahl, S., and Schmid, R. D. (2004) Microbial P450 enzymes in biotechnology, *Appl. Microbiol. Biotechnol. 64*, 317–325.
110. Bistolas, N., Wollenberger, U., Jung, C., and Scheller, F. W. (2005) Cytochromes P450 biosensors – a review, *Biosens. Bioelectron. 20*, 2408–2423.
111. Guengerich, F. P., Ballou, D. P., and Coon, M. J. (1975) Purified liver microsomal cytochrome P-450. Electron-accepting properties and oxidation–reduction potential, *J. Biol. Chem. 250*, 7405–7411.
112. Scheller, F., Renneberg, R., Strnad, G., Pommerening, K., and Mohr, P. (1977) Electrochemical aspects of cytochrome P-450 system from liver microsomes, *Bioelectrochem. Bioenerg. 4*, 500–507.
113. Krishnan, S., Dhanuka, W., Zhao, L., Schenkman, J., and Rusling, J. (2011) Efficient bioelectronic actuation of the natural catalytic pathway of human metabolic cytochrome P450s, *J. Am. Chem. Soc. 133*, 1459–1465.

8

MOLECULAR PROPERTIES AND REACTION MECHANISM OF MULTICOPPER OXIDASES RELATED TO THEIR USE IN BIOFUEL CELLS

EDWARD I. SOLOMON, CHRISTIAN H. KJAERGAARD AND DAVID E. HEPPNER
Department of Chemistry, Stanford University, Stanford, CA, USA

8.1 INTRODUCTION

The multicopper oxidases (MCOs) are a ubiquitous family of proteins that carry out four single-electron oxidations of substrates coupled to the four-electron reduction of dioxygen to water (Eq. 8.1) [1]. These enzymes carry out a diverse set of oxidation reactions and based on their reactivity can be placed into one of two categories: those that oxidize organic molecules (organic oxidases) and those that oxidize metal ions (metalloxidases) (Table 8.1). The metalloxidases are specific toward their substrate, while the organic oxidases generally exhibit broad substrate specificity:

$$4H^{+} + 4\text{Substrate} + O_2 \rightarrow 2H_2O + 4\text{Substrate}^{+} \qquad (8.1)$$

For MCOs, the substrate can be replaced with an electrode where the reducing potential is varied for reduction of dioxygen at low overpotential. This has been a motivating factor in the design of membrane-less biofuel cells where an MCO would be an ideal catalyst on the cathode.

The MCOs contain at least four copper atoms arranged in a mononuclear type 1 (T1) site where substrate is oxidized, and this electron is transferred to a trinuclear

Electrochemical Processes in Biological Systems, First Edition. Edited by Andrzej Lewenstam and Lo Gorton.

TABLE 8.1 Members of the MCO family separated into two substrate classes: organic oxidases and metalloxidases.

Organic oxidases	Metalloxidases
Plant and fungal laccases (Lc)	Fet3p
Ascorbate oxidase	Ceruloplasmin
CotA	CueO
Phenoxazinone synthase (PHS)	Mnxg
Bilirubin oxidase (BOD)	

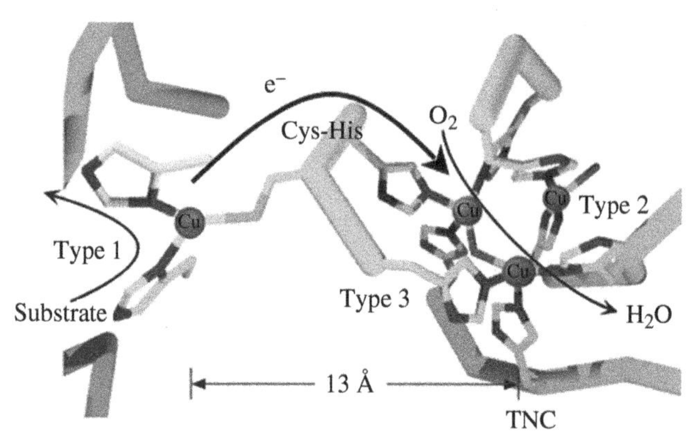

FIGURE 8.1 The structure of the MCO active site with arrows marking the flow of substrate, electrons (e^-), and O_2. Reproduced with permission from Ref. [2]. © 2008 RSC Publishing.

copper cluster (TNC) where dioxygen is reduced to water (*vide infra*). In this review, the mechanism of substrate oxidation, intramolecular electron transfer (ET), and dioxygen reduction to water will be presented. Then studies of the MCOs immobilized on electrodes will be summarized. Finally, parallels and differences between electrochemical and solution studies will be considered as relevant to fuel cells.

8.2 MCOs IN SOLUTION: STRUCTURE AND MECHANISM

8.2.1 Copper Sites and Mechanism of the MCOs

The general topology of the active sites in the MCOs (Fig. 8.1) consists of a T1 [3, 4] center where substrate is oxidized and electrons are transferred across approximately 13 Å through protein to the TNC [5, 6] where dioxygen is reduced to water. The MCOs contain at least four copper atoms and are characterized on the basis of their spectral features in the context of the fully oxidized resting form. The T1 or blue copper site exhibits an intense absorption band (Fig. 8.2, bottom) near 600 nm (~5000 $M^{-1}cm^{-1}$) assigned to a Cys—Sπ to Cu-d($x^2 - y^2$) charge transfer (CT) transition that is responsible

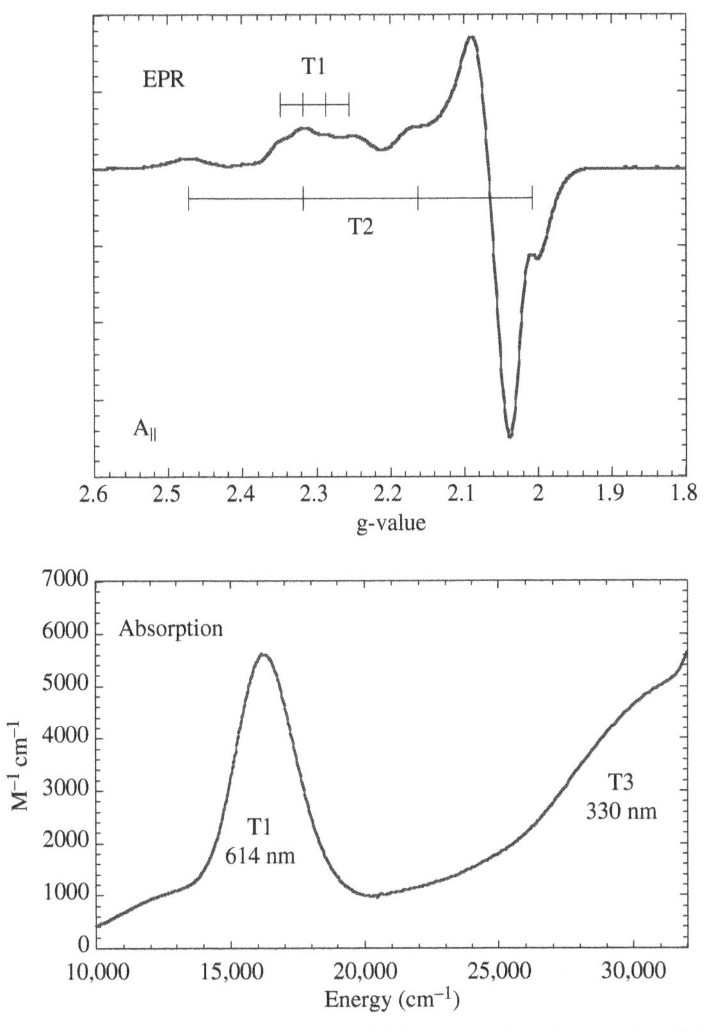

FIGURE 8.2 EPR and absorption spectra of *Rhus vernicifera* laccase highlighting the representative spectral features of copper sites in the MCOs.

for the blue color of the MCOs. The EPR spectrum (Fig. 8.2, top) of the T1 exhibits a small Cu(II) $A_{\parallel}$ hyperfine coupling constant ($40-95 \times 10^4\,cm^{-1}$), which has been shown to be due to a highly covalent Cu—S(Cys) bond (*vide infra*). The TNC is composed of a mononuclear, normal type 2 (T2) and a coupled binuclear type 3 (T3) site. The T2 center exhibits an EPR spectrum with a normal Cu(II) $A_{\parallel}$ hyperfine coupling constant ($158-201 \times 10^4\,cm^{-1}$) but lacks strong absorption features (Fig. 8.2). The T3 site has an intense CT band at approximately 330 nm (Fig. 8.2, bottom) originating from a μ_2-hydroxo ligand bridging the two coppers, which provides a superexchange pathway resulting in antiferromagnetic (AF) coupling between the coppers, leading to an $S=0$ ground state that results in the lack of an EPR signal.

TABLE 8.2 Comparison of the redox potential of T1 copper sites with and without an axial methionine ligand. Entries in bold are T1 sites in MCOs.

Protein	E° (mV vs. NHE)	Reference
T1 copper sites with an axial methionine ligand		
Achromobacter cycloclastes nitrite reductase	240	Suzuki et al. [7]
Pseudomonas aeruginosa azurin	310	Pascher et al. [8]
Cucumber ascorbate oxidase	344	Kawahara et al. [9]
Spinach plastocyanin	370	Katoh et al. [10]
***Rhus vernicifera* laccase**	434	Reinhammar et al. [11]
Human ceruloplasmin (redox-active T1 sites)	448	Machonkin et al. [12]
***Myrothecium verrucaria* bilirubin oxidase**	670	Christenson et al. [13]
Thiobacillus ferrooxidans rusticyanin	680	Ingledew et al. [14]
***Polyporus pinsitus* laccase F463M**	680	Xu et al. [15]
T1 copper sites without an axial methionine ligand		
Pseudomonas aeruginosa azurin M121L	412	Pascher et al. [8]
***Saccharomyces cerevisiae* Fet3p**	427	Machonkin et al. [16]
***Myceliophthora thermophila* laccase**	465	Xu et al. [17]
***Coprinus cinereus* laccase**	550	Ducros et al. [18]
***Trachyderma tsunodae* bilirubin oxidase**	660	Christenson et al. [13]
***Trametes versicolor* laccase**	778	Malmstrom et al. [19]
***Trametes hirsuta* laccase**	780	Shleev et al. [20]
Thiobacillus ferrooxidans rusticyanin M148L	798	Hall et al. [21]

T1 copper sites are found in a number of proteins (Table 8.2) in addition to the MCOs [22]. The geometry of the T1 site is either found as 4-coordinate with two histidines, a cysteine, and an axial methionine in a trigonally elongated tetrahedral geometry, where the z-axis is along the Cu—S_{Met} bond, as found in tree laccase [23] (Lc), ascorbate oxidase [24], and human ceruloplasmin [25] (hCp) or as 3-coordinate trigonal planar, where the site lacks the axial methionine. In the latter T1 sites, the methionine is replaced in the protein sequence by a noncoordinating phenylalanine or leucine as in Fet3p [26] and the fungal Lcs (Fig. 8.3). These sites exhibit wide variance in electrochemical potential [16] (Table 8.2), which is an important consideration in the design of biofuel cells (*vide infra*). T1 sites without the axial methionine generally have higher potentials than the ones with the ligand [27]. It has been shown that eliminating the methionine ligand in the same protein environment raises the T1 potential approximately 100 mV. Additionally, protein electrostatics and hydrogen bonds also significantly contribute to the redox potential of a T1 center in a protein matrix [22, 28].

The T2 and T3 sites comprise the TNC [5, 6] where dioxygen is reduced (Fig. 8.4). The T2 site is coordinated by two histidines and an aquo-derived hydroxo ligand exterior to the TNC, while the T3 pairs are both coordinated by three histidines and bridged by a μ_2-OH ligand [30]. All of the histidines of the TNC are in His–X–His sequence motifs where the two histidines bridge two coppers. All three of the coppers have coordinatively unsaturated positions oriented toward the center of the cluster ready to bridge dioxygen and its intermediates as evident from its electronic structure

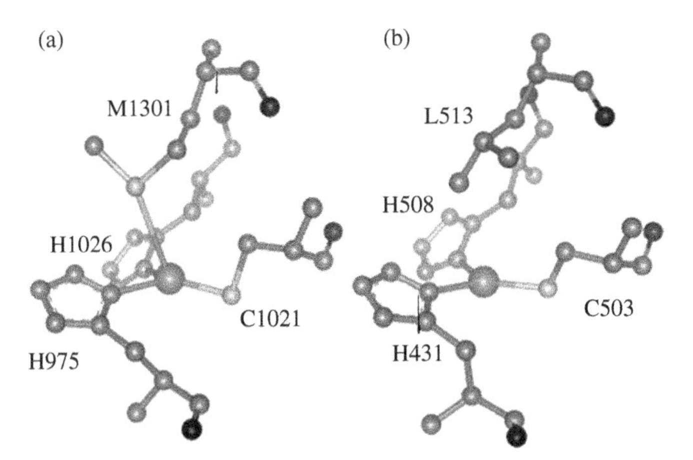

FIGURE 8.3 Comparison of the T1 Cu site in hCp (PDB: 1KCW) with an axial methionine that is redox active near the substrate binding site (a) with that of *M. albomyces* fungal laccase (PDB: 3FU8) that does not have a methionine (b).

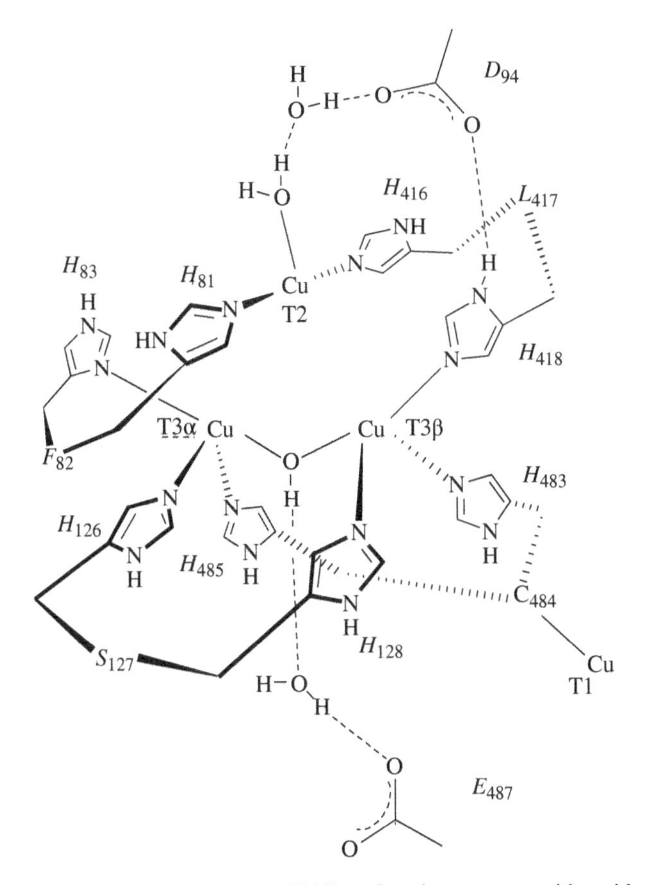

FIGURE 8.4 Trinuclear copper cluster (TNC) active site structure with residue numbering according to the Fet3p sequence. Reproduced with permission from Ref. [29]. © 2010 the American Chemical Society.

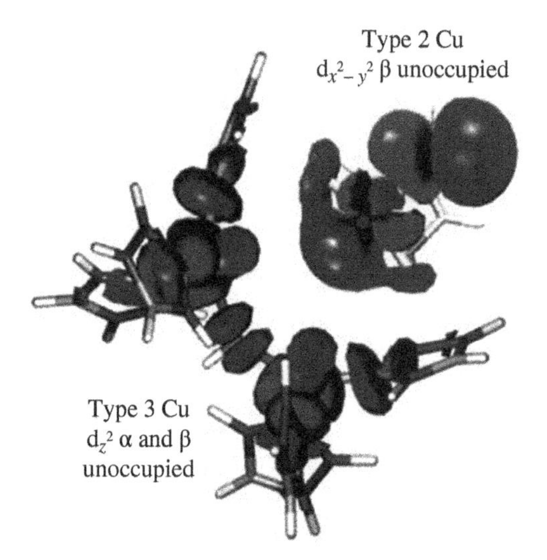

FIGURE 8.5 Electronic structure of the resting TNC. Reproduced with permission from Ref. [30], © 2005 the American Chemical Society.

of the resting site in Figure 8.5. This 4^+ cluster remains coordinatively unsaturated due to the presence of 4^- carboxylates within approximately 10 Å of the TNC that prevent the binding of aquo solvent species as OH^- or O^{2-} in the center of the TNC. Two of these carboxylates (Fig. 8.4), D94 and E487 in the Fet3p sequence [26], contribute to the dioxygen reactivity of this cluster (*vide infra*).

8.2.2 Dioxygen Reactivity of the T3 Sites: Hemocyanin versus Lc

An important point of comparison is the reactivity of the T3 site in the MCOs relative to that of the T3 site in hemocyanin (Hc), which contains only a coupled binuclear copper site and functions as the reversible dioxygen-binding protein in arthropods and mollusks [1]. In order to isolate only the T3 site in Lc, the T2 center in the MCOs can be removed to give a "T2-depleted form" (T2D), which only contains the reduced T3 (deoxy-T3) and T1 sites [31]. Cu K-edge X-ray absorption spectroscopy (XAS) can be used to quantify the amount of reduced copper via the intensity of a peak at 8984 eV peak [32]. Exposing dioxygen to reduced Hc (deoxy-Hc) results in a change in the XAS spectrum showing the oxidation of 2Cu(I) to 2Cu(II) in the formation of oxy-Hc (Fig. 8.6a). However, exposure of dioxygen to reduced T2D (deoxy-T3) does not change the XAS spectrum, demonstrating that the T3 site does not react with dioxygen in the absence of the T2 (Fig. 8.6b). The origin of this different dioxygen reactivity has been evaluated with potential energy surface (PES) scans [33]. These show that while dioxygen binding to the deoxy-T3 Cu(I)—Cu(I) site in Lc is uphill by 6 kcal/mol, this binding in deoxy-Hc is 4 kcal/mol downhill (Fig. 8.7). This 10 kcal/mol difference in dioxygen-binding energy derives from differences in electrostatic repulsion of the reduced T3 Cu's. With proper protein constrains, the

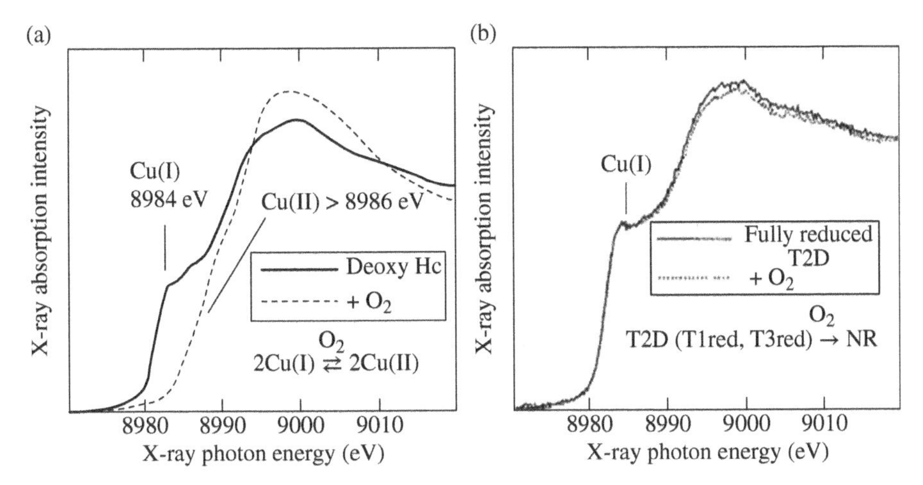

FIGURE 8.6 Cu K-edge X-ray absorption spectra of deoxy-Hc (a) and T2D laccase (b) and their reactions with O_2. Reproduced with permission from Ref. [1]. © 1996 the American Chemical Society.

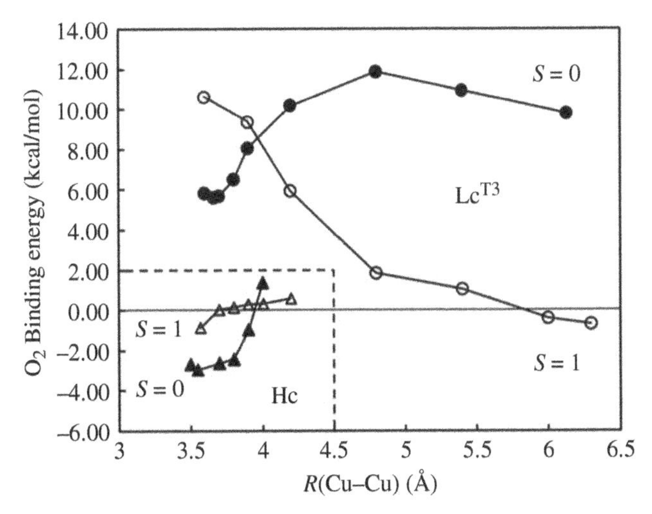

FIGURE 8.7 Energy of O_2 binding to the reduced T3 site with $S = 0$ in closed circles and $S = 1$ in open circles and Hc site with $S = 0$ in closed triangles and $S = 1$ in open triangles. Reproduced with permission from Ref. [33]. © 2009 the National Academy of Sciences of the United States of America.

optimized Cu(I)—Cu(I) distance in deoxy-Hc is approximately 4 Å, which is much shorter than the approximately 6 Å Cu(I)—Cu(I) distance optimized in the deoxy-T3 center (Fig. 8.8). The shorter distance for deoxy-Hc leads to electrostatic destabilization of the reduced coppers and drives the binding of dioxygen. This distance

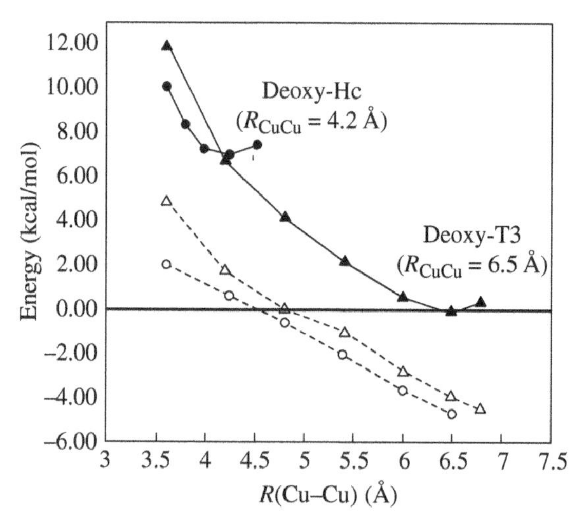

FIGURE 8.8 Potential energy surfaces for deoxy-Hc and deoxy-T3 sites as a function of Cu—Cu distance (dashed lines show electrostatic interactions between the two Cu(I) centers). Reproduced with permission from Ref. [33]. © 2009 the National Academy of Sciences of the United States of America.

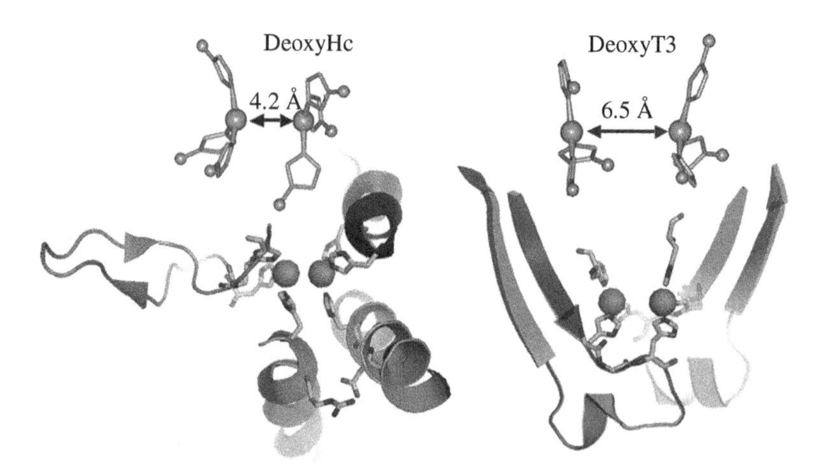

FIGURE 8.9 Comparison of the constrained structures of deoxy-Hc (PDB code 1JS8) and deoxy-T3 (PDB code 1GYC). Reproduced with permission from Ref. [33]. © 2009 the National Academy of Sciences of the United States of America.

difference originates from differences in the histidine protein constrains (Fig. 8.9). In Hc, the T3 coppers are coordinated by His ligands on α-helices held together by salt bridges, which is more rigid compared to the T3 site in Lc where the coppers are coordinated by His–X–His motifs on loops from β-sheets.

8.2.3 Mechanism of the MCOs

The catalytic mechanism of the MCOs will be considered in three steps. First, the oxidation of both metal and organic substrates near the T1 site will be explored followed by the intramolecular ET over 13 Å from the T1 to the TNC (Fig. 8.1). Then the mechanism of dioxygen reduction to water will be presented first by the characterization of the peroxy intermediate (PI) and then the native intermediate (NI), followed by the reductive cleavage of the O—O bond, and summarized by a discussion of the overall catalytic cycle [2, 34].

8.2.4 Substrate Binding and ET to the T1

Substrate oxidation occurs via ET to the T1 site [22]. This is important when considering how electrode surfaces or mediators transfer electrons to the MCOs on the cathode of biofuel cells (*vide infra*). For the organic oxidases, which exhibit broad oxidative specificity, the catalytic efficiency (k_{cat}/K_m) of a variety of organic oxidases is directly proportional to the difference between the reduction potentials of the substrate employed and that of the T1 site, implying outer-sphere ET in these proteins (Fig. 8.10) and that this is the rate-determining step in turnover [35]. The nature of the substrate binding site of these proteins comes from crystal structures of *Trametes versicolor* [36] and *Melanocarpus albomyces* [37] Lc with substrate and analogues bound in a shallow hydrophobic pocket near the T1 site. In the structure of 2,6-dimethylphenol (DMP) bound to *M. albomyces* Lc (Fig. 8.11a), only two protein–substrate interactions are evident: a hydrogen bond to one of the histidine ligands of the T1 and to a nearby carboxylate. From this structure, an ET pathway can be drawn from the oxygen of the phenol to the T1 Cu through the bound histidine where the carboxylate can act as a proton acceptor (Fig. 8.12, left). This is consistent with the broad oxidase activity of these enzymes since these interactions can accommodate a diverse array of substrates.

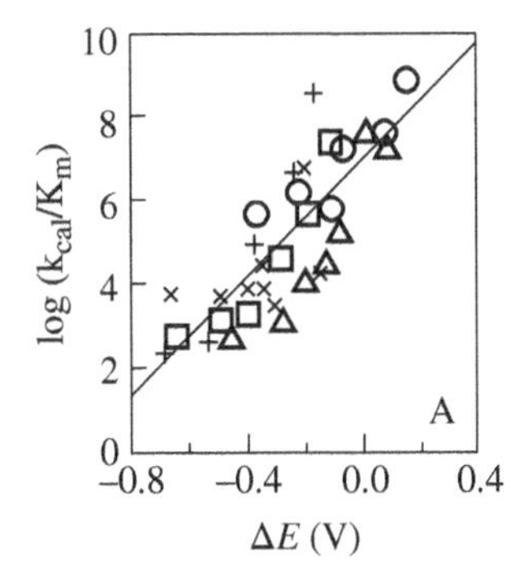

FIGURE 8.10 Correlation of log(kcat/Km) with ΔE ($= E_{T1site} - E_{Substrate}$) at pH = 5.0. Reproduced with permission from Ref. [35]. © 1996 the American Chemical Society.

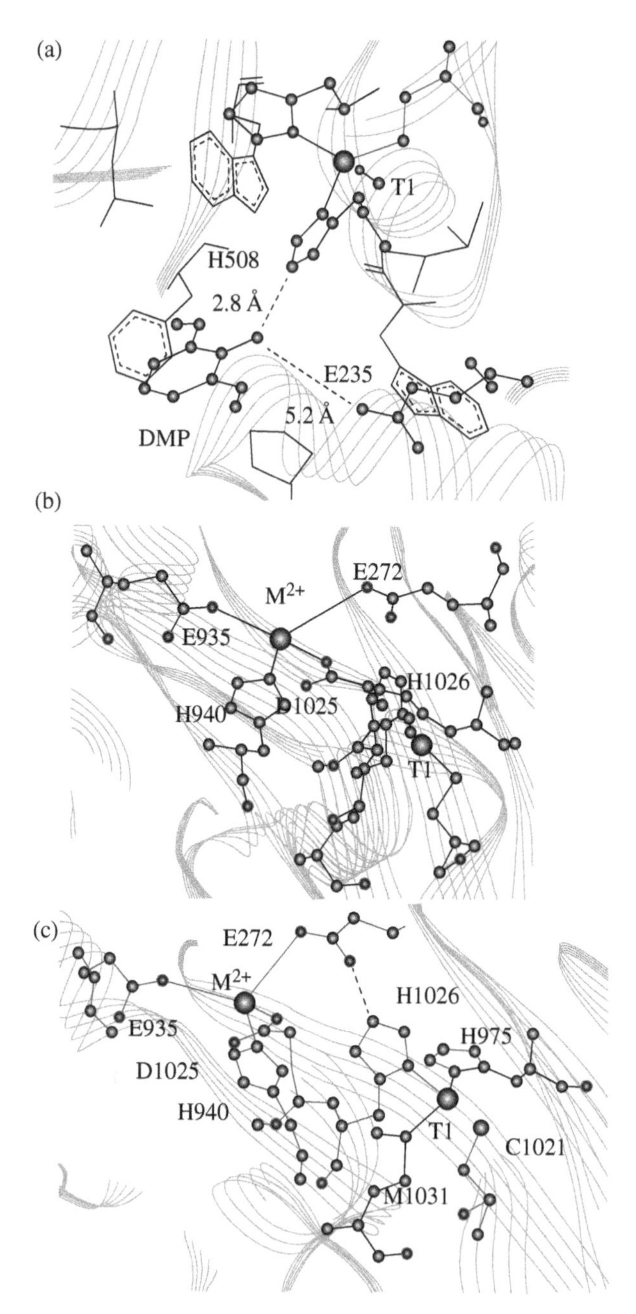

FIGURE 8.11 Crystal structures of (a) Ma Lc with DMP bound in the substrate binding sites near the T1 Cu (PDB, 3FU8) and (b) hCp with dication bound to substrate site (PDB, 1KCW) and another view highlighting the ET pathway from substrate to T1 through the hydrogen bond of E272 and H1026 (c).

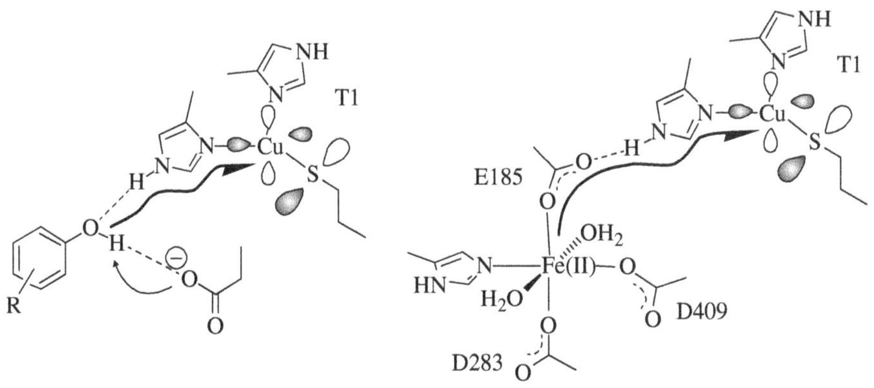

FIGURE 8.12 Single-electron transfer to the T1 from phenolic substrate in the organic oxidases (left) and for the Fe(II) the ferroxidases (right) with Fet3p sequence numbers.

For the metalloxidases, which oxidize specific substrates, significant insight into substrate binding and oxidation has come from spectroscopic and kinetic studies on Fe(II) bound to the substrate binding site in the ferroxidases Fet3p and hCp [38]. Variable temperature, variable field (VTVH) magnetic circular dichroism (MCD) spectroscopy can be used to probe nonheme Fe(II) active sites [39, 40] and confirms that Fe(II) binds in an identical fashion to both hCp and Fet3p in a 6-coordinate geometry with a binding constant greater than $10^5\,M^{-1}$, indicating that these proteins contain sophisticated tight binding sites for the Fe(II) substrate. A crystal structure of a divalent metal ion binding site is available in hCp composed of three carboxylates and a histidine ligand (Fig. 8.11b) and two aquo-derived ligands verified by MCD, but not evident in crystal structure [25]. Additionally, mutating carboxylate E185 in Fet3p, analogous to E272 in hCp (Fig. 8.11), near the T1 site to Asp (a shorter carboxylate residue) results in no change to the coordination number or binding constant of the Fe(II) substrate site, but lowers the T1 rate by an order of magnitude implying that the mutation affects the superexchange pathway for ET from the Fe(II) to the T1 [38]. Mutating this residue to a neutral alanine diminishes the binding constant by three orders of magnitude and lowers the ET rate by two orders of magnitude. From the crystal structure of hCp, an ET pathway from the substrate to the T1 is evident from substrate bound to E272, which is hydrogen bonding with H1026 of the T1 site (Fig. 8.11c). Therefore, E185 in Fet3p and E272 in hCp function to both bind Fe(II) tightly and supply a specific superexchange pathway for ET to the T1 through a hydrogen bond to one of the histidines coordinating the T1 (Fig. 8.12, right). This binding site also functions to lower the potential of Fe(II) to less than 190 mV (from 420 mV of aqueous Fe(II)) in order to achieve a large driving force for ET from Fe(II) to the T1 [38]. Additionally, a general feature of the crystal structure of the ferroxidases (hCp—Fig. 8.11b) is that the T1 site is buried in the protein by approximately 7–10 Å, greatly limiting access to the site from solvent. Therefore, an inaccessible T1 site in combination with sophisticated metal substrate binding and an

ET pathway to the T1 site define the origin of the substrate specificity of these enzymes and the limited reactivity of the metalloxidases with organic substrates.

8.2.5 ET from the T1 to the TNC

Once an electron has been transferred to the T1 from substrate, the electron is then transferred from the T1 over 13 Å to the TNC site via a protein-derived Cys—His pathway (Fig. 8.1). The T1 site has been studied in detailed and has been shown to have a highly covalent Cu—S_{Cys} bond [3]. The electronic coupling matrix element (H_{DA}) in Marcus theory (*vide infra*), which is a key contributor to fast ET rates, is directly proportional to ligand covalency, and therefore, the Cys—His pathway is activated for ET to the TNC [41]. This is evident from Figure 8.13a where the contour of the T1 redox-active molecular orbital (RAMO) is modeled into the active site showing the delocalization of the T1 onto the Cys—S ligand toward the TNC [3]. Additionally, Figure 8.13b shows the molecular orbital delocalization of spin densities between the Cu—S π donor bond of the T1 and the Cu—N σ bond to the T3. The molecular orbital contour (Fig. 8.13c) shows two superexchange pathways from the π bond of the T1 that extends through the protein backbone and exits σ to the remote T3 copper of the TNC and a second pathway involving a hydrogen bond through a backbone carbonyl, where the H-bond pathway is dependent on distance. Resonance Raman (rR) studies have shown that the resonance enhanced vibrations of the T1 copper site change depending on the redox state of the TNC (Fig. 8.14) [42]. This change reflects the Cu—S stretch and its coupling to the vibrations of the cysteine and to the protein backbone, showing that the protein backbone is affected by the redox states of the TNC. Changes in the TNC can also affect the C=O—H distance of the H-bond shunt pathway, where small changes in this distance can impact H_{DA} through constructive or destructive interference. This implies that redox states and intermediates of the TNC can change the backbone H bond and regulate ET from the T1 to the TNC.

8.2.6 Dioxygen Reduction at the Trinuclear Cluster of the MCOs

Key insights into the intermediates in this process have been defined from studies on wild type (WT) and derivatives of these enzymes (Fig. 8.15) [2]. The native enzyme containing a fully reduced TNC and T1 reacts with dioxygen to form the NI at a rate of $1.7 \times 10^6\ M^{-1}s^{-1}$ [43]. Alternatively, the T1 copper can be replaced chemically with a redox and spectroscopic inert mercuric ion (T1Hg) or, in an expressed enzyme, eliminated by mutating the T1 cysteine to a serine known as the T1-depleted (T1D) form [44, 45]. Without the T1 copper, the enzyme lacks one of its electrons to transfer to dioxygen. In both of these forms, the TNC is still intact. The reaction of either T1-eliminated form produces the PI [46] at a rate ($2.2 \times 10^6\ M^{-1}s^{-1}$) that is comparable with the formation of NI [43]. Therefore, PI is kinetically competent to precede NI and it has been found that PI decays to NI. The conversion of PI to NI in the native enzyme is immeasurably fast (>350 s^{-1}) but much slower (~$10^{-3}\ s^{-1}$) in the T1D/T1Hg derivatives allowing for the study of the reductive cleavage of the O—O bond (*vide infra*) [47, 48].

(a)

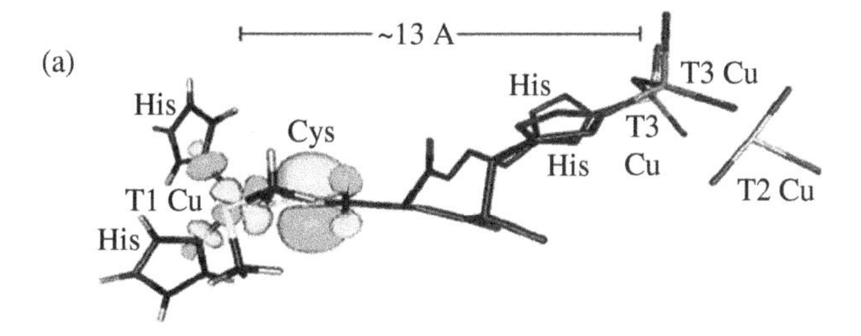

Cys-His superexchange pathway for ET

(b)

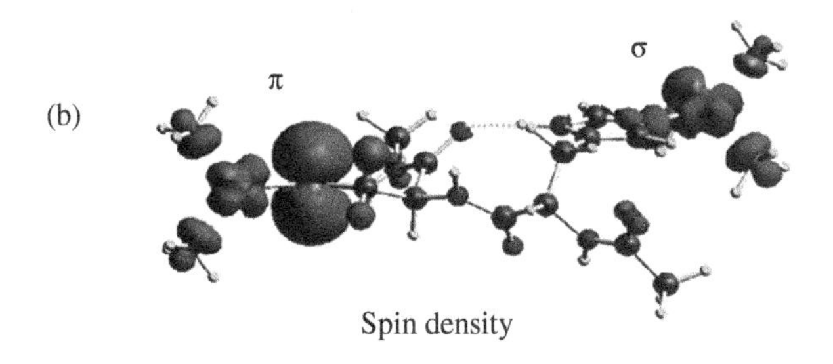

Spin density

(c)

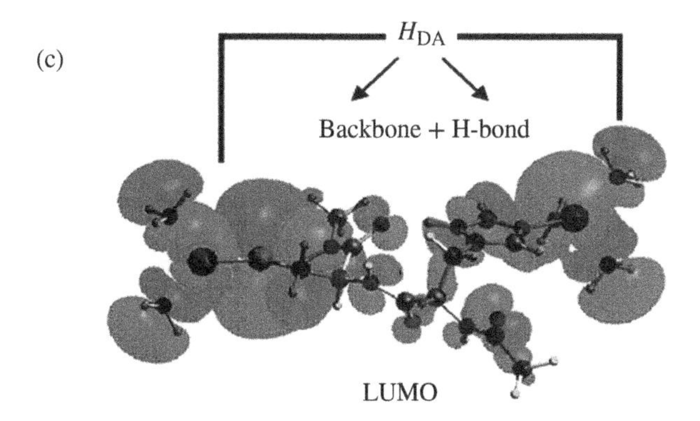

LUMO

FIGURE 8.13 Superexchange pathway of the blue copper site: (a) RAMO superimposed on part of the crystal structure of a MCO (b) π-to-σ hole delocalization between the T1 and the T3 site (c) superexchange pathways through a Cys—His molecular orbital. Reproduced with permission from Ref. [3]. © 2006 the American Chemical Society.

PI has been extensively studied and spectroscopically characterized to be an internal bridging peroxo in the TNC with two oxidized AF-coupled coppers and one reduced [49]. An optimized model of PI with dioxygen bound in the center of the TNC yielded a structure with peroxide bound as a side-on bridge between oxidized T3 pairs (Fig. 8.16a). However, a comparison of the peroxo to Cu(II) CT absorption spectrum of PI to that of the side-on peroxo formed in oxy-Hc clearly eliminates a side-on peroxo as a possible structure of PI (Fig. 8.17). These spectra are very

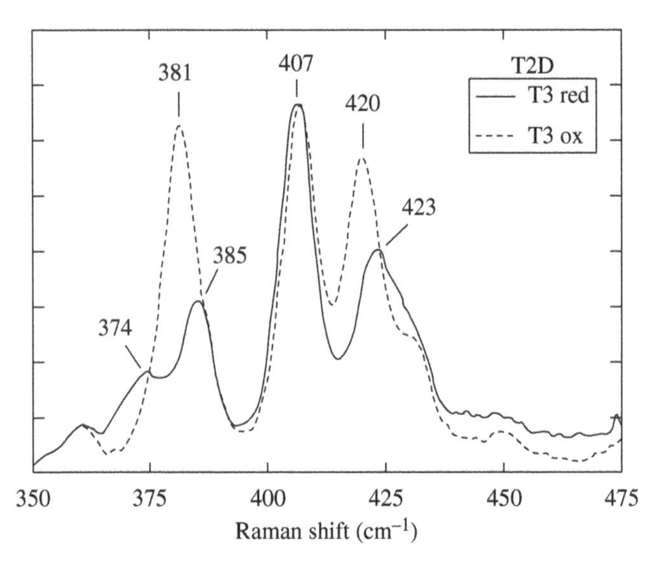

FIGURE 8.14 The Raman spectrum of T1 Cu of T2D Lc with the T3 site in both oxidized and reduced states. Reproduced with permission from Ref. [2]. © 2008 RSC Publishing.

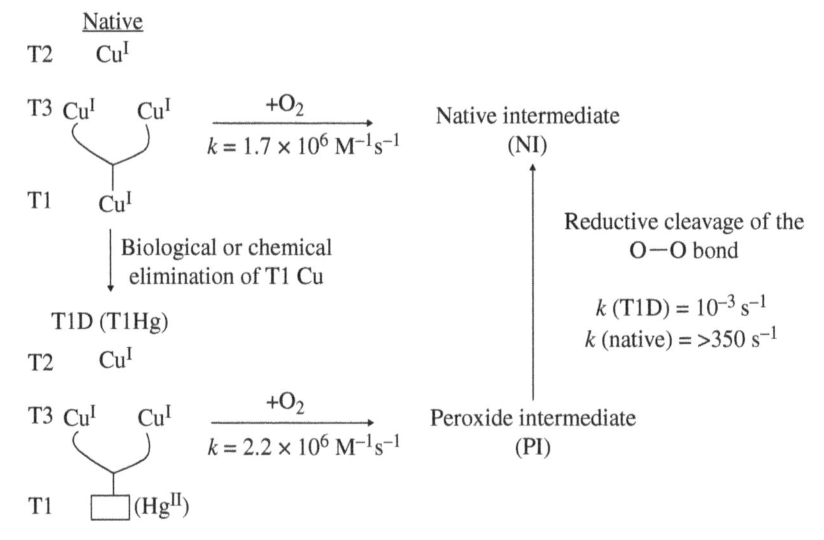

FIGURE 8.15 Schematic of the O_2 reactivity of both native and T1D/T1Hg forms of the MCOs and the cleavage of the O—O bond. Reproduced with permission from Ref. [2]. © 2008 RSC Publishing.

different where the peroxy to Cu(II) CT transitions in PI are weaker and shift to lower energy, indicating a different electronic and geometric structure. Extending the computational model in Figure 8.16a to include a highly conserved aspartate residue (D94 in Fet3p; Fig. 8.4) near the T2 and T3β, which is found to be essential for dioxygen

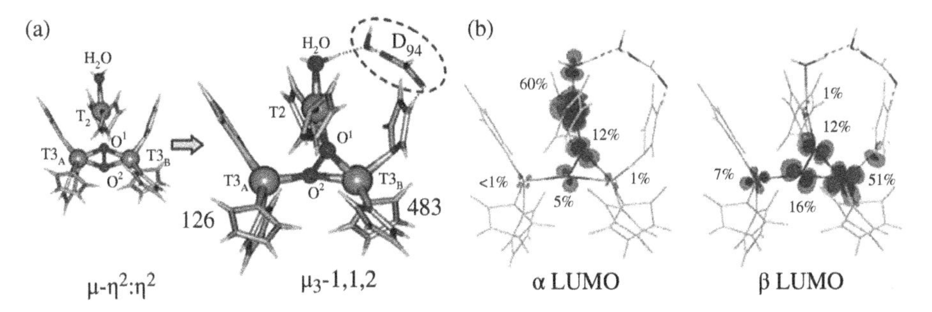

FIGURE 8.16 (a) Calculated geometric structures of PI with and without D94. The PI structure without D94 (left) has both T3 Cu's oxidized and the T2 Cu reduced, while in the structure with D94 (right), the T3B and T2 are oxidized and the T3A is reduced. (b) Contours of the α-LUMO (based on the T2 $d_{x^2-y^2}$) and the β-LUMO (based on the T3B $d_{x^2-y^2}$) of PI + D94. Reproduced with permission from Ref. [49]. © 2007 the American Chemical Society.

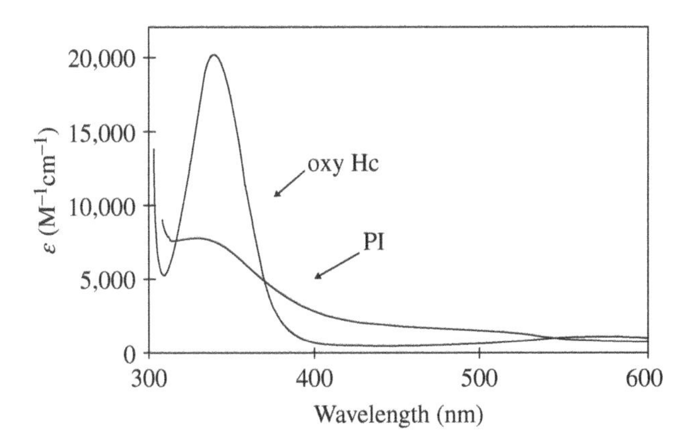

FIGURE 8.17 Peroxide to Cu(II) charge transfer absorption spectrum of oxy-Hc and PI. Reproduced with permission from Ref. [2]. © 2008 RSC Publishing.

reactivity (mutating D94 to an alanine eliminates dioxygen reactivity) [50], stabilizes a different structure consisting of a μ_3-1,1,2 peroxo (Fig. 8.16a, right), which is more stable than the side-on peroxo. This optimized structure includes the peroxide bridging all three coppers of the TNC, with the T2 and T3β oxidized and T3α reduced, and is consistent with spectroscopy [49]. Figure 8.16b includes both of the half-occupied d-orbitals that overlap the $\pi*_\sigma$ of the peroxo, leading to AF coupling of the coppers. Therefore, the proximity of the negative charge of D94 lowers the redox potentials of the T2 and T3β coppers for electron donation in the binding of dioxygen. Additional support for this comes from inner-sphere mutants in Fet3p. Mutation of H126 (Fig. 8.16a, right), which is bound to the T3α copper, to a glutamine produces a TNC capable of reacting with dioxygen to form PI [29]. However, mutation of

H483 to glutamine (Fig. 8.16a, right), which is bound to the T3β copper, leads to a TNC that does not react with dioxygen. Therefore, mutation to the T3α copper does not affect dioxygen reactivity, while mutation to T3β does, which further support the aforementioned model. These studies also imply an intrinsic asymmetry of the TNC where each T3 copper contributes differently to dioxygen binding. This originates from the lower potential of the T2 and T3β due to their proximity to D94 and to a δ-coordinated histidine (H83 in Fet3p) on the T3α side that imposes a trigonal planar geometry further stabilizing its reduced state (Fig. 8.4).

As mentioned earlier, NI is formed from the reaction of the fully reduced holoenzyme with dioxygen (Fig. 8.15). NI is transient but decays slowly in Lc, which makes it trappable for spectroscopic study [51]. Extensive spectroscopic studies performed on NI elucidated its key diagnostic spectral features and lead to the assignment of its geometric and electronic structure. The absorption spectrum (Fig. 8.18a) consists of a pair of CT transitions at the TNC at 318 and 365 nm accompanied by the 614 nm CT of the oxidized T1. Cu K-edge XAS of NI (Fig. 8.18b) showed that the TNC and the T1 are fully oxidized, as in the fully oxidized resting state of the enzyme, thus dioxygen is fully reduced to the water level oxidation state. However, this intermediate is unlike the resting form as shown by its different low-temperature (LT) EPR spectra. Figure 8.17d shows the normal T2 Cu(II) EPR spectra of the resting TNC (black). Alternatively, NI exhibits a broad EPR signal with g-values below 2.0. The latter is very unusual for Cu(II), indicating a significant change at the TNC. The absorption CT features show an intense derivative-shaped feature in the corresponding LT MCD spectrum (Fig. 8.18a bottom), which is called a pseudo-A term. The LT magnetic saturation data for these CT features (Fig. 8.18c) fit to a Brillouin function for an $S = ½$ spin state confirming that this is the spin of the ground state with the unusual EPR g-values. Importantly, the MCD intensity at high field does not decrease in intensity with varying temperature as 1/T, the Curie law behavior predicted for an isolated $S = ½$ ground state (Fig. 8.18d). For instance, the intensity at 25,000 cm^{-1} first decreases and then increases in intensity (Fig. 8.18e), consistent with the Boltzmann population of a low-lying excited state at approximately 150 cm^{-1} above the $S = ½$ ground state and with a different MCD signal (Fig. 8.18f). This low-lying excited state requires all three coppers of the TNC are close to equally AF coupled (~500 cm^{-1}). This situation results in a spin-frustrated ground state for NI. Figure 8.19a shows that AF coupling of Cu_1 with Cu_2 and Cu_2 with Cu_3 leads to a parallel alignment of the spins on Cu_3 and Cu_1, although they want to be AF coupled due to the superexchange pathway associated with a Cu_1–Cu_3 bridging ligand. Therefore, all three coppers must be bridged by the products of complete dioxygen reduction. Additionally, the unique $g < 2.0$ values, which originate from a phenomenon known as antisymmetric exchange that is associated with this spin frustration [52], and the sign and nature of the temperature dependence of the pseudo-A term in the MCD spectra in (Fig. 8.18a, bottom) unambiguously lead to the structural assignment of NI as containing a μ_3-oxo ligand bridging all three of the coppers of the TNC and a μ_2-OH between the T3s (Fig. 8.19b) where the two bridging oxygen ligands originate from dioxygen reduction [53, 54].

Having defined the electronic and geometric structures of PI and NI, the mechanism of the 2nd two-electron reductive cleavage of the O—O bond was explored [49].

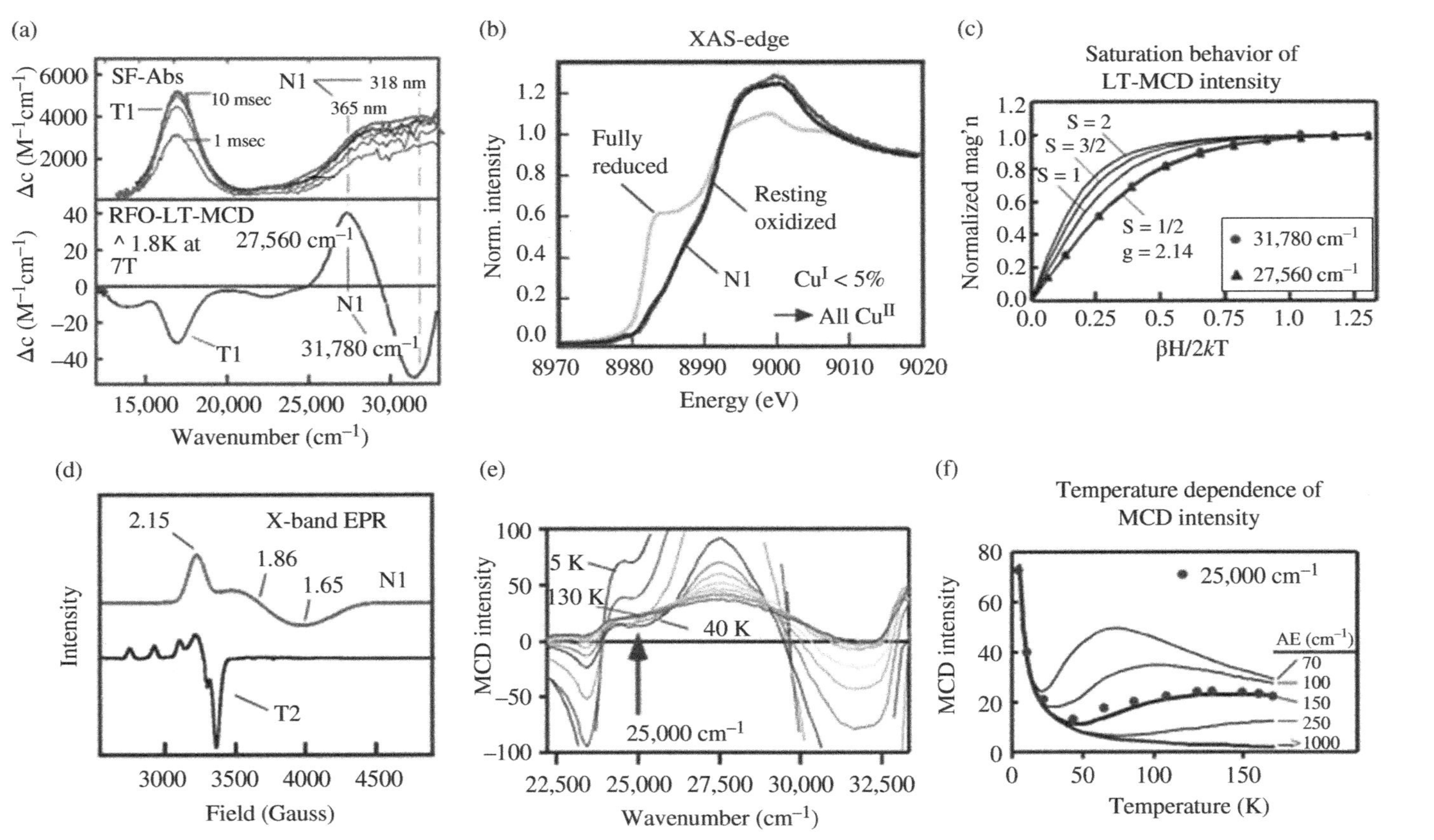

FIGURE 8.18 Stopped flow absorption spectra (top) and rapid freeze quench MCD spectra (a), Cu K-edge XAS spectrum (b), variable field behavior at low T of the MCD spectra (c), low-temperature X-band EPR (d), temperature dependence of the MCD spectra at high field (e), and plot of temperature dependence of MCD intensity at 25,000 cm^{-1} of NI. Reproduced with permission from Refs. [2, 51]. © 2002 the American Chemical Society and © 2008 RSC Publishing.

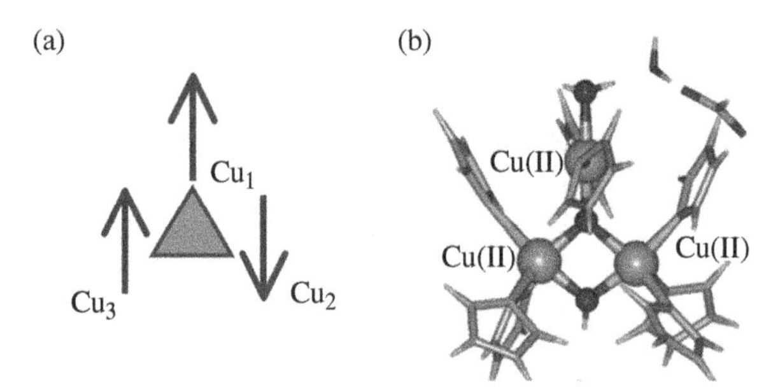

FIGURE 8.19 Orientation of electron spins in the spin-frustrated ground state of NI (a) and the calculated geometric structure of NI (b). Reproduced with permission from Ref. [34]. © 2011 RSC Publishing.

Starting from PI (Fig. 8.20a, left), where the T2 and T3β are oxidized and the T3α is reduced, ET occurs from the T1 to reduce PI (PI + e^-), where the T2 is now reduced. With two reduced coppers, the TNC is set up to facilitate the two-electron reduction and O—O bond cleavage. This has been modeled in a two-dimensional PES (Fig. 8.20b) where one dimension is the O—O elongation and the other is the proton donation from a carboxylate (E487 in Fet3p) distal to the TNC (Fig. 8.4), as confirmed by experiments on mutants at this position in Fet3p [48]. Two pathways are evident where either protonation occurs prior to or after O—O cleavage. In pathway 2 (low pH), the proton is transferred prior to the transition state for O—O cleavage consistent with the experimentally observed inverse solvent kinetic isotope effect [47]. The other pathway (high pH) consists of protonation after O—O bond cleavage. Both of these pathways have low barriers of approximately 5–6 kcal/mol, as consistent with experiment, and the reaction is found to be highly exothermic. The low barrier reflects the fact that the triangular topology of the TNC is highly efficient at performing O—O bond cleavage due to its frontier molecular orbitals (Fig. 8.20c) [49]. The reduced T3α and T2 have significant overlap with the σ* of the peroxo to donate an α,β pair directly into this orbital. Additionally, the oxidized T3β functions as a Lewis acid, similar to a proton, by lowering the energy of the σ* orbital and thus assisting in facile O—O bond cleavage with a low barrier.

Once the O—O bond has been cleaved to form NI, this decays slowly to the fully oxidized resting form of the MCOs. The conversion of NI to the resting TNC has been modeled and found to proceed first via protonation of the μ_3-oxo and then the μ_2-hydroxo (Fig. 8.21, top) [54]. Then the internal hydroxo takes a proton from the water bound to the T3β (to restore the μ_2-hydroxo bridge) and rotates through the T3α–T2 edge to replace the solvent ligand at the T2 with a barrier consistent with the decay kinetics of NI (Fig. 8.21, bottom). Importantly, the rate of this process (0.05 s^{-1}) [55] is considerably slower than the turnover rate for the enzyme (560 s^{-1}) [56], which eliminates it from the catalytic cycle. Therefore, NI is the catalytically relevant fully oxidized form of the MCOs and undergoes fast reduction back to the reduced form in turnover.

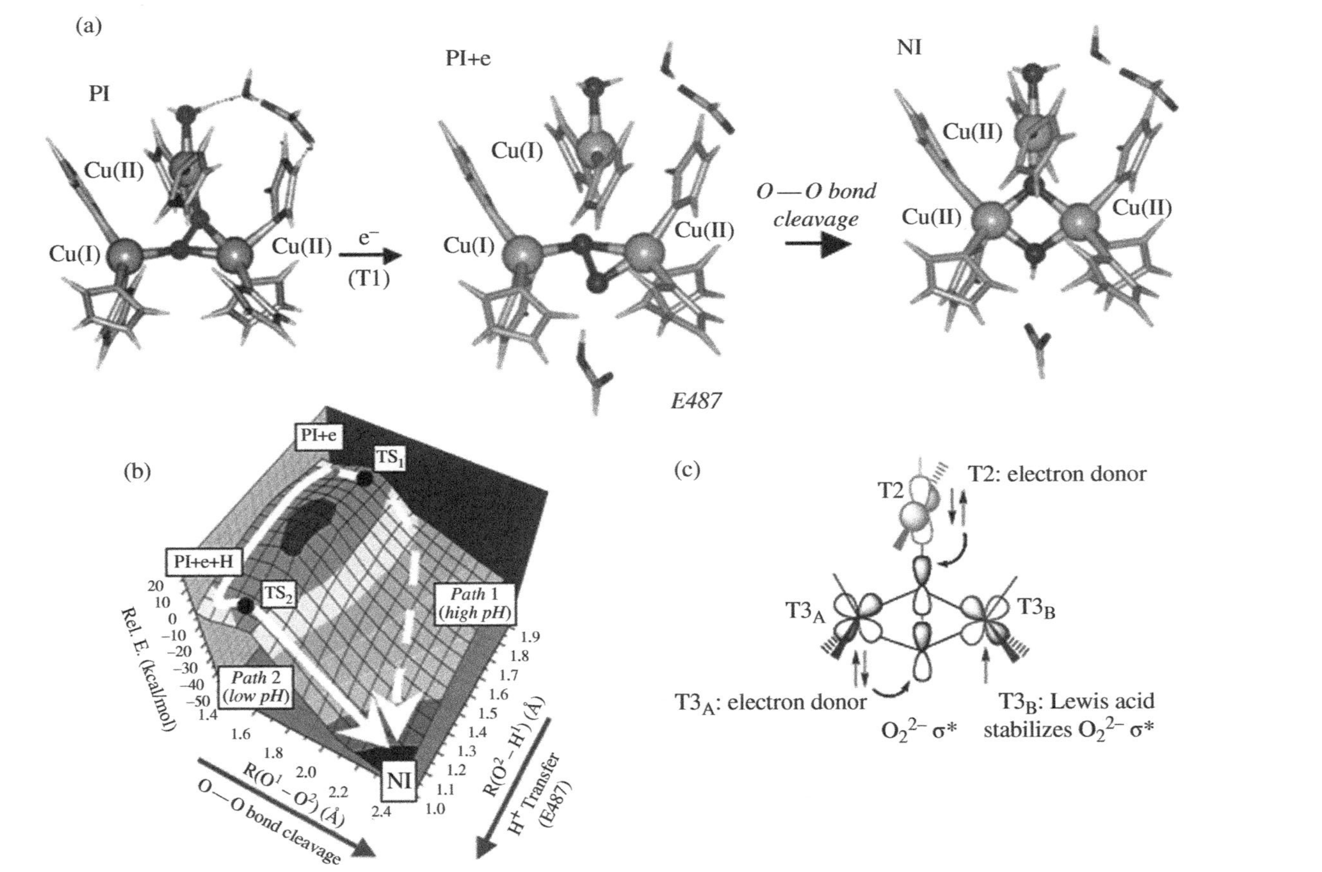

FIGURE 8.20 Calculated structures of PI, PI+e, and NI (a). 2D potential energy surface of the reductive cleavage of the O—O bond (b). Schematic of triangular topology of the TNC and depiction of the frontier molecular orbitals relevant in O—O cleavage. Note that T2 and T3α transfer two electrons into the peroxide σ∗ orbital, while T3β is oxidized and acts as a Lewis acid in stabilizing the σ∗ orbital. Reproduced with permission from Ref. [34].

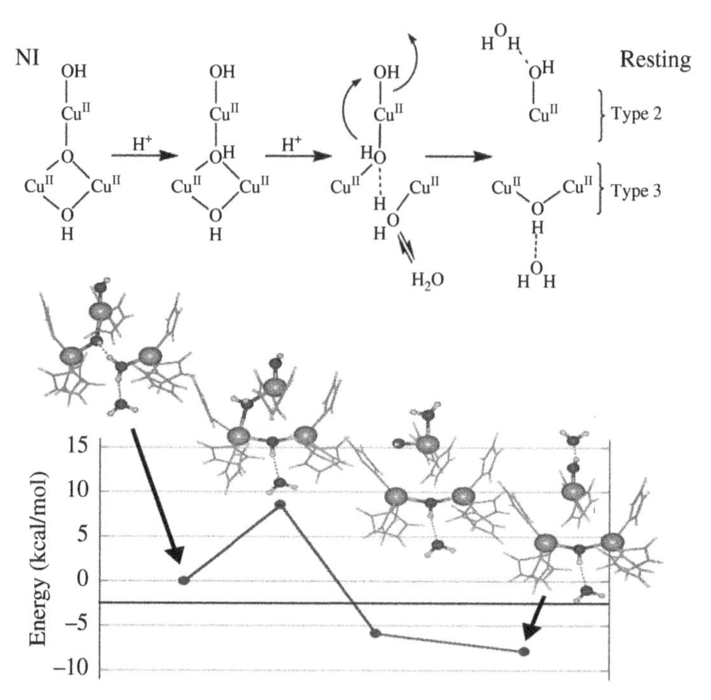

FIGURE 8.21 Schematic showing the conversion of NI to resting oxidized TNC marking the position of the oxygen atoms from dioxygen (top) and geometric structures and energies for the doubly protonated form of NI rotating the μ_3OH from inside to outside the cluster (bottom). Reproduced with permission from Ref. [54]. © 2007 the National Academy of Sciences of the United States of America.

In summary, the mechanism of dioxygen reduction by the TNC of the MCOs has been determined to proceed through 2 two-electron steps (Fig. 8.22). The 1st two-electron step is the reduction of dioxygen to peroxide in the formation of the bridged PI. This binds in an asymmetric structure to promote for this reduction. The O—O bond of PI is cleaved via the 2nd two-electron reduction to the water level in the formation of the NI. This is facilitated by the FMOs of the TNC. In the presence of substrate, NI is rapidly reduced back to the catalytically relevant fully reduced form of the enzyme, whereas if substrate is not in excess, NI will decay to the fully oxidized resting form of the enzyme that will require reduction from substrate for activation to return to the catalytic cycle.

8.3 MCOs IN ELECTROCHEMISTRY

8.3.1 Direct ET from Electrodes to MCOs

MCOs have attracted significant interest as candidates for cathodic enzymes in biofuel cells and implantable devices. This is primarily due to their ability to operate at low overpotentials in O_2 reduction [57, 58]. High stability and efficient electrode contact

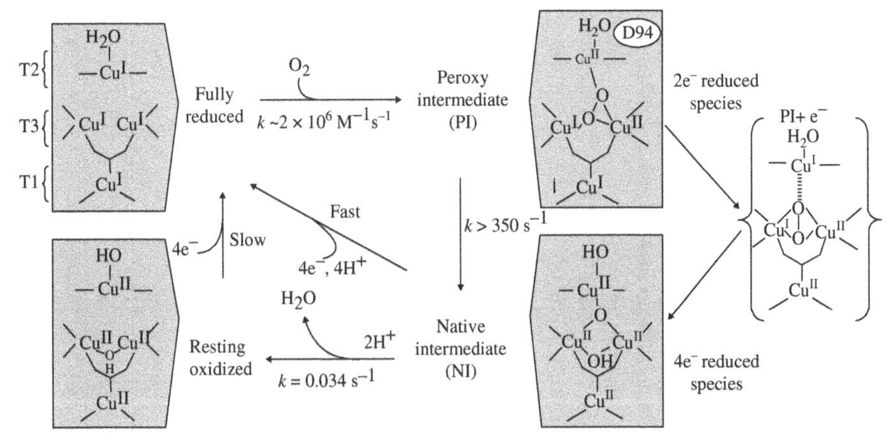

FIGURE 8.22 Mechanism of dioxygen reduction to water by the multicopper oxidases. Bold arrows indicate the steps that take place in the catalytic cycle. Black arrows indicate steps that can be experimentally observed but are not a part of catalysis. The dashed arrows at the right indicate the transfer of an electron from the T1Cu to the T2 to generate PI+e and the fascicle cleavage of the O—O bond due to the FMOs of the TNC. Reproduced with permission from Ref. [2]. © 2008 RSC Publishing.

are also important parameters in this respect [59, 60]. Two methods of ET are utilized for contact between the electrode and the enzyme: mediated ET (MET) and direct (DET) ET. In MET, the mediator shuttles electrons between the electrode and the redox-active T1 site of the enzyme. This often leads to high currents but with significant drawbacks including instability and toxicity of the mediator. Also, when a mediator is employed, it is the redox potential of the mediator, not the enzyme, that dictates the thermodynamic efficiency of catalysis, often leading to increased overpotential [60]. Alternatively, in DET, electrons are transferred directly from the electrode to the redox active site of the enzyme. Although this may result in lower currents than MET due primarily to inefficient ET pathways, the advantages compared to MET make it an attractive approach for bioelectrocatalysis [61]. Also, it allows for more fundamental molecular studies of MCOs compared to MET.

In the following sections, common approaches to and results from immobilizing MCOs for DET are presented (summarized in Table 8.3). This is not an exhaustive treatment, but rather an overview of the different approaches, emphasizing their advantages and disadvantages, based on selected literature. Also, a brief description of the coimmobilization of MCOs and redox hydrogels is given at the end of Section 8.3.5.

8.3.2 Unmodified Electrodes

The simplest method for immobilizing enzymes onto electrodes is by adsorption to unmodified electrodes. Some of the first examples of DET to MCOs on unmodified electrodes were reported by Lee et al. [62] and Thuesen et al. [63], both observing electrocatalytic O_2 reduction with the high-potential fungal Lc, *T. versicolor*

TABLE 8.3 Immobilization method and performance of MCO-modified electrodes (in DET contact) described in Section 8.3.2.

	Immobilization method	MCO	Max. onset potential (pH)	Current density (rpm, electrode potential)	Reference
Adsorption on unmodified electrodes	Pyrolytic graphite electrode	*Trametes versicolor* Lc	805 mV (3.8)	0.193 mAcm^{-2} (900 rpm, 240 mV)	Lee et al. [62]
	Pyrolytic graphite electrode	*T. versicolor* Lc	~875 mV (3.8)	4.5 μA[b] (no rotation dependence, 360 mV)	Thuesen et al. [63]
	Spectrographic graphite electrode	*Myrothecium verrucaria* BOD	~800 mV (4.0)	30 μA cm^{-2} (no rotation rate reported, 300 mV)	Shleev et al. [64]
	Spectrographic graphite electrode	*Trachyderma tsunodae* BOD	~700 mV (7.0)	40 μA cm^{-2} (no rotation rate reported, 150 mV)	Ramirez et al. [65]
SAM	Covalent attachment to MPA[a] SAMs	*Rhus vernicifera* Lc	~450 mV (7.0)	~6 μA[b] (no rotation rate reported, 150 mV)	Johnson et al. [66]
	Covalent attachment to different functional group SAMs	*M. verrucaria* BOD	~700 mV (6.8)	~10 μA cm^{-2} (no rotation rate reported, 300 mV)	Tominaga et al. [67]
	Covalent attachment to different thiol functional group SAMs	*Trametes hirsuta* Lc	~800 mV (5.0)	~0.5 μA cm^{-2} (no rotation rate reported, 400 mV)	Shleev et al. [68]
	Covalent attachment to MPA[a] SAMs	*T. tsunodae* BOD	N/A	~0.5 μA cm^{-2} (no rotation rate reported, 500 mV)	Ramirez et al. [65]
Diazonium coupling	Anthracene-2-diazonium-modified electrode	*Pycnoporus cinnabarinus* Lc	~850 mV (4.0)	~0.6 mAcm^{-2} (2500 rpm, 450 mV)	Blanford et al. [69]
	Chrysene-2-diazonium-modified electrode	*T. versicolor* Lc	~900 mV (4.0)	~20 μA[b] (3000 rpm, 450 mV)	Blanford et al. [70]
	Amino-2-naphtonic acid-modified electrode	*M. verrucaria* BOD	750 mV (6.0)	~.7 mAcm^{-2} (2500 rpm, 350 mV)	dos Santos et al. [71]
	Attachment to anthracene- and anthraquinone-modified electrodes	*T. hirsuta* Lc	~800 mV (5.0)	<0.5 μA cm^{-2} (no rotation rate reported)	Sosna et al. [72]
	Attachment to 2-aminophenol-modified electrodes	*T. hirsuta* Lc	~800 mV (4.2)	~25 μA[b] (no rotation rate reported, 400 mV)	Vaz-Dominguez et al. [73]

3D electrodes	Cross-linked with glutaraldehyde to multiwalled CNTs	*T. versicolor* Lc	~850 mV (6.0)	~70 μA cm^{-2} (no rotation rate reported, 400 mV)	Zheng et al. [74]
	Cross-linked with glutaraldehyde to multiwalled CNTs	*M. verrucaria* BOD	~750 mV (7.0)	~35 μA cm^{-2} (no rotation rate reported, 400 mV)	Zheng et al. [74]
	Cross-linked with glutaraldehyde to single-walled CNTs	*T. versicolor* Lc	~770 mV (6.0)	~25 μA cm^{-2} (no rotation rate reported, 650 mV)	Zheng et al. [75]
	Covalent attachment to PBSE-modified multiwalled CNTs	*T. versicolor* Lc	~780 mV (5.8)	~300 μA cm^{-2} (no rotation rate reported, 600 mV)	Ramasamy et al. [76]
	Covalent attachment to PBSE-modified multiwalled CNTs	*M. verrucaria* BOD	~780 mV (5.8)	~300 μA cm^{-2} (no rotation rate reported, 600 mV)	Ramasamy et al. [76]
	Immobilized in single-walled CNT forests	*Trametes* sp. Lc	~800 mV (5.0)	~4 $mAcm^{-2}$ (stirred solution, 600 mV)	Miyake et al. [77]
	Adsorbed on carbon aerogel-modified electrodes	*Trametes* sp. Lc	~800 mV (5.0)	~10 $mAcm^{-2}$ (8000 rpm, 0 mV)	Tsujimura et al. [78]
	Adsorbed on carbon aerogel-modified electrodes	*M. verrucaria* BOD	~700 mV (7.0)	~6 $mAcm^{-2}$ (4000 rpm, 200 mV)	Tsujimura et al. [78]
	Adsorbed on carbon aerogel-modified electrodes	*E. coli* CueO	~600 mV (5.0)	~13 $mAcm^{-2}$ (8000 rpm, 200 mV)	Miura et al. [79]
	Adsorbed on Ketjenblack-modified carbon paper (air-diffusion electrode)	*E. coli* CueO	~570 mV (5.0)	2.7 $mAcm^{-2}$ (no rotation, 200 mV)	Kontani et al. [80]

[a] Mercaptopropionic acid.
[b] No area adjusted current reported.

(TvL), by DET from unmodified pyrolytic graphite electrodes. Onset potentials were observed to be greater than 800 mV in both cases, consistent with the redox potential of approximately 780 mV for the T1 site determined by poised potential methods [19]. This indicates that electrons are transferred via the T1 Cu site to the TNC where dioxygen is reduced, with very little overpotential. Also, in the study by Thuesen et al., quasireversible peaks were observed under anaerobic conditions, with an estimated midpoint potential of 790 mV, also consistent with the observed onset potential of electrocatalysis. Observation of voltammetric peaks under anaerobic conditions is important, both for determination of redox potentials and as a method to estimate the amount of active immobilized enzyme.

In recent years, relatively few papers have been concerned with DET to MCOs on unmodified electrodes. Shleev and coworkers have reported DET on unmodified graphite electrodes to fungal Lcs and bilirubin oxidase (BOD). In a study from 2004, Shleev et al. investigated DET from a spectrographic graphite electrode (SPGE) to BOD from *Myrothecium verrucaria* (MvBOD), adsorbed on the surface of the electrode [64]. Electrocatalytic O_2 reduction was observed over a wide pH range with the onset potential of approximately 800 mV observed at pH 4.0. At pH 7.4, the onset potential had dropped to approximately 700 mV, indicating a pH dependence of approximately 30 mV, which cannot be ascribed simply to the pH dependence of O_2 reduction (59 mV pH^{-1}). Instead, the authors proposed that the observed pH dependence was related to the T1 Cu site. Based on this, the T1 Cu site was assumed to be the entry point of electrons in the electrocatalytic O_2 reduction. Under anaerobic conditions, a quasireversible peak was observed with a midpoint potential of 515 mV at pH 7.4. This peak, however, was only observed when a permselective membrane was applied to the electrode, indicating weak adsorption of the enzyme to the electrode.

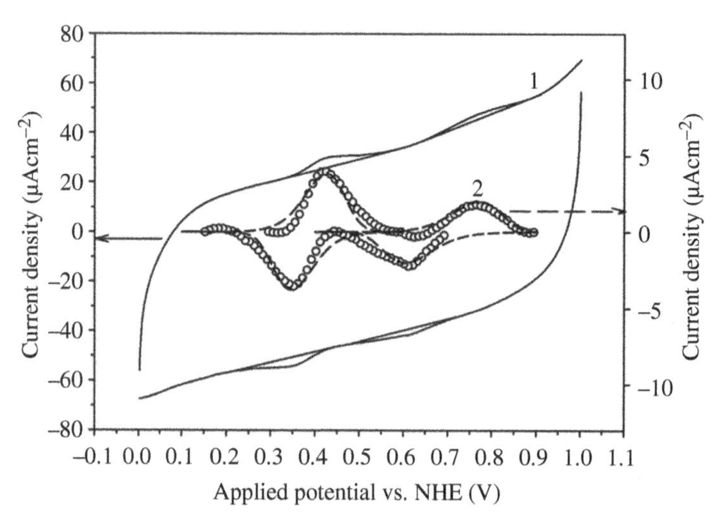

FIGURE 8.23 Anaerobic CV of TtBOD immobilized on an SPGE electrode. Circles represent enzymatic current obtained by subtraction of background spectrum. pH 7.0, 0.1 M phosphate, 100 mV s^{-1}. Reproduced with permission from Ref. [65]. © 2008 Elsevier.

In a subsequent study, similar experimental conditions were applied to BOD from *Trachyderma tsunodae* (TtBOD) [65]. Again, electrocatalytic O_2 reduction was observed when BOD was adsorbed on an SPGE electrode, with a similar onset potential as that observed previously (~700 mV at pH 7). In contrast to the previous study, symmetrical peaks were observed in CV conducted under anaerobic conditions (Fig. 8.23), with two separate midpoint potentials of 390 and 690 mV, respectively. The peak at 690 mV was ascribed to ET to the T1 Cu site, estimated at approximately 660 mV in solution [13]. The low-potential peak, on the other hand, was ascribed to ET directly to the T2 Cu site of the TNC, consistent with the few reported solution potentials of this site in MCOs [1]. The authors proposed that two different orientations of the enzyme caused the ET to different sites of the immobilized enzyme. It should also be mentioned that DET to BOD adsorbed on unmodified gold electrodes was investigated in the same study, but no current was observed under aerobic or anaerobic conditions.

8.3.3 Self-Assembled Monolayer Immobilization

The immobilization of enzymes to self-assembled monolayers (SAMs) on gold electrodes is a well-established technique, aimed at securing a stabile attachment and/or site-directed orientation of the enzyme allowing for efficient ET from the electrode to the active redox center [81, 82]. In one of the first reports on MCOs, Johnson et al. attached *Rhus vernicifera* Lc (RvL) to a 3-mercaptopropionic acid linker on gold sheets [66]. Subsequent CV, under anaerobic conditions, gave rise to well-defined anodic and cathodic peaks with a midpoint potential of 410 mV, close to that of the T1 Cu site measured in solution [11]. Interestingly, the ratio of the anodic and cathodic peaks, i_a/i_c, increased from 1:1 at less than 1 mV s^{-1} scan rate to 4:1 at 200 mV s^{-1} (Fig. 8.24), indicating that the TNC is oxidized, via the T1 Cu, more rapidly than it is reduced, at least under nonturnover conditions. Electrocatalytic O_2 reduction was also observed in aerobic CV experiments, with an onset potential consistent with the anaerobically determined midpoint potential.

One of the advantages of SAMs is the capability of attaching different functional groups at the end of the linker. In a study by Tominaga et al., only COOH- and SO_3H-derivatized SAMs provided reliable DET between electrode and MvBOD, whereas no electrocatalytic activity was observed with NH_2, CH_3, and OH groups [67]. In addition to different functional groups, the authors also investigated the effect of the length of the linker and found that catalytic current decreased from C2 to C5 and C7 and no current was identified for a C10 linker. Assuming that the attachment of the enzyme is independent of the length of the linker, this illustrates the ET dependence on pathway length, as described in standard ET theory [83]. Shleev and coworkers have also investigated the efficiency of different functional groups, in their case to high-potential *Trametes hirsuta* Lc [68]. Out of 16 different functional groups, only 4-aminothiophenol resulted in DET to the enzyme with irreversible anaerobic peaks observed at 380 and 800 mV. The authors ascribed these peaks to DET to the TNC and T1 Cu, respectively, similar to studies performed on unmodified electrodes

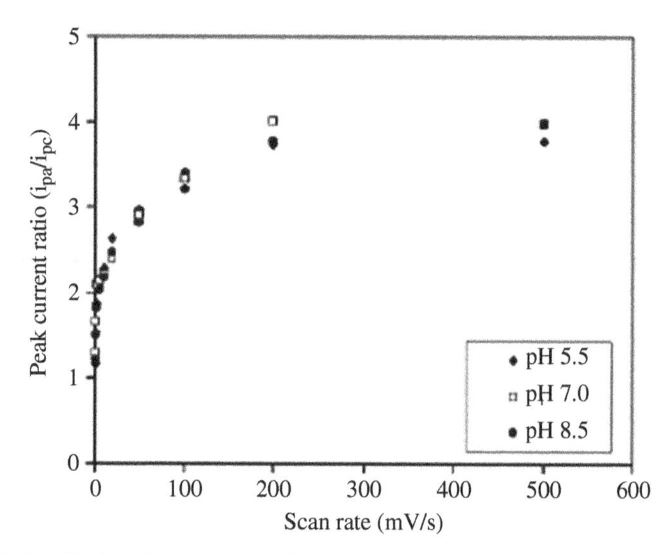

FIGURE 8.24 Ratio of peak-current dependence on scan rate for RvL under anaerobic conditions. Reproduced with permission from Ref. [66]. © 2003 the American Chemical Society.

(*vide supra*) by the same authors [64]. However, under aerobic conditions, the O_2 reduction current was extremely small, with a maximal onset potential around 800 mV. In a subsequent study, the authors immobilized TtBOD on gold electrodes via mercaptopropionic functional groups, again with ambiguous results [65]. Under anaerobic conditions, a single quasireversible peak with a midpoint potential of 360 mV was observed; no ET at high potentials was observed, indicating limited connection to the T1 Cu site. Electrocatalytic current was observed in chronoamperometric experiments with the potential held at 500 mV. The current reached less than 0.5 $\mu A\,cm^{-2}$ but did decrease upon the addition of fluoride, indicating "normal" activity of the enzyme (*vide infra*).

8.3.4 Immobilization by Diazonium Coupling

In a different approach to site-directed immobilization of MCOs, diazonium coupling to carbon/graphite electrodes has been investigated. The first report was presented by Blanford et al., who used a previously developed modification technique [84] to attach polyaromatic functional groups to a graphite electrode [69]. The technique involves generating a diazonium salt ($Ar\text{-}N_2^+$) from an aryl diamine, followed by covalent attachment of the diazonium salt to the electrode surface by electrochemical reduction, releasing N_2. Depending on the nature of the aryl, different functionalities can be attached, targeted for optimal binding and orientation of a given enzyme. Blanford et al. investigated a selection of arylamines for site-specific attachment of *Pycnoporus cinnabarinus* Lc, similar in sequence to TvL, via hydrophobic interaction to the putative substrate pocket [69], identified in the crystal structure of TvL [85]. Lc immobilized on 2-anthacene-modified electrodes showed the highest electrocatalytic activity, with an onset potential of greater than 800 mV. Furthermore, more than 50% of the activity was retained 8 weeks after immobilization. In comparison, electrocatalytic activity of the unmodified graphite electrode with physi-adsorbed fungal Lc showed a lower onset potential and significantly reduced currents, which quickly diminished due to desorption of the enzyme. Later studies, utilizing the same immobilization technique, have been presented by Blanford and coworkers. In these, TvL showed similar electrocatalytic behavior to that of *P. cinnabarinus*, although a slightly different aryl unit, 2-aminochrysene, resulted in the highest activity [70]. MvBOD activity, on the other hand, was optimized by attachment to a 6-amino-2-naphthoic acid-derivatized electrode [71]. Whether this optimization was a result of the difference in enzyme structure or simply because the acid derivative was not investigated in the fungal Lc studies is uncertain. The onset potential of the BOD activity was found to depend on pH, with higher potentials observed at low pH. By analyzing this potential dependence with the "dispersion" model developed by Leger [86], the authors determined that the attachment of different enzyme molecules was highly uniform, with a variation in ET distance to the entry site of the enzyme of only 6–7 Å [71].

Other groups have reported DET to diazonium-coupled MCOs. Similar to Blanford et al., Sosna et al. used anthracene to immobilize *T. hirsuta* Lc to a glassy carbon electrode, but only very limited currents were observed ($<0.5\ \mu A\ cm^{-2}$) [72]. The authors ascribed this to a different immobilization strategy where Boc-protected amines were utilized in the attachment of the linker molecules, proposed to ensure only monolayer formation. In Blanford et al., the unprotected linkers could potentially lead to the generation of multilayer structures, resulting in higher currents. It should be noted, however, that in the presence of ABTS mediator, currents increased by more than 30-fold, indicating that poor connection between the enzyme and the electrode caused the low current and not the monolayer structure.

Finally, Vaz-Dominguez et al. reported on a different linker strategy, where aminophenyl groups were utilized to link *T. hirsuta* Lc via amide or ester bonds to surface-exposed enzyme residues [73]. Again, high onset potentials for electrocatalytic reduction of O_2 were observed, close to the redox potential of the T1 Cu site [20]. Although the orientation is assumed to be somewhat random, it is interesting to observe that the reduction current only increased twofold when a mediator was added to the solution.

8.3.5 Three-Dimensional Electrodes

An often encountered limitation for use in biofuel cells, common to the aforementioned immobilization methods, is low catalytic current density, related to small total surface areas. If MCOs are to be incorporated into biofuel cells and other biodevices, much higher currents are desired. This has been addressed in a number of studies, where modification with different carbon materials allows for immobilization of enzyme in three dimensions. Carbon nanotubes (CNTs), carbon aerogels (CAs), and various other nanoparticles have been investigated for electrode modification with

MCOs, and in this section, some of these approaches will be described, highlighting advantages compared to monolayer approaches described previously but also pointing out some of the obstacles that are encountered.

CNTs have attracted a great deal of interest in a variety of bioelectrical applications, due to their unique properties, including high conductivity, high total surface area, and efficient electric contact to immobilized enzymes [87]. The first report of DET to a CNT-immobilized MCO was by Zheng et al. in 2006 [74]. In this study, the authors investigated DET to immobilized TvL as well as MvBOD and observed significant catalytic activity of both enzymes on multiwalled CNTs (MWCNTs). As pointed out by the authors, unmodified CNTs tend to aggregate in solution. This can, however, be overcome by either covalent or noncovalent modification [74]. In this case, the CNTs were noncovalently modified with different cellulose derivatives, which significantly increased their solubility. The catalytic activities of TvL and MvBOD were investigated with both unmodified and modified CNTs, but only unmodified CNTs resulted in catalytic current. The onset of catalysis was observed at approximately 850 and 750 mV for TvL and BOD, respectively, consistent with ET to the T1 Cu's (*vide supra*). For TvL, catalytic currents of approximately 45 $\mu A\ cm^{-2}$ were obtained at 200 mV without rotation. It should be mentioned that no voltammetric peaks were observed in CVs under anaerobic conditions.

In a subsequent study, the same group investigated DET to TvL on single-walled CNTs (SWCNTs) and again observed significant catalytic current [75]. The maximum current and onset potential were similar to those obtained on the MWCNTs. Furthermore, by employing differential pulse voltammetry techniques, the authors were able to observe redox peaks with a formal potential of 780 mV, consistent with the redox potential of the T1 site in TvL.

As for the monolayer electrodes described previously, CNTs can be modified with different functional groups, in order to obtain better contact between the CNTs and the enzyme. One example was presented by Ramasamy et al., where MWCNTs were modified with 1-pyrenebutanoic acid, succinimidyl ester (PBSE) [76]. The aromatic pyrenyl moiety forms an irreversible π–π interaction directly with the CNTs, whereas the succinimidyl ester is proposed to form amide bonds with amines on the protein surface [88]. When TvL and MvBOD were immobilized on CNT–PBSE-treated carbon paper electrodes, both enzymes showed catalytic currents in excess of 300 $\mu A\ cm^{-2}$, both with onset potentials consistent close to their respective T1 Cu potentials. Interestingly, the effect of PBSE treatment was significantly higher for TvL-modified electrodes, where catalytic current was absent on unmodified CNTs, whereas the catalytic current for BOD immobilized on unmodified CNTs was approximately 80% of that observed for the modified electrode. This illustrates the inherent differences in MCOs and emphasizes the need to optimize conditions for the desired MCO.

A report on DET to CNT-immobilized MCOs, by Miyake and coworkers, deserves attention. In this study, the authors present an interesting strategy to *in situ* modify the CNT structure to match the dimensions of a given protein [77]. Normally, the CNTs are immobilized on the electrode, before the addition of the redox-active enzyme, resulting in poor correlation between the enzyme and CNT dimensions [77]. In their approach, Miyake et al. used approximately 1 mm-long SWCNTs, known as carbon

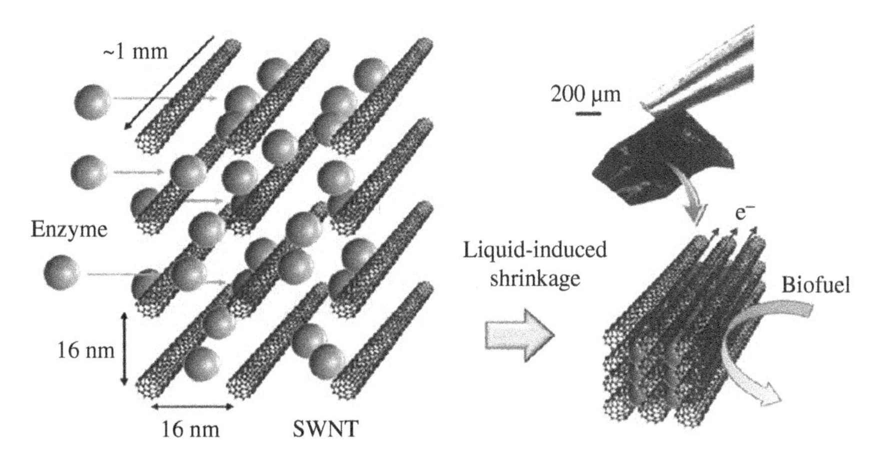

FIGURE 8.25 Schematic representation of liquid-induced shrinkage of CNTFs resulting in efficient entrapment of enzyme. CNTF-modified electrodes can be handled with tweezers, as illustrated in top right corner. Reproduced with permission from Ref. [77]. © 2011 the American Chemical Society.

nanotube forests (CNTFs), that shrink upon the addition of liquid followed by drying. With an enzyme-containing liquid, the CNTFs shrink to match the size of the enzyme, as illustrated in Figure 8.25. With this approach, the authors designed a biofuel cell, incorporating commercially available *Trametes* sp. Lc (TspL) on the cathode and fructose dehydrogenase on the anode. The evaluation of the catalytic properties of the cathode showed currents as high as 4 mA cm^{-2} (under stirred conditions) and an onset at approximately 800 mV. The stability of the enzyme was relatively good, with more than 80% activity retained after 24 h.

Another approach for DET to MCOs on three-dimensional (3D) electrodes is immobilization in the so-called CAs, primarily investigated by Kano and coworkers. CAs are porous nanoparticle carbon structures, which can be synthesized with different pore sizes. In addition to their porosity, CAs are highly conductive and have large surface areas, all valuable properties for enzymatic electrodes [89]. For electrode attachment, poly(vinylidene difluoride) is used as binder, followed by drying on either glassy carbon or carbon paper electrodes. Subsequent adsorption of enzyme is performed by immersion in enzyme solution or simply adding enzyme solution to the electrode. In a study from 2007 by Tsujimura et al., TspL and MvBOD were immobilized on CAs with electrocatalysis of O_2 onset at approximately 800 and 700 mV, respectively. High catalytic currents were observed for both enzymes, limited by the diffusion of O_2 as evaluated on rotating disk electrodes [78]. For TspL, the current density dependence on rotation rate showed an almost linear correlation with $\omega^{1/2}$, up to $\omega = 8000$ rpm, where a current of more than 10 mA cm^{-2}, at 0 mV, was obtained (Fig. 8.26). For MvBOD, the linearity was less pronounced, but again, high currents of greater than 6 mA cm^{-2} were observed at 0 mV.

In a subsequent study, Miura et al. used a similar approach to immobilize CueO (a bacterial Lc) in a CA [79]. O_2-limited current densities as high as 13 mA cm^{-2}

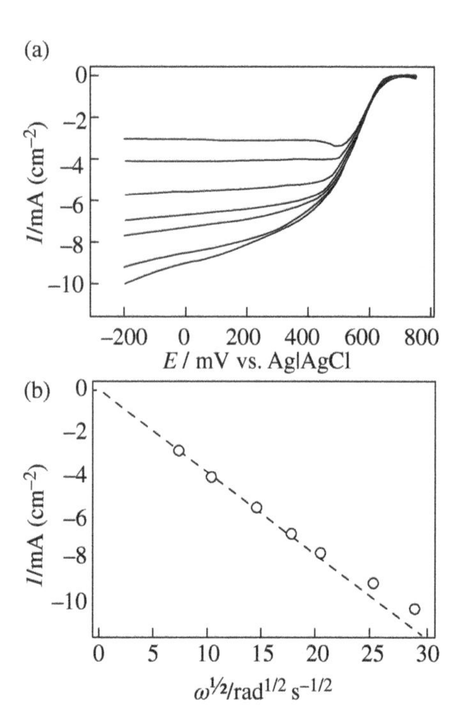

FIGURE 8.26 (a) Rotation rate dependence on catalytic current density for immobilized CueO. 500 (top curve), 1000, 2000, 3000, 4000, 6000, and 8000 rpm (bottom curve). (b) Current density versus square root of angular velocity (at 0 mV), fit to a four-electron transfer process. Reproduced with permission from Ref. [78]. © 2007 John Wiley and Sons.

at 8000 rpm were observed. However, the onset potential was approximately 600 mV, significantly lower than those observed for TspL and MvBOD (*vide supra*). This correlated with the low T1 Cu potential observed in solution [79], indicating that DET occurs to this site. Subsequently, upon mutation of selected T1 Cu residues, the onset potential was found to shift, parallel to shifts in T1 redox potential observed in solution [79].

One significant challenge with 3D MCO bioelectrodes is slow diffusion of O_2, resulting in limited current in the absence of rotation. mA currents are only obtained at relatively high rotation rates. This is impractical if the electrodes are to be employed in biofuel cells. One approach to address this is the so-called air-diffusion electrode, where O_2 is provided from the air directly to the electrode, thereby limiting the effects of slow diffusion [90]. Kontani et al. designed an air-diffusion electrode, where CueO was immobilized with Ketjenblack as support [80]. This electrode showed significantly higher activity, relative to solution-type electrodes under the same conditions. In fact, the currents were comparable to solution-type electrodes in air-saturated buffer, rotated at 10,000 rpm.

As a final example of immobilized 3D MCO cathodes, the so-called redox hydrogels deserve attention, realizing, however, that this technique involves MET. Redox hydrogels, incorporating enzymes, were pioneered by Adam Heller, Nicolas Mano, and coworkers [91, 92] and have been applied to a range of enzymes, most prominently in a functional glucose sensor, utilizing glucose oxidase to detect blood serum levels of glucose in diabetes [93]. The general structure consists of an electrode-immobilized organic polymer to which various redox-active metal complexes, most commonly osmium complexes, are covalently attached. The conducting properties arise from electron self-exchange between individual metal ions. When redox-active enzymes are coimmobilized within the hydrogel, electrical contact can be established between the Os ion and the relevant redox center of the enzyme. Os complexes with a range of redox potentials have been synthesized, allowing for optimized potential differences between the redox-active metal ion and the enzyme, minimizing overpotentials in catalysis while still allowing for facile ET [60]. Attachment of the Os complex to long flexible linkers allows for facile ET to the enzyme, circumventing the requirement for site-directed attachment of the enzyme, as well as allowing for electrical contact to several enzyme layers [94]. The efficiency of MCOs immobilized in redox hydrogels has been demonstrated in several studies by Mano and coworkers and others [95–98]. An example is the electrocatalytic reduction of O_2 observed for TvL, in a study where the electrocatalytic properties were compared directly to those of an inorganic platinum catalyst [94]. As observed in Figure 8.27, the onset potentials of the enzymatic cathodes (1 and 2) are close to optimal and superior to that of the platinum cathode (3). Furthermore, with the Os complex attached via a long flexible linker (1), the current density is comparable to that of the platinum catalyst.

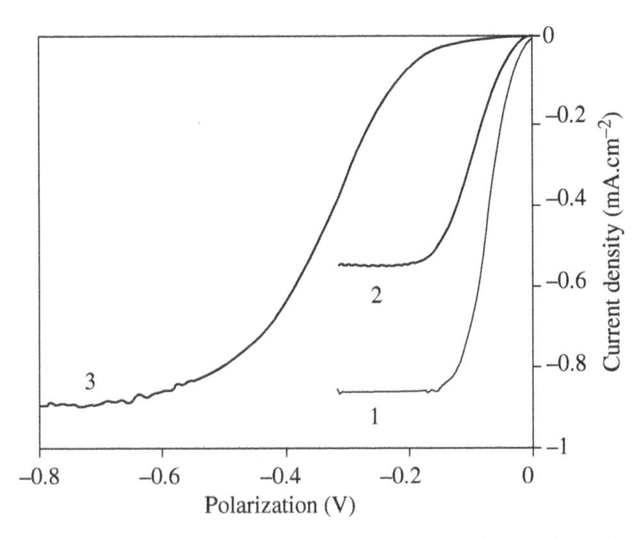

FIGURE 8.27 Polarizations of TvL-modified redox hydrogel with Os redox complex via long (1) and short (2) linkers. Platinum electrode under similar conditions is incorporated for comparison (3). Reproduced with permission from Ref. [94]. © 2006 the American Chemical Society.

8.4 FUNCTIONALITY OF IMMOBILIZED AND SOLUBILIZED MCOs

8.4.1 ET from Substrate to Enzyme

As evident from Section 8.2, much is known about the intermediates and mechanism of the MCOs in solution. Section 8.3 highlights how these enzymes, by immobilization on electrode surfaces, can be employed as electrocatalysts. At this point, it is important to compare and contrast the chemistry of the MCOs under these different conditions. First, the processes by which electrons are delivered to the enzyme in solution or enzyme immobilized on electrodes will be explored. Differences and similarities with respect to the terms in the Marcus equation for ET will be considered, followed by a discussion of the possibility of DET to the TNC in MCOs. Then, inhibition of the MCOs by high pH and chloride concentration, both critical for application in implantable devices, will be addressed.

The differences between solution and immobilized MCOs are evident in comparing the variables in the expression for the rates of ET from Marcus theory ([83] Equation 8.2), which has been applied to solution ET from the Fe(II) substrate to the T1 site in Fet3p and hCp [38]:

$$k_{\mathrm{ET}} = \sqrt{\frac{4\pi^3}{h^2\lambda kT}}[H_{\mathrm{DA}}]^2 \exp\left(\frac{-(\Delta G^\circ + \lambda)^2}{4\lambda kT}\right) \tag{8.2}$$

The variables that directly affect the rate of solution ET include the reorganization energy (λ), which is the energy required to structurally reorganize the ligands and protein environment upon change in oxidation state and the free energy (ΔG°, thermodynamic driving force), which reflects the difference between reduction potentials in the donor and acceptor for the ET reaction. The electronic coupling matrix element (H_{DA}) quantifies the efficiency of the overlap between the donor and acceptor through the protein superexchange pathways.

In general, between solution phase and immobilized enzymes, the driving force in solution originates from the difference in redox potentials of the *bound* substrate and the T1 site implying faster rates with lower-potential substrates. This is analogous for the immobilized enzymes where the reducing capability of the electrode can be controlled by the applied potential, and therefore, the driving force is the difference between the T1 potential (the site of electron entry into the MCOs) and the applied potential of the electrode. Therefore, faster ET rates can be accessed via lower applied potentials.

Considering reorganization energy, ET in solution involves transferring an electron from the localized donor state of the substrate to the localized acceptor state of the T1, meaning that the total reorganization energy (λ) is the sum of the reorganization energies of both the substrate and T1 center. For an immobilized MCO, electrons are delivered to the enzyme from the electrode surface. In this case, the electron originates from a state where the change in charge upon loss of an electron is rapidly delocalized in the conduction band on a time scale that is faster than structural and solvent rearrangements [99, 100]. Therefore, the reorganization energy of the electrode is limited, and ET in the immobilized systems is mostly impacted by the reorganization energy of the T1 site.

Major differences, however, are expected when considering the pathways for the delivery of the electrons to the enzyme. In solution, the enzyme has to bind substrate prior to ET. A prefactor (SK_A) is included in Equation 8.2, where S reflects the sterics of the substrate interaction with the T1 copper of the enzyme, where only specific orientations of substrate are optimal for ET, and K_A is the equilibrium constant for binding of substrate in the bimolecular ET rate [83]. However, once substrate is bound in solution, efficient superexchange pathways exist for ET from the substrate to the T1, which give the H_{DA}. Since, in many cases, the enzyme immobilization on the electrode surface is not controlled, there can be various pathways and a range of H_{DA}'s for ET to the T1 site. Since H_{DA} decreases exponentially as a function of distance between the donor and acceptor, the longer the distance of the T1 to the electrode surface, the less likely is ET from the electrode to the enzyme. This depends on the type of electrode and/or the method of immobilization and the choice of MCO enzyme. The aforementioned study by Tominaga et al. observes diminishing and eventual termination of catalytic current with longer organic linkers to a SAM electrode, which is proposed to be due to the distance dependence on ET [67]. Additionally, depending on the MCO employed, the T1 site can be more or less buried in the protein as in the metalloxidases or organic oxidases, respectively. Therefore, it would be expected that enzymes with exposed T1 sites immobilized closer to the electrode surface would produce a maximally efficient fuel cell. Finally, specific superexchange pathways can be very efficient for ET. This comes from the covalency of bonds and could be particularly efficient for MCOs with specific modifications that allow direct orientation of a superexchange pathway from the electrode surface to the T1 site.

8.4.2 DET to TNC of Immobilized MCOs

As described previously, ET to the TNC in MCOs occurs via the T1 Cu. This is the case for homogeneous (i.e., solution) and generally for heterogeneous (i.e., electrochemical) turnover (*vide supra*). In addition to ensuring efficient ET to the TNC, the T1 Cu dictates the potential at which electrons can be extracted from a substrate for oxygen reduction by MCOs (*vide supra*). In fuel cells, the redox potential of the T1 Cu determines the energy, in the absence of a redox mediator, that can be extracted from a given fuel, compared to the thermodynamic limit for oxygen reduction [57]. The difference between the two is the overpotential of the cathodic reaction. Although high-potential Lcs have been demonstrated to operate with much lower overpotentials than the most efficient inorganic catalysts [94], electron shuttling via the T1 Cu will inevitably result in the loss of energy. This raises the question of the intrinsic overpotential of MCOs in oxygen reduction, that is, the potential at which the TNC can directly transfer electrons to dioxygen [71]. In that respect, T1-depleted MCOs (T1D or T1Hg (*vide supra*)) should be evaluated. In these derivatives of the MCOs, oxygen reduction is retained, as evident from solution experiments (*vide supra*), but catalytic activity is impaired as a result of having only three electrons available for reduction of O_2 (to PI), as compared to the four-electron reduction that occurs in the presence of the T1 Cu (*vide supra*). In order to reestablish facile four-electron reduction in T1-depleted enzymes, efficient ET from the substrate directly to the TNC must be established. Through numerous experiments, this has not proved possible in solution. The only molecule thus far found to be capable of reducing TNC Cu's in the absence of the T1 is dithionite, and despite its low potential and small size, this is still a slow process. Alternatively, ET directly to the TNC may be possible for T1-depleted immobilized MCOs. In this case, the onset potential of catalytic current would reveal the intrinsic overpotential of the TNC. ET to immobilized T1-depleted MCOs has not been reported, but several studies on holoenzyme suggest that the connection between electrode and TNC may already have been observed.

In the study by Ramirez et al. [65] (Section 8.3.2), two CV peaks were observed under anaerobic conditions with E° of approximately 390 and 690 mV, respectively, for TtBOD immobilized on an SPGE electrode. In the presence of O_2, catalytic onset was observed at the two different potentials corresponding to the anaerobic peaks. Whereas the 690 mV peak can be assigned to T1 Cu reduction, consistent with the solution determined potential of this Cu (*vide supra*), the origin of the second peak is unclear. The authors assign it to DET to the TNC, based on the literature value of the redox potential of this site (~400 mV), determined by anaerobic redox titration (*vide supra*), and propose that an uphill ET from the T1 to the T2 Cu occurs in catalysis. While a low-potential TNC Cu may be present under anaerobic conditions, the fact that it is also observed under aerobic conditions must be considered in the context of the generally accepted model for catalytic O_2 reduction of MCOs (*vide supra*). Under anaerobic conditions, the TNC is in the resting state where the T2 Cu is uncoupled from the binuclear T3 Cu's, which leads to slow reduction of the TNC [2]. Under turnover conditions, however, reduction of the NI of the TNC is facile, consistent with its all-bridged structure (*vide supra*). This rules out the presence of a low-potential TNC Cu under turnover. A possible explanation for the observation of two separate onset potentials for electrocatalysis of BOD has been provided by a recent study. Here, two distinct onset potentials, similar to those observed by Ramirez et al., were observed with immobilized BOD from *Magnaporthe oryzae* [101]. However, the onset potentials were correlated to the enzyme having two different resting forms of the TNC, identified by spectroscopic methods, with the low-potential form converted to the high-potential form when electrochemically activated at approximately 400 mV.

DET to the TNC has also been proposed in other MCOs. In a similar study to Ramirez et al., Frasconi and coworkers observed two redox peaks, under anaerobic conditions, for *T. versicolor* and *T. hirsuta* in a polymeric film of an anionic exchange resin immobilized on gold electrodes [102]. Consistent with the T1 Cu potential of these enzymes, one redox peak was observed at approximately 800 mV, while the second peak was observed at approximately 400 mV. In contrast to Ramirez et al., however, only one onset potential, corresponding to the high-potential peak under anaerobic conditions, was observed for catalytic current, consistent with the mechanism for O_2 reduction by MCOs.

As indicated earlier, DET to the TNC of an immobilized MCO have been proposed. However, confirmation will have to come from electrochemical studies, for example, of the T1-depleted enzyme. In the holoenzyme (with the T1 Cu present), the T1 Cu is connected via superexchange pathways to the T3α and T3β, respectively, ensuring optimal delivery of electrons to the TNC, both in conversion of PI to NI and in the rereduction of the TNC following NI formation. In the conversion of PI to NI, the electron delivered by the T1 Cu cannot enter the TNC via the T3α Cu, as this is reduced (*vide supra*). Similarly, the correct sequence of TNC Cu reduction (NI to fully reduced TNC) may be required for facile turnover. For immobilized T1-depleted enzyme, a given pathway providing direct ET between electrode and the TNC may not be optimized for electron delivery, in which case redox peaks observed under anaerobic conditions will not result in an observable electrocatalytic current in the presence of O_2. This should be considered in the investigation of T1-depleted MCOs immobilized at electrodes.

8.4.3 Inhibition of MCOs at High pH

It is generally accepted that catalytic activity in solution of MCOs is inhibited at high pH (>7). A study by Xu showed that the catalytic activities of *Trametes villosa* (high-potential) and *Myceliophthora thermophila* (low-potential) Lcs were significantly inhibited at pH above 7 [103]. Importantly, this was observed with aprotic as well as protic substrates. Similarly, for MCOs immobilized on electrodes, inhibition at high pH is generally observed. While studies of Mv and TtBODs, immobilized on redox polymers, have indicated that these enzymes retain their catalytic activity up to pH ~10 [96, 97], these results have proven somewhat ambiguous. First, there was little effect on catalytic current upon fluoride addition, inconsistent with the high affinity of the TNC for fluoride established as a general trait for MCOs (*vide supra*). Second, a later study with a similar preparation of a *T. tsunodae*-immobilized electrode showed significant inhibition at pH > 7 [104].

In general, MCO activity is inhibited at high pH, whether in solution or immobilized on an electrode. The explanation for this has been somewhat elusive, partly because the mechanism for catalysis has only recently been elucidated. Considering the different reaction steps in this mechanism, substrate to T1 Cu ET, formation of NI from fully reduced enzyme, and T1 to TNC ET, inhibition at high pH is likely associated with T1 ET to the TNC. First, the redox potential of the T1 Cu has been shown to have little pH dependence, with facile ET from substrate observed even at pH >8 [105]. Second, the reaction of fully reduced enzyme and O_2 to form PI is pH independent, and while the reduction of PI to NI can be either H^+ assisted or unassisted, both are significantly faster than the overall turnover rate [2, 49]. This leaves the T1 reduction of the TNC, in either the resting form or as NI, both consisting of three T1 to TNC ET's, as the likely source of inhibition at high pH.

Some existing data support this option. In the case of reduction of the TNC under anaerobic conditions (i.e., reduction of resting oxidized enzyme), Andreasson and Reinhammar observed a significant decrease in reduction rate of the T2 Cu, as monitored by EPR, with more than 60% oxidized Cu remaining after 50 s, with 30-fold excess of hydroquinone substrate at pH 8.4 [105]. This was initially interpreted as deprotonation of a T2 Cu-coordinated water ligand at high pH, resulting in a lowering of the redox potential of this site. This, however, is inconsistent with spectroscopic results where Quintanar et al. showed that the water ligand of the T2 Cu in the resting oxidized T1-depleted RvL remains deprotonated over the functional pH range (4–10) of this enzyme [30]. Regardless of the mechanistic origin, if a subpopulation of the TNC Cu's of the resting oxidized enzyme remains oxidized at high pH under turnover, this will significantly decrease the observed turnover number.

With respect to inhibition of TNC reduction (in the NI state) at high pH under turnover, a recent study by dos Santos et al. deserves attention. Here, the researchers investigated the pH dependence of catalytic activity of MvBOD, immobilized on an electrode in DET contact with the T1 Cu site [71]. At low pH, the turnover current was observed to linearly increase when the electrode potential was lowered, indicating that interfacial (i.e., electrode to T1) ET is rate limiting, whereas at high pH, the turnover current showed a sigmoidal increase with lower electrode potential, indicating a shift in the rate-limiting step to intramolecular (T1 to TNC) ET. This is consistent with pH-dependent rereduction of NI to fully reduced enzyme. Whereas the formation of NI by four-electron reduction of O_2 has been studied in detail [51], relatively little is known about the involvement of protons in the reduction of NI. It is known that NI decay to the resting oxidized form is accelerated at low pH in the absence of reducing electrons [106]. Whether this will translate to more facile reduction of NI at low pH is yet to be determined.

8.4.4 Chloride Inhibition of MCOs

Interaction between halide and other anions and MCOs has long been studied in solution. In a study from 1996, Xu investigated the catalytic tolerance, with ABTS as the substrate, of five different Lcs and MvBOD toward fluoride, chloride, and bromide and found a general trend of decreasing tolerance with smaller anion [35]. Among the different MCOs, all were inhibited at low fluoride concentrations, whereas chloride and bromide inhibition varied significantly, with RvL showing the highest degree of inhibition.

For implantable biodevices incorporating MCOs as the cathode enzyme, chloride inhibition is of particular interest, due to the high concentrations in the body. Early studies conducted by Heller and coworkers found very low Cl^- inhibition of BODs, when these were immobilized in Os redox polymers [96, 97, 107]. Later studies, by other groups, found similar low inhibition with *T. hirsuta* fungal Lc [98] and CotA bacterial MCO [108] immobilized in redox polymers. In the case of CotA, however, the tolerance was found to be dependent on the type of Os complex employed. For immobilized MCOs, in DET contact with the electrode, different results have been obtained with respect to Cl^- inhibition. In a study by Blanford et al., *T. versicolor* Lc lost more than 50% of its activity in the presence of 25 mM Cl^- when immobilized on an anthracene-modified graphite electrode [70]. In contrast, Vaz-Dominguez et al., found practically no inhibition of *T. hirsuta* Lc, covalently attached to an aminophenyl-derivatized electrode, with Cl^- levels of greater than 350 mM (Fig. 8.28a) [73]. In the same study, when ABTS was added as a mediator, a significant increase in catalytic current was observed. Interestingly, the additional mediated current was

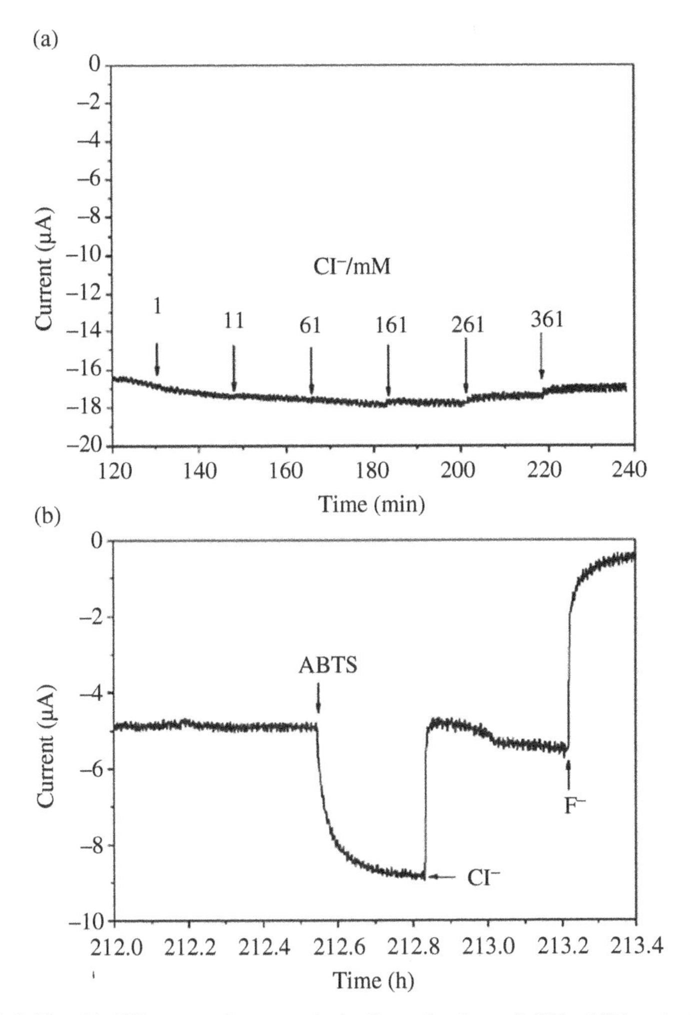

FIGURE 8.28 (a) Effect on electrocatalytic O_2 reduction of Cl^- addition to *T. hirsuta* laccase in DET with the electrode. Arrows indicate the addition of chloride ions with the final concentration in mM listed above. (b) Effect of Cl^- addition to ABTS-mediated electrocatalytic O_2 reduction. Background DET current is approximately 5 μA and is seen to be inhibited by the addition of F^-. Reproduced with permission from Ref. [73]. © 2008 Elsevier.

completely inhibited by the addition of 150 mM Cl^- (Fig. 8.28b). A similar result was reported by Beneyton et al., with no inhibition of enzymatic current, with CotA enzyme adsorbed directly on a graphite electrode, whereas additional current, in the presence of ABTS, was completely inhibited by Cl^- addition [108]. Overall, ABTS-mediated catalysis is inhibited by Cl^- regardless of whether the MCO is in solution or immobilized on electrodes, whereas inhibition of electrocatalytic current with enzyme in direct contact with the electrode, or mediated via Os complexes in a polymer, seems to be more ambiguous.

The aforementioned results raise the issue of how chloride inhibition occurs mechanistically and whether a single model for this inhibition is sufficient to explain the observed behaviors. In that respect, it is relevant to compare Cl^- interaction to interactions of MCOs with other anions, fluoride and azide in particular, for which spectroscopic data are available and perturbations are observed upon binding to the TNC [5, 6, 45]. The trinuclear cluster has an unusual high affinity for fluoride, where the addition of stoichiometric concentrations results in superhyperfine splitting of the T2 Cu signal, observable by EPR [109]. A μ_3-F^- all-bridged fully oxidized TNC structure was developed by Quintanar et al. (Fig. 8.29), where, based on spectroscopic and computational evaluation, the high affinity was ascribed to electrostatic attractions between the positive environment of a fully oxidized TNC and the negative fluoride ion [30]. For azide, the affinity of the TNC was evaluated by UV–Vis absorption and LTMCD methods, revealing high- and low-affinity binding to the TNC [5, 6, 45]. Interestingly, the high-affinity binding was ascribed to a partially reduced TNC, where only the T2 Cu was in the oxidized form [45]. In concert, studies on fluoride and azide and their competition led to the first proposal of a trinuclear Cu site in biology [5, 6], which was later confirmed by crystallography [24]. Whereas fluoride and azide binding to the TNC result in observable spectroscopic perturbations of the TNC, no definitive interactions between this site and chloride have been spectroscopically defined. Crystal structures of MCOs do however indicate that Cl^- has access to the TNC. In a structure of *M. albomyces* Lc, Cl^- is refined to the position of the water-derived ligand at the T2 Cu [110]. Alternatively, in studies of CueO, Cl^- has been observed to bridge the T3 Cu's in a similar way as OH^- [111]. Whether the redox states of the TNC involved in these structures are consistent with the Cl^- inhibition observed under solution or mediated turnover is yet to be determined. Interactions between chloride and the T1 Cu of MCOs may also occur, although no spectroscopic or crystallographic evidence for this exists and there is no obvious Cl^- binding site

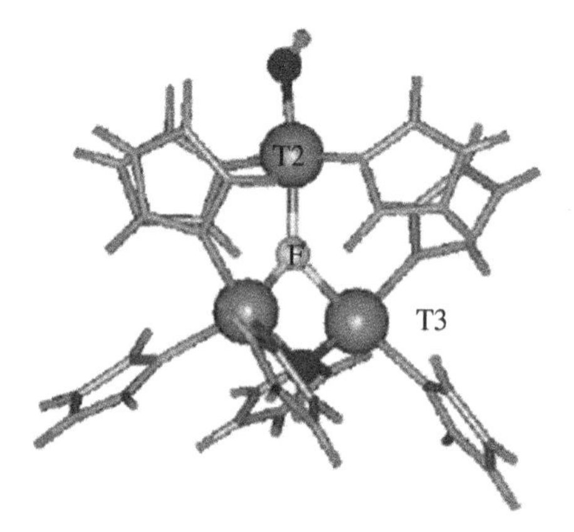

FIGURE 8.29 Optimized structure of fluoride binding to a fully oxidized resting TNC. Reproduced with permission from Ref. [30]. © 2005 the American Chemical Society.

near the T1 Cu. In order to elucidate the mechanism for chloride inhibition, a systematic approach investigating different steps in catalysis correlated with a spectroscopic evaluation of the different states of the individual Cu sites is required. In this respect, it is important to keep in mind the affinity dependence on redox state of the TNC toward anions (*vide supra*), especially considering observations with MCOs in solution compared to immobilized on electrodes. It may be envisioned that the different environments alter the timing of electron delivery to the T1 Cu and subsequently to the TNC, thereby generating different intermediates with varying Cl^- affinities, for immobilized versus solution enzyme.

As evident from earlier, the inhibitions of MCO activity at high pH and high Cl^- concentration are not yet fully understood. Whereas inhibition at high pH is a general feature of MCOs, inhibition by Cl^- varies from enzyme to enzyme, and in a given enzyme, with immobilization strategy to an electrode. Studies aimed at systematically investigating the proposed models may lead to a better understanding of these inhibition processes and how they can be avoided, which will be of great importance especially for implantable devices incorporating MCOs on the cathode.

8.5 CONCLUDING COMMENTS

Significant attention has been devoted toward understanding the unique chemistry of the MCOs, the oxidation of high-potential substrates coupled to the facile four-electron reduction of dioxygen to water. These properties have generated much interest in the application of MCOs as oxygen-reducing biocathodes. The combination of solution and electrochemical experiments has the potential to enhance our understanding of the MCOs in general and for biofuel cell applications in particular. Ambiguous redox peaks observed electrochemically might be difficult to assign without direct chemical insight from solution studies. Likewise, chemical effects on MCO activity, including high potential and chloride inhibition, are difficult to interpret only by electrochemistry. In parallel, electrochemistry can provide important insight into solution observations. The rate-determining step in the organic oxidases is the oxidation of substrate (ET from substrate to T1), which limits the elucidation of fast kinetic processes including intramolecular ET from the T1 to the TNC. Since the applied potential of the electrode can be varied to access higher driving forces and therefore faster ET rates from the electrode to the T1, it may be possible to probe mechanistic events not accessible in solution. Also, most organic oxidizing MCO substrates are protic, which prevents the evaluation of proton-dependent steps at the TNC. An electrode-mediated process should eliminate the substrate pH effect and reflect proton-coupled ET process. Another interesting possibility, which may be accessible on an electrode and not in solution, is DET to the TNC in catalysis. If achieved, and the enzyme reduces dioxygen catalytically, information will be obtained about the intrinsic overpotential of the TNC. Furthermore, if DET to the TNC can be established, this could circumvent the Cys—His ET pathway from the T1 resulting in "nonnative" delivery of electrons to the TNC. Compared to the native enzyme, this could result in new chemistry at the active site that would be inaccessible in solution.

ACKNOWLEDGMENTS

This research is supported by NIH grant DK-31450.

We thank Dr. Federico Tasca for his insightful evaluation of this manuscript.

REFERENCES

1. Solomon, E. I., Sundaram, U. M., Machonkin, T. E., *Chemical Reviews* 1996, 96(7), 2563–2605.
2. Solomon, E. I., Augustine, A. J., Yoon, J., *Dalton Transactions* 2008, (30), 3921–3932.
3. Solomon, E. I., *Inorganic Chemistry* 2006, 45(20), 8012–8025.
4. Solomon, E. I., Hadt, R. G., *Coordination Chemistry Reviews* 2011, 255(7–8), 774–789.
5. Allendorf, M. D., Spira, D. J., Solomon, E. I., *Proceedings of the National Academy of Sciences of the United States of America* 1985, 82(10), 3063–3067.
6. Spirasolomon, D. J., Allendorf, M. D., Solomon, E. I., *Journal of the American Chemical Society* 1986, 108(17), 5318–5328.
7. Suzuki, S., Kohzuma, T., Deligeer, Yamaguchi, K., Nakamura, N., Shidara, S., Kobayashi, K., Tagawa, S., *Journal of the American Chemical Society* 1994, 116(24), 11145–11146.
8. Pascher, T., Karlsson, B. G., Nordling, M., Malmstrom, B. G., Vanngard, T., *European Journal of Biochemistry* 1993, 212(2), 289–296.
9. Kawahara, K., Suzuki, S., Sakurai, T., Nakahara, A., *Archives of Biochemistry and Biophysics* 1985, 241(1), 179–186.
10. Katoh, S., Shiratori, I., Takamiya, A., *Journal of Biochemistry* 1962, 51(1), 32–40.
11. Reinhamm B. R., Vanngard, T. I., *European Journal of Biochemistry* 1971, 18(4), 463–&.
12. Machonkin, T. E., Solomon, E. I., *Journal of the American Chemical Society* 2000, 122 (50), 12547–12560.
13. Christenson, A., Shleev, S., Mano, N., Heller, A., Gorton, L., *Biochimica Et Biophysica Acta-Bioenergetics* 2006, 1757(12), 1634–1641.
14. Ingledew, W. J., Cobley, J. G., *Biochimica Et Biophysica Acta* 1980, 590(2), 141–158.
15. Xu, F., Palmer, A. E., Yaver, D. S., Berka, R. M., Gambetta, G. A., Brown, S. H., Solomon, E. I., *Journal of Biological Chemistry* 1999, 274(18), 12372–12375.
16. Machonkin, T. E., Quintanar, L., Palmer, A. E., Hassett, R., Severance, S., Kosman, D. J., Solomon, E. I., *Journal of the American Chemical Society* 2001, 123(23), 5507–5517.
17. Xu, F., Shin, W. S., Brown, S. H., Wahleithner, J. A., Sundaram, U. M., Solomon, E. I., *Biochimica Et Biophysica Acta-Protein Structure and Molecular Enzymology* 1996, 1292 (2), 303–311.
18. Ducros, V., Brzozowski, A. M., Wilson, K. S., Brown, S. H., Ostergaard, P., Schneider, P., Yaver, D. S., Pedersen, A. H., Davies, G. J., *Nature Structural Biology* 1998, 5(4), 310–316.
19. Reinhamm BR, *Biochimica Et Biophysica Acta* 1972, 275(2), 245–259.
20. Shleev, S., Christenson, A., Serezhenkov, V., Burbaev, D., Yaropolov, A., Gorton, L., Ruzgas, T., *Biochemical Journal* 2005, 385, 745–754.
21. Hall, J. F., Kanbi, L. D., Strange, R. W., Hasnain, S. S., *Biochemistry* 1999, 38(39), 12675–12680.

22. Quintanar, L., Stoj, C., Taylor, A. B., Hart, P. J., Kosman, D. J., Solomon, E. I., *Accounts of Chemical Research* 2007, 40(6), 445–452.
23. Nitta, K., Kataoka, K., Sakurai, T., *Journal of Inorganic Biochemistry* 2002, 91(1), 125–131.
24. Messerschmidt, A., Rossi, A., Ladenstein, R., Huber, R., Bolognesi, M., Gatti, G., Marchesini, A., Petruzzelli, R., Finazziagro, A., *Journal of Molecular Biology* 1989, 206(3), 513–529.
25. Zaitseva, I., Zaitsev, V., Card, G., Moshkov, K., Bax, B., Ralph, A., Lindley, P., *Journal of Biological Inorganic Chemistry* 1996, 1(1), 15–23.
26. Taylor, A. B., Stoj, C. S., Ziegler, L., Kosman, D. J., Hart, P. J., *Proceedings of the National Academy of Sciences of the United States of America* 2005, 102(43), 15459–15464.
27. Palmer, A. E., Randall, D., Xu, F., Solomon, E. I., *Journal of Inorganic Biochemistry* 1999, 74(1–4), 259–259.
28. Marshall, N. M., Garner, D. K., Wilson, T. D., Gao, Y. G., Robinson, H., Nilges, M. J., Lu, Y., *Nature* 2009, 462(7269), 113–U127.
29. Augustine, A. J., Kjaergaard, C., Qayyum, M., Ziegler, L., Kosman, D. J., Hodgson, K. O., Hedman, B., Solomon, E. I., *Journal of the American Chemical Society* 2010, 132(17), 6057–6067.
30. Quintanar, L., Yoon, J. J., Aznar, C. P., Palmer, A. E., Andersson, K. K., Britt, R. D., Solomon, E. I., *Journal of the American Chemical Society* 2005, 127(40), 13832–13845.
31. Graziani, M. T., Morpurgo, L., Rotilio, G., Mondovi, B., *Febs Letters* 1976, 70(1), 87–90.
32. Kau, L. S., Spirasolomon, D. J., Pennerhahn, J. E., Hodgson, K. O., Solomon, E. I., *Journal of the American Chemical Society* 1987, 109(21), 6433–6442.
33. Yoon, J., Fujii, S., Solomon, E. I., *Proceedings of the National Academy of Sciences of the United States of America* 2009, 106(16), 6585–6590.
34. Solomon, E. I., Ginsbach, J. W., Heppner, D. E., Kieber-Emmons, M. T., Kjaergaard, C. H., Smeets, P. J., Tian, L., Woertink, J. S., *Faraday Discussions* 2011, 148, 11–39.
35. Xu, F., *Biochemistry* 1996, 35(23), 7608–7614.
36. Bertrand, T., Jolivalt, C., Briozzo, P., Caminade, E., Joly, N., Madzak, C., Mougin, C., *Biochemistry* 2002, 41(23), 7325–7333.
37. Kallio, J. P., Auer, S., Janis, J., Andberg, M., Kruus, K., Rouvinen, J., Koivula, A., Hakulinen, N., *Journal of Molecular Biology* 2009, 392(4), 895–909.
38. Quintanar, L., Gebhard, M., Wang, T. P., Kosman, D. J., Solomon, E. I., *Journal of the American Chemical Society* 2004, 126(21), 6579–6589.
39. Solomon, E. I., Brunold, T. C., Davis, M. I., Kemsley, J. N., Lee, S. K., Lehnert, N., Neese, F., Skulan, A. J., Yang, Y. S., Zhou, J., *Chemical Reviews* 2000, 100(1), 235–349.
40. Solomon, E. I., Pavel, E. G., Loeb, K. E., Campochiaro, C., *Coordination Chemistry Reviews* 1995, 144, 369–460.
41. Lowery, M. D., Guckert, J. A., Gebhard, M. S., Solomon, E. I., *Journal of the American Chemical Society* 1993, 115(7), 3012–3013.
42. Augustine, A. J., Kragh, M. E., Sarangi, R., Fujii, S., Liboiron, B. D., Stoj, C. S., Kosman, D. J., Hodgson, K. O., Hedman, B., Solomon, E. I., *Biochemistry* 2008, 47(7), 2036–2045.
43. Cole, J. L., Tan, G. O., Yang, E. K., Hodgson, K. O., Solomon, E. I., *Journal of the American Chemical Society* 1990, 112(6), 2243–2249.

44. Moriebebel, M. M., Morris, M. C., Menzie, J. L., McMillin, D. R., *Journal of the American Chemical Society* 1984, 106(12), 3677–3678.
45. Cole, J. L., Clark, P. A., Solomon, E. I., *Journal of the American Chemical Society* 1990, 112(26), 9534–9548.
46. Shin, W., Sundaram, U. M., Cole, J. L., Zhang, H. H., Hedman, B., Hodgson, K. O., Solomon, E. I., *Journal of the American Chemical Society* 1996, 118(13), 3202–3215.
47. Palmer, A. E., Lee, S. K., Solomon, E. I., *Journal of the American Chemical Society* 2001, 123(27), 6591–6599.
48. Augustine, A. J., Quintanar, L., Stoj, C. S., Kosman, D. J., Solomon, E. I., *Journal of the American Chemical Society* 2007, 129, 13118–13126.
49. Yoon, J. J., Solomon, E. I., *Journal of the American Chemical Society* 2007, 129, 13127–13136.
50. Quintanar, L., Stoj, C., Wang, T. P., Kosman, D. J., Solomon, E. J., *Biochemistry* 2005, 44 (16), 6081–6091.
51. Lee, S. K., George, S. D., Antholine, W. E., Hedman, B., Hodgson, K. O., Solomon, E. I., *Journal of the American Chemical Society* 2002, 124(21), 6180–6193.
52. Yoon, J., Mirica, L. M., Stack, T. D. P., Solomon, E. I., *Journal of the American Chemical Society* 2004, 126(39), 12586–12595.
53. Yoon, J., Solomon, E. I., *Inorganic Chemistry* 2005, 44(22), 8076–8086.
54. Yoon, J., Liboiron, B. D., Sarangi, R., Hodgson, K. O., Hedman, B., Solomona, E. I., *Proceedings of the National Academy of Sciences of the United States of America* 2007, 104(34), 13609–13614.
55. Huang, H. W., Zoppellaro, G., Sakurai, T., *Journal of Biological Chemistry* 1999, 274(46), 32718–32724.
56. Petersen, L. C., Degn, H., *Biochimica Et Biophysica Acta* 1978, 526(1), 85–92.
57. Cracknell, J. A., Vincent, K. A., Armstrong, F. A., *Chemical Reviews* 2008, 108(7), 2439–2461.
58. Soukharev, V., Mano, N., Heller, A., *Journal of the American Chemical Society* 2004, 126(27), 8368–8369.
59. Heller, A., *Physical Chemistry Chemical Physics* 2004, 6(2), 209–216.
60. Barton, S. C., Gallaway, J., Atanassov, P., *Chemical Reviews* 2004, 104(10), 4867–4886.
61. Ghindilis, A., *Biochemical Society Transactions* 2000, 28, 84–89.
62. Lee, C. W., Gray, H. B., Anson, F. C., Malmstrom, B. G., *Journal of Electroanalytical Chemistry* 1984, 172(1–2), 289–300.
63. Thuesen, M. H., Farver, O., Reinhammar, B., Ulstrup, J., *Acta Chemica Scandinavica* 1998, 52(5), 555–562.
64. Shleev, S., El Kasmi, A., Ruzgas, T., Gorton, L., *Electrochemistry Communications* 2004, 6(9), 934–939.
65. Ramirez, P., Mano, N., Andreu, R., Ruzgas, T., Heller, A., Gorton, L., Shleev, S., *Biochimica Et Biophysica Acta-Bioenergetics* 2008, 1777(10), 1364–1369.
66. Johnson, D. L., Thompson, J. L., Brinkmann, S. M., Schuller, K. A., Martin, L. L., *Biochemistry* 2003, 42(34), 10229–10237.
67. Tominaga, M., Ohtani, M., Taniguchi, I., *Physical Chemistry Chemical Physics* 2008, 10(46), 6928–6934.
68. Shleev, S., Pita, M., Yaropolov, A. I., Ruzgas, T., Gorton, L., *Electroanalysis* 2006, 18 (19–20), 1901–1908.

69. Blanford, C. F., Heath, R. S., Armstrong, F. A., *Chemical Communications* 2007, (17), 1710–1712.
70. Blanford, C. F., Foster, C. E., Heath, R. S., Armstrong, F. A., *Faraday Discussions* 2008, 140, 319–335.
71. dos Santos, L., Climent, V., Blanford, C. F., Armstrong, F. A., *Physical Chemistry Chemical Physics* 2010, 12(42), 13962–13974.
72. Sosna, M., Chretien, J.-M., Kilburn, J. D., Bartlett, P. N., *Physical Chemistry Chemical Physics* 2010, 12(34), 10018–10026.
73. Vaz-Dominguez, C., Campuzano, S., Rudiger, O., Pita, M., Gorbacheva, M., Shleev, S., Fernandez, V. M., De Lacey, A. L., *Biosensors & Bioelectronics* 2008, 24(4), 531–537.
74. Zheng, W., Li, Q. F., Su, L., Yan, Y. M., Zhang, J., Mao, L. Q., *Electroanalysis* 2006, 18(6), 587–594.
75. Zheng, W., Zho, H. M., Zheng, Y. F., Wang, N., *Chemical Physics Letters* 2008, 457(4–6), 381–385.
76. Ramasamy, R. P., Luckarift, H. R., Ivnitski, D. M., Atanassov, P. B., Johnson, G. R., *Chemical Communications* 2010, 46(33), 6045–6047.
77. Miyake, T., Yoshino, S., Yamada, T., Hata, K., Nishizawa, M., *Journal of the American Chemical Society* 2011, 133(13), 5129–5134.
78. Tsujimura, S., Kamitaka, Y., Kano, K., *Fuel Cells* 2007, 7(6), 463–469.
79. Miura, Y., Tsujimura, S., Kurose, S., Kamitaka, Y., Kataoka, K., Sakurai, T., Kano, K., *Fuel Cells* 2009, 9(1), 70–78.
80. Kontani, R., Tsujimura, S., Kano, K., *Bioelectrochemistry* 2009, 76(1–2), 10–13.
81. Wink, T., vanZuilen, S. J., Bult, A., vanBennekom, W. P., *Analyst* 1997, 122(4), R43–R50.
82. Song, S., Clark, R. A., Bowden, E. F., Tarlov, M. J., *Journal of Physical Chemistry* 1993, 97(24), 6564–6572.
83. Marcus, R. A., Sutin, N., *Biochimica Et Biophysica Acta* 1985, 811(3), 265–322.
84. Bourdillon, C., Delamar, M., Demaille, C., Hitmi, R., Moiroux, J., Pinson, J., *Journal of Electroanalytical Chemistry* 1992, 336(1–2), 113–123.
85. Piontek, K., Antorini, M., Choinowski, T., *Journal of Biological Chemistry* 2002, 277(40), 37663–37669.
86. Leger, C., Jones, A. K., Albracht, S. P. J., Armstrong, F. A., *Journal of Physical Chemistry B* 2002, 106(50), 13058–13063.
87. Jacobs, C. B., Peairs, M. J., Venton, B. J., *Analytica Chimica Acta* 2010, 662(2), 105–127.
88. Chen, R. J., Zhang, Y. G., Wang, D. W., Dai, H. J., *Journal of the American Chemical Society* 2001, 123(16), 3838–3839.
89. Kamitaka, Y., Tsujimura, S., Setoyama, N., Kajino, T., Kano, K., *Physical Chemistry Chemical Physics* 2007, 9(15), 1793–1801.
90. Barton, S. C., *Electrochimica Acta* 2005, 50(10), 2145–2153.
91. Degani, Y., Heller, A., *Journal of the American Chemical Society* 1989, 111(6), 2357–2358.
92. Heller, A., *Journal of Physical Chemistry* 1992, 96(9), 3579–3587.
93. Gregg, B. A., Heller, A., *Analytical Chemistry* 1990, 62(3), 258–263.
94. Mano, N., Soukharev, V., Heller, A., *Journal of Physical Chemistry B* 2006, 110(23), 11180–11187.
95. Barton, S. C., Kim, H. H., Binyamin, G., Zhang, Y. C., Heller, A., *Journal of the American Chemical Society* 2001, 123(24), 5802–5803.

96. Mano, N., Kim, H. H., Zhang, Y. C., Heller, A., *Journal of the American Chemical Society* 2002, 124(22), 6480–6486.
97. Mano, N., Kim, H. H., Heller, A., *Journal of Physical Chemistry B* 2002, 106(34), 8842–8848.
98. Beyl, Y., Guschin, D. A., Shleev, S., Schuhmann, W., *Electrochemistry Communications* 2011, 13(5), 474–476.
99. Marcus, R. A., *Journal of the Electrochemical Society* 1959, 106(3), C71–C72.
100. Marcus, R. A., *Canadian Journal of Chemistry-Revue Canadienne De Chimie* 1959, 37 (1), 155–163.
101. Kjaergaard, C. H., Durand, F., Tasca, F., Qayyum, M. F., Kauffmann, B., Gounel, S., Suraniti, E., Hodgson, K. O., Hedman, B., Mano, N., Solomon, E. I. *Journal of the American Chemical Society* 2012, 134, 5548–5551.
102. Frasconi, M., Boer, H., Koivula, A., Mazzei, F., *Electrochimica Acta* 2010, 56(2), 817–827.
103. Xu, F., *Journal of Biological Chemistry* 1997, 272(2), 924–928.
104. Hyosul, S., Sangeun, C., Heller, A., Chan, K., *Journal of the Electrochemical Society* 2009, 156(6), F87–F92.
105. Andreasson, L. E., Reinhammar, B., *Biochimica Et Biophysica Acta* 1979, 568(1), 145–156.
106. Branden, R., Deinum, J., *Biochimica Et Biophysica Acta* 1978, 524(2), 297–304.
107. Mano, N., Heller, A., *Journal of the Electrochemical Society* 2003, 150(8), A1136–A1138.
108. Beneyton, T., Beyl, Y., Guschin, D. A., Griffiths, A. D., Taly, V., Schuhmann, W., *Electroanalysis* 2011, 23(8), 1781–1789.
109. Braenden, R., Malmstrom, B. G., Vanngard, T., *European Journal of Biochemistry* 1973, 36(1), 195–200.
110. Hakulinen, N., Kiiskinen, L. L., Kruus, K., Saloheimo, M., Paananen, A., Koivula, A., Rouvinen, J., *Nature Structural Biology* 2002, 9(8), 601–605.
111. Roberts, S. A., Wildner, G. F., Grass, G., Weichsel, A., Ambrus, A., Rensing, C., Montfort, W. R., *Journal of Biological Chemistry* 2003, 278(34), 31958–31963.

9

ELECTROCHEMICAL MONITORING OF THE WELL-BEING OF CELLS

KALLE LEVON, QI ZHANG, YANYAN WANG, AABHAS MARTUR AND RAMYA KOLLI

Department of Chemical and Biomolecular Engineering, New York University Polytechnic School of Engineering, Six Metrotech Center, Brooklyn, USA

The beauty of our life can be a fascinating scale-down evaluation of the well-being of our cells. The diversity of human cells takes care of us with their effective nutrition control, energy production, and waste management with other extravagant metabolic events. The cells are comfortable and functional in microenvironment (in the near)—because extracellular matrix (ECM) not only provides structural support but also promotes physical and chemical cues for them to behave well in the complex environment. The interactions of the cells with ECM occur with minute domains of proteins, such as the binding of integrin's RGD segment on cell surface to ECM components like fibronectin (FN) and laminin. ECM with its nanoscale components controls the softness/elasticity, support/rigidity, and free volume for motility and also manages the release of growth factors as biochemical cues. Understanding ECM is of most importance when analyzing cell growth environment, and for this reason, the nanostructure of ECM is the guiding feature.

The interactions of cells with ECM through ligands like integrins are essential as the guidance for proliferation, differentiation, and cell–cell interactions. The interaction of cell membrane receptors forms signaling possibilities with gap junctions and other membrane domains. Often, the interactions result in conformational changes in membrane proteins so that the components on the cytoplasmic side change their function and initiate signaling paths with additional phosphorylation as an example.

Electrochemical Processes in Biological Systems, First Edition. Edited by Andrzej Lewenstam and Lo Gorton.

FIGURE 9.1 SEM pictures of PC12 cells on PCL fibers secreting thin fibers as new components of the extracellular matrix [1].

Signaling between cells is essential for cells to maintain their functionalities in the ever-changing (entropic) environment, and it is important that the cell/ECM interactions do not cause too strong enthalpic linkages to increase the free energy for the event. Figure 9.1 shows PC12 cells secreting already very thin fibers, maybe glycoproteins, as they have been interacting with each other while getting a minimal support from the scaffold fibers [1]. The thermodynamic ratio between enthalpy and entropy again proves right—well-being is optimized with the proper ratio: cell interaction with the scaffold does not inhibit the cell–cell interactions.

It is thus understandable that the biocompatibility of materials is strongly dependent on the initial protein adhesion. Electrostatic, hydrophobic, and van der Waals interactions along with hydrogen bonding influence the concentrations, conformations, and orientations of adsorbed proteins and their physical characteristics on the surface. These adsorbed proteins mediate cell adhesion and cell attachment with their proper tertiary structures [2]. Subsequent cell adhesion depends strongly on the adsorbed protein: FN is secreted quickly and adheres and its conformation affects the eventual cell adhesion depending on the RGD (arginylglycylaspartic acid) sequence in FN as exposed for the formation of the focal points [3, 4]. Vinculin is an important molecule connecting the integrin-mediated network and is used as a marker in the formation of focal adhesion [5]. Thus, protein–substrate interaction is a major factor determining cell–substrate interaction, which is mediated by the adsorbed protein.

Although the substrate chemistry is a major factor affecting the protein/cell adhesion, substrate curvature has also been shown to influence the tertiary structure of the

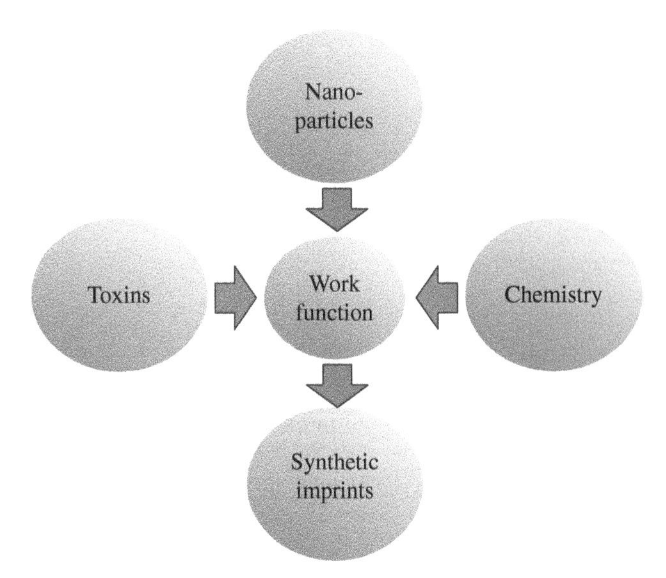

SCHEME 9.1 Effects of toxins, nanoparticles, and covalent bonds on the property of gold as presented with the work function, which then is utilized for effective surface imprinting.

bounded proteins [6]. In studies comparing nanophase to conventional ceramics including hydroxyapatite, Webster et al. found that the nanoscale surface structure improved cell function, which indicated the potentiality of nanophase ceramics for implant formulations due to the enhanced improved osseointegrative behavior [7].

Scheme 9.1 presents the different parts we discuss: effects of cytotoxicity, nanoparticles (NPs), and chemical interactions measured as a function of work function (WF) in gold and then utilizing the information to optimizing surface imprinting process. It is our goal to relate the adsorption behavior by electrochemical monitoring of toxin, curvature, and covalent anchoring effects on protein and/or cell adhesion and then to discuss how the substrate can reflect the information using gold electrodes. Eventually, we relate the monitoring methods to evaluate a protein surface imprinting.

9.1 ELECTROCHEMICAL MONITORING

In general, cells drift down to the bottom of cell culture dish when seeded out to the medium. The attachment to the surface is a requirement for cells to grow and divide, although cancer cells are able to multiply in suspension. This is called the anchorage dependence, and the initially spherical cells start to spread into a thin layer on the surface after the anchoring. And if the conditions are favorable, the cells keep crawling and thinning on the surface, eventually even copying their chromosomes and dividing. In cell culture experiments, the cells must be detached, harvested, and reseeded to

SCHEME 9.2 Visual presentation of cell addition, adhesion, anchorage, spreading, and mitosis.

different dishes in order to keep them in log phase (growth phase) of the cell cycle. Scheme 9.2 gives an illustrative presentation of the processes. The process of cell growth is our focus when using electrochemical methods to monitor effects of toxins, topography, and covalent chemical bonds in addition to the local confinement by surface imprinting.

These conditions can be monitored electrochemically without additional labels and extensive conjugation chemistry. **Amperometry** monitors current changes when constant potential is applied: the changes can be related to the enzymatic redox reactions in the electrolyte solution in proximity of the working electrode. Although enzyme efficiency certainly would be related to the environment the cells are in, we do not discuss amperometry, as these reactions apply immobilized enzymes on the reactive substrate. **Impedance measurements** rely on the capacitive interface between cells and electrodes and in essence the attached healthy cell monolayer on the electrode surface that inhibits the measurement of the conductance between the electrolyte solutions with the electrodes acting as the cells that act as insulating particles. In cytotoxicity measurements, various toxins are added to the electrolyte solutions/cell culture media, cells leave the surface, and as they detach, the resistivity decreases abruptly [8]. The Electric Cell–Substrate Impedance Sensing (ECIS) method has been widely applied to cell screening due to its simplicity: the capacitance is basically a function of the electrical double layer and thus frequency dependent as the resistance is frequency independent [9–15]. Potentiometry is in essence same measurements, but with no potential applied, the circuit remains open when potential difference between the two electrodes is measured. Field effect with transistors is also similar to potentiometry—potential applied from gate electrode over on capacitive interface lying on top of a semiconducting channel between source and drain electrodes [16, 17]. Potentiometry can measure potential changes during redox reactions but can also monitor ion concentration gradients through ion-selective membranes with pH measurement using glass membrane as the best example. When the membrane is modified with ion-selective components, ligands like ionophores, the ion-sensitive electrodes (ISE) result in quantitative determination of not only charge changes during redox reaction but also ion concentration profile alterations during adsorption processes. The potential change is measured between two electrodes, a reference electrode, commonly Ag/AgCl electrode, and a working electrode, which can be modified for the selectivity enhancement. The requirement for stable, well-quantified reference electrode is obvious as the working electrode with the selectivity factors should alone respond to changes. The high-impedance voltmeter used in the measurements ensures that current flow doesn't affect the measurement. As the potential measurement is done under zero flux conditions, the potential difference, which can be attributed to the redox species concentration by the Nernst equation, is an indicator for free energy change of the reaction.

Chemical modulation of the WF (in metals equals to ionization energy, electron affinity) can also be measured potentiometrically. WF can be divided into two components, the Fermi level and the surface potential, and it has been shown that long-range intermolecular dipole interactions can induce a cooperative effect to the layer influencing the net surface dipole. We shall discuss the use of open-circuit potentiometry (OCP) in various situations when monitoring the well-being of cells.

As we concentrate on electrochemical tools, we apply the surface modifications to the working electrodes, especially with gold electrodes, that are easy to modify. For instance, the gold nanoparticle (AuNP) formation is convenient on the surface of the electrode with gold salt compound reduction. The size of the NP can be controlled by the electrochemical condition including potential range, cycling number, salt concentration, and electrolyte strength and pH.

9.2 CELL DEATH: ELECTROCHEMICAL CYTOTOXICITY MEASUREMENTS

Cytotoxicity deals with conditions, mainly the effects of toxins, which cause cell death. ECIS monitors resistance changes during cell death but also during attachment, spreading, and proliferation of fibroblastic cells. Usually, the gold electrodes are first coated with FN, gelatin, or L-cysteine as a means to improve the adhesion and the potential ligand immobilization and to stabilize the interface. A large variety of toxins have been used for these cytotoxicity studies [8]. Figure 9.2 clearly reveals that no attachment happens when the surface has been pretreated with the toxin (c), and the attachment occurs clearly without any toxin (b) with the time-dependent increase in the resistance and then a decrease in the resistance as the toxin was added (d).

The use of potentiometry for cell screening is somewhat more simplified as no closed circuit with applied current is involved. Hydroquinone toxicity effect by OCP was first presented by Wang et al. [16]. The profile of potential change as cells attach to the working electrode is shown in Figure 9.3 [18]. This negative change in the potential can be assigned to the anionic charge of the cell membrane and will be related later to carboxylic acid group interactions with gold surface, as will later be discussed in Section 9.4. Initially, the system needed about 3 h to get stabilized after medium was added. The initial response, after adding 6×10^5 cells suspended with 400 μl medium, was due to the medium, and the later drop in potential overnight (or increase in negative potential) was due to the cell attachment. There was significant instability in potential due to the addition of 100 μg/ml Triton X-100. After 40 min, the negative potential decreased, indicating cell death as shown. The flat control line was the parallel measurement of blank medium without cells. Both ECIS and OCP rely on the deadhesion of the cell monolayer due to the presence of toxins.

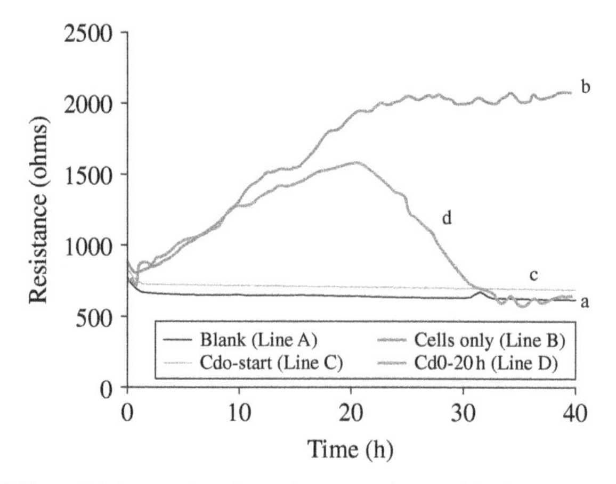

FIGURE 9.2 EIS results (measuring the resistance only): (a) blank control, (b) cell adhesion, (c) toxin added from the beginning of the experiments (no cell adhesion), and (d) toxin added after 20 h (cell adhered).

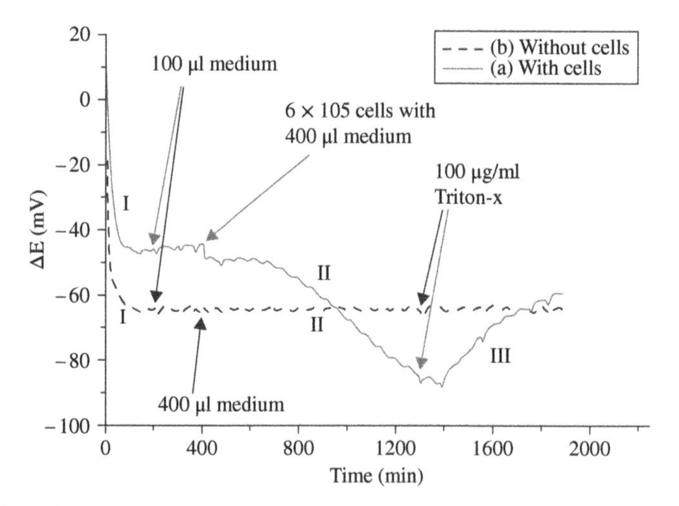

FIGURE 9.3 Open-circuit potentiometry results for addition of the medium (I), cell adhesion (II), and cells leaving the surface (III) due to the addition of the Triton X-100 toxin. The dotted line is a control without cells. Also, pH was monitored to be stable during the I–III events.

The cells have adhered in an assembly manner, cooperatively forming the monolayer, and thus, the disappearance of the cells is neither a function of chemical reaction nor interactions.

9.3 CURVATURE/SIZE EFFECT ON DENATURATION OF PROTEINS

Foreign substance and synthetic surfaces cause immunoresponses mainly because the proteins that interact with the surfaces denature and become nonfunctional. AuNPs have become important due to their great potential with imaging as well as the possible simultaneous NP toxicity effects. Although their behavior in solutions is different from that of surface bound, their interactions with proteins causing denaturation can provide the guidance on the nanosize effect. Teichroeb et al. investigated the effect of the size of the NPs on the denaturizing process by monitoring the change in the peak wavelength of the optical extinction spectrum of BSA-coated nanospheres with surface plasma resonance (SPR) method [19]. This wavelength expresses the changes in the adsorbed protein conformation caused by the denaturation process. The measurements were done isothermally in the 60–70°C temperature range, which was found to be the temperature range where the denaturation mainly occurs, while nanospheres of 5, 10, 20, 30, 40, 50, 60, and 100 nm sizes were used. As the denaturing started to happen almost immediately with larger NPs, the process was strongly delayed with NPs of 20 nm and smaller, as shown in Figure 9.4. Actually, the denaturing of the BSA on the 60 and 100 nm spheres was already completed before the 70°C was reached. As it is clear that denaturation occurs faster on larger particles, the surface coverage changes similarly from end-on adsorption to side-on adsorption.

Further on with similar studies but using calorimetry, they showed that the denaturing of large (>20 nm) particles was not different from the bulk solution studies. Consequently, the behavior of BSA adsorbed on NPs of size less than 20 nm is different from that of the larger ones; the authors propose the possibility of their sensitivity to the local surface curvature [20]. In addition to monitoring individual proteins and their denaturation, interaction effects have also been measured. Linse et al. have related the NP size to the nucleation of protein fibrillation [21]. Such surface-assisted nucleation increases the risk for amyloid formation in Alzheimer's disease

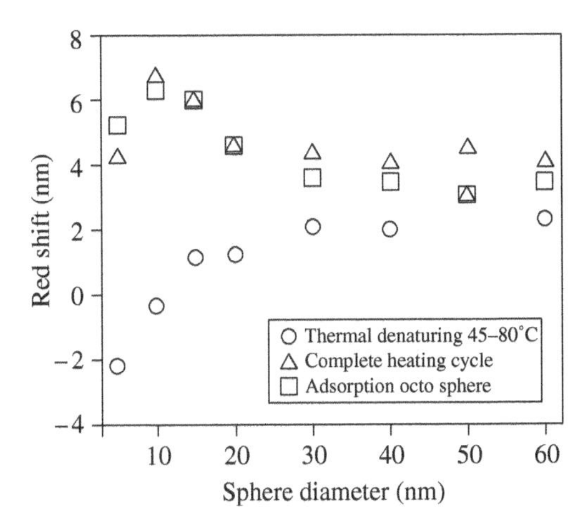

FIGURE 9.4 Redshift observed at various stages in experiment. The squares show the redshift upon protein adsorption to the nanosphere in phosphate buffer solution, the circles show the redshift (relative to room temperature) as the sample is heated from 45 to 80°C, and the triangles show the difference between room temperature measurements before and after a complete heating cycle.

development. Kaur and Forrest followed protein–protein interactions and saw remarkable surface NP size effects: for instance, IgG adsorbed on particles larger than 20 nm did not bind with protein A. They indicate the reason being the particle-bound proteins being either inaccessible due to steric constrains or inactive due to changes in tertiary structures [22].

Roach et al. compared BSA binding with fibrinogen by varying the chemical nature of the silica NP substrate. The amide I band for C=O stretching vibration in infrared spectroscopy with an attenuated total reflectance (ATR) was used for the evaluation of the denaturing process as this band is sensitive to conformational changes in the secondary structure. The main distinguished feature was that 30 nm size was an important turning point (Fig. 9.5): proteins on larger spheres didn't show changes in amide I band, indicating that the denaturing process had come to a completion. Both proteins became more denatured on hydrophobic surfaces, but BSA followed such trend also with increased size of the sphere and at the same the flatness of the surface [23].

The incubation of a gold surface in albumin solution causes a decrease in the substrate open-circuit potential and eventually a potential plateau is reached. The concomitant variation of the ellipsometric angles has confirmed gold modification with the albumin layer, with a thickness of approximately 1.5 nm. The application of negative (−100 mV) potentials promotes the albumin adsorption on gold surfaces, indicating the flexibility and ability of protein conformational change on the surface [24].

Similar use of OCP along with quartz crystal microbalance (QCM) has been done by Lori and Hanawa [25]. The use of electrochemical QCM allowed them to monitor the adsorption of bovine serum albumin (BSA, Fig. 9.6) on gold and titanium surfaces and to compare simultaneously the frequency change measured by QCM and the potential drop by OCP. The obtained concentration profiles gave comparative information on the adsorption conditions, such as ion concentration.

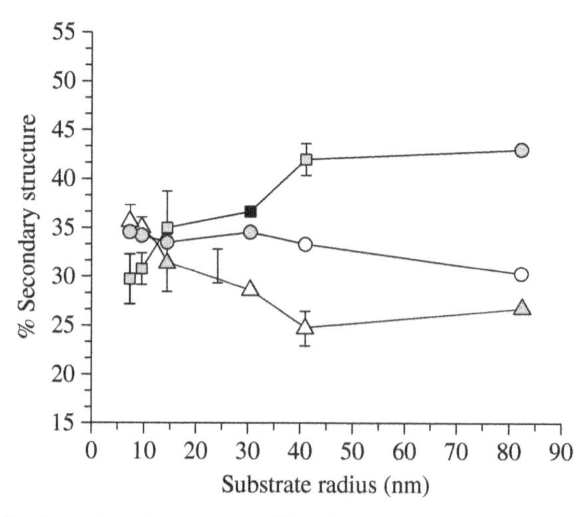

FIGURE 9.5 Conformational assessment if surface-bound BSA onto hydrophilic surface. ■, random/extended; Δ, helix; O, sheet/turn.

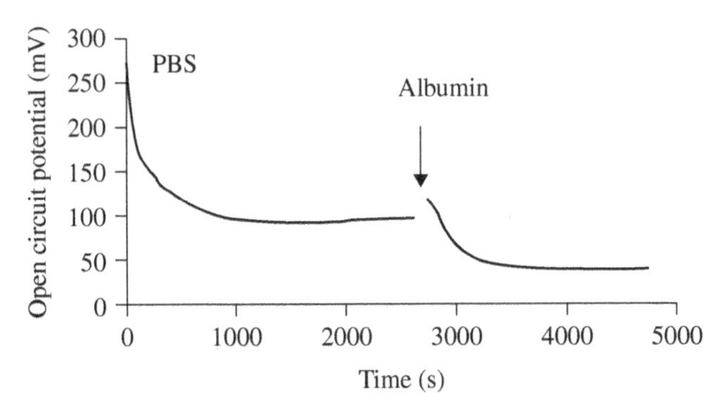

FIGURE 9.6 Open-circuit potential of gold surface in PBS and albumin solution. Albumin concentration: 0.1 mg mL^{-1}.

9.4 COVALENT ANCHORING: CHEMISTRY ON GOLD

Although surface roughness and morphology are important parameters for cell interactions with surfaces, also chemical reactions affect cell development. Covalent linkages, electrostatic interactions, hydrogen bonding, and even van der Waals interactions all anchor proteins and influence with their changes in 3D structure and their functions. The chemical interactions can be monitored based on the molecular level reactions. Direct electron transfer between an enzyme and gold surface can also be monitored amperometrically. Gold electrodes with adsorbed recombinant HRP forms exhibited high and stable current response to H_2O_2 due to its bioelectrocatalytic reduction through direct electron transfer between gold and HRP. This immobilization of HRP restricted the enzyme activity [26].

Electron transfer between an organic molecule and the metal has been monitored in terms of WF change. A controlled molecular donor layer was shown to decrease gold WF from 5.5 to 3.3 eV by Broker et al. UV photoelectron spectroscopy (UPS) was used to follow a bipyridinylidene derivative deposition on Au(111) surface. The reduced WF gold by the 2.2 eV is a good example of reducing the injection barriers between metal electrodes and organic semiconducting materials. The low Au WF surface introduces low electron transfer barriers to the deposited organic electron transfer materials [27].

Covalent reactions between organic molecules and AuNPs on surfaces can be monitored by the change of gold WF. Thiolation of AuNP is a well-known reaction for ligand immobilization and development of protection layer against nonspecific binding, both on NPs and on gold surfaces:

$$R-SH + Au(s) \rightarrow R-S-Au + \tfrac{1}{2} H_2$$

Cysteine is the amino acid most commonly used for the substation reaction. Such redox reaction had been followed by OCP, and the decrease of the potential is indicative of increased negative charge on Au surface [28]. Paik et al. followed the reaction with open-circuit potential monitoring the decrease in the potential during the anodic thiol reaction with gold surface. The gold surface is reduced to show negative potential shifts (Fig. 9.7a). Disulfide compound was shown to react on gold surface with an opposite reaction: the increase of the potential was due to the now cathodic reaction with the disulfide bond and gold (Fig. 9.7b) [29]. Also, weaker forces, such as carboxylic acids, amine groups, and alcohols, have been shown to bind with gold surfaces, but as we focus on the well-being of the cells, we assume that such interactions do not play a role in the buffered solutions [30].

Paik et al. were also able to relate with the QCM and OCP experiments that compounds with carboxylic acids bind to gold surface and with negative potential changes as with the thiol groups. They indicated that common for all the reactions was the surface oxidation of the substrate metal, gold [30]. This reaction was the one we related to

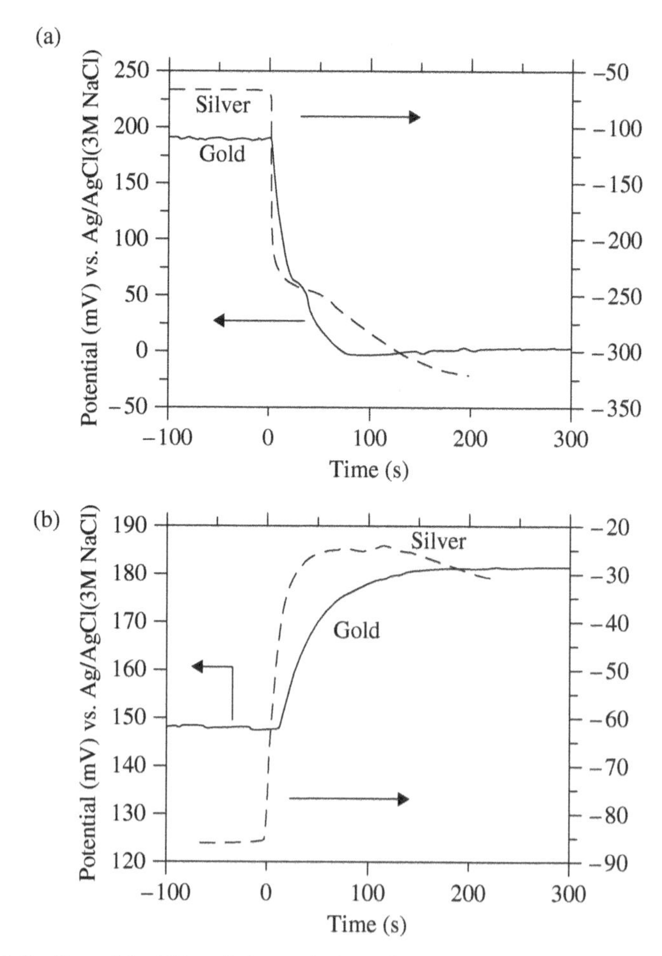

FIGURE 9.7 Potential shifts of Au and Ag substrates under open-circuit situation in acetonitrile 0.1 M $LiClO_2$ solution after addition of decanethiol (a) or dipropyl disulfide (b).

earlier when the cell adhesion was shown to cause negative shift in the oxidation potential due to the anionic cell membrane containing phosphate groups.

Same interactive forces apply to proteins. When focusing on BSA only, as in the aforementioned work, BSA has been shown to interact with gold nanospheres with pH-induced conformational changes depending on the size of the NPs using dynamic light scattering experiments in addition to the spectroscopic methods. The strong affinity was accounted for the thiolation of AuNP as BSA has a free exterior cysteine/thiol available for the bonding. This adsorption interaction induces conformational changes in BSA secondary structures with less alpha-helix structures and increased amount of beta-sheet, random, or expanded structures. With the smallest, 10 nm, particles, the size difference between BSA-coated and BSA-noncoated AuNPs was largest, assuming that the protein is able to maintain its 3D structure. The largest, 60 nm, AuNPs seem to provide wide enough surface as the protein loses tertiary structures and starts to lay flat [31]. As the aforementioned authors considered only a potential cysteine/gold reaction, Sen et al. extended the possibilities to defined cysteines using tryptophans with surface energy transfer (SET) method and concluded that AuNPs react through Cys53–Cys62 disulfide bonds located at subdomain IA of HAS [32].

Saturated hydrocarbons on the surface are related to the formation of the interface dipole, which shifts the Fermi energy levels and lowers the WF of the metal. As alkanes have been shown to change the WF of gold, it cannot result from charge transfer effects as alkanes do not possess electron charge donation properties [33]. With noble gases, Pauli repulsion has been shown to cause perturbation of the electronic structure. But due to the physical interactions, the gas atoms are pulled toward the surface via van der Waals interactions resulting in the overlap of the electronic wave functions of those of the gas and the metal [34]. Although potential hydrophobic interactions might be able to result in dipole orientation of polarization, Bagus et al. related alkane chemisorption on gold surface to charge density differentiation instead of the aforementioned traditional explanation [35].

9.5 COOPERATIVITY ON GOLD: MEASURING WF

The aforementioned reactions present activities of individual molecules. Very large research area has focused on cooperative forces, with a focus on weak interactions during self-assembly processes such as the formation of self-assembled monolayer (SAM). Common ways to monitor their effect are Kelvin probes (KP) and UPS, both monitoring the changes in the WF of gold, the latter though giving directly quantitative analysis and KP with contact potential difference only after calibration with UPS method. UPS, also known as electron spectroscopy for chemical analysis (ESCA), follows the kinetic energies of ejected electrons, and the information can be related to WF, which is the minimum energy needed for an electron to move from the Fermi level to vacuum as determined by the chemical and morphological conditions of the surface. In UPS, Fermi level is positioned at 0 eV and UPS measures only occupied states giving the E_F value. Extensive work has been done on parameter control during SAM formation so that the surface potential, carrier density, electron affinity, and WF can be correlated,

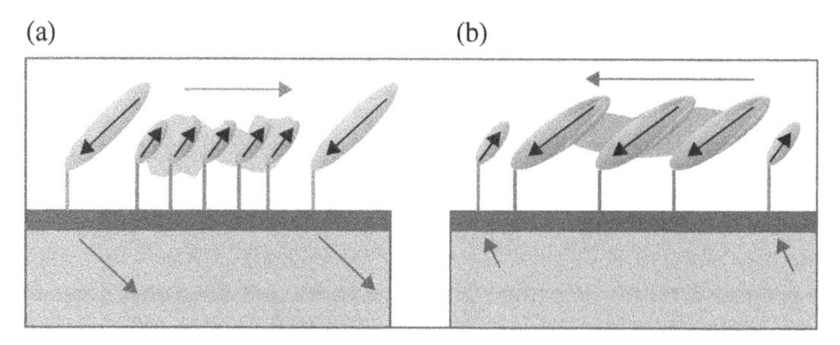

FIGURE 9.8 Cooperative effect through a mixed organic layer. (a) Small dipoles pointing away from the surface and (b) large dipoles pointing toward the surface.

remembering that for metals electron affinity (ionization energy) equals WF [36]. Yitzhak et al. showed that a cooperative performance arises from the use of mixed dipole monolayer with parallel and antiparallel organized dipoles. The intermolecular interaction of neighboring dipoles induces a cooperative effect influencing the net surface charge. The dipoles behave as individual dipoles when far away from each other in a layer that is not dense. But in a highly ordered monolayer, shown in Figure 9.8, the dipole–dipole interactions induce a molecular cooperative effect with a depolarization that suppresses the contribution to the net surface dipole [36].

Cahen's group investigated similarly monolayers on Si/SiO_2 surfaces using silanes with systematic change in their chain length ($n = 3$, 6 or 11) and the chemical nature of the end group [37]. The effect of the dipole moment on electron affinity was measured, and band bending (BB) at metal/semiconductor interface was further estimated with eventual presentation of the controlled WF. The effective dipole moment on gold surface can be tuned using a binary SAM system as shown by Lee et al. [38]. The electron-donating and electron-accepting pendant groups were the dipole moment controls (amine- and carboxylic acid-bearing alkylthiols). UPS was applied for the measurements and the WF was observed to increase to 5.8 eV. This increase was because of the inward dipole moments due to lower electronegativity of gold. Oppositely, amine groups caused the decrease of the WF to 5.1 eV. As novel electronic technologies depend on the metal (or semiconductor)/organic interfaces and the tuning of the WF of metal (or semiconductor) with molecular adsorbates has become effective, Armstrong et al. were able to tune the WF over a range of 1.8 eV [39]. They used hexadecylthiol with methyl terminated and alike with fluorinated tails (C16F2) to adjust the electronegativity with this two-component SAM, and the formed interface dipole from the deposition of the adsorbate impacted the current/voltage properties remarkably as needed in organic field effect transistors. Figure 9.9 shows the effect of controlled inductive effect on the potential change.

9.6 COOPERATIVE CELL IMPRINTS

The beginning of molecular imprinting can be dated to 1972 when Gunter Wulff used covalent binding approach to prepare the molecularly imprinted organic polymers

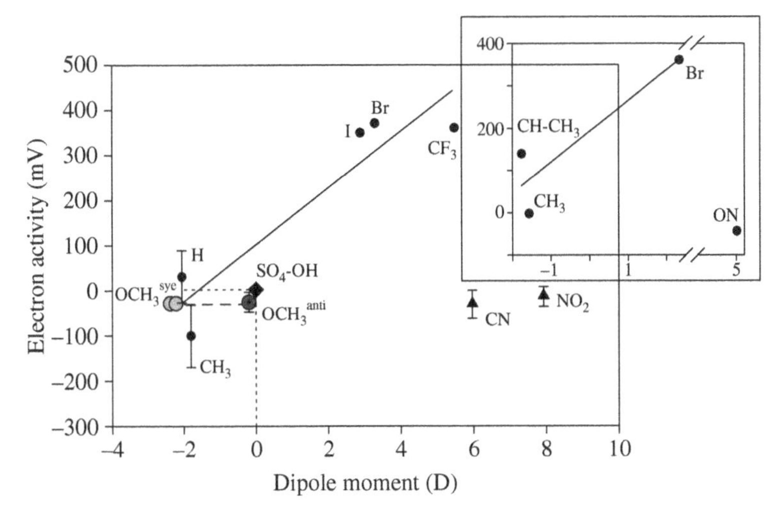

FIGURE 9.9 Measured contact potential differences (plotted as electron affinity) versus the dipole moments of the free molecules. The end groups of the monolayer compounds indicated in the figure.

(MIP) for differentiation between enantiomers of glyceric acid [40]. The concept of MIP was eventually popularized by Mosbach and coworkers who created molecular imprints using a noncovalent binding with the template to form the imprints/molds [41].

An improvement in quantitating the binding with the imprint was achieved when the surface confinement was included [42]. Initially, functional silanes were used to form a polysiloxane layer around the analyte, which then was removed for the cavity formation. Later, the approach was modified for imprinting with functional thiols on gold surfaces, the sequences shown in Figure 9.10 [43–45]. The cancer biomarkers were imprinted with a formation of a partial SAM around the analyte with additional interactive forces between the analyte and the hydroxyl functional groups on thiols. The possibility for obtaining presumptive knowledge of disease has made the imprints popular for immunoassay-type applications including even the efforts for epitope and cell receptor imprintings [46–49]. A very recent extension has been the application of the imprints on the surfaces of spherical particles for drug delivery-type application; only now, the delivery concept attains the possibility of inhibition and/or activation of cell membrane receptors [49, 50].

The interest with the imprints is the interaction of the protein with the cavity or even the interaction of a cell with the cavity. We have learned that toxicity, denaturing, and unfavorable interactions relate to the conditions proteins and cells do not easily survive. But weak interactions, small nanoscale domains, and cooperativity enhance the well-being. So although covalent bonding with cysteines and the bottom of the gold cavity is possible, results have been obtained where cysteines are not available as with myoglobin. The detection of myoglobin with its cavity is noticed with a 40 mV

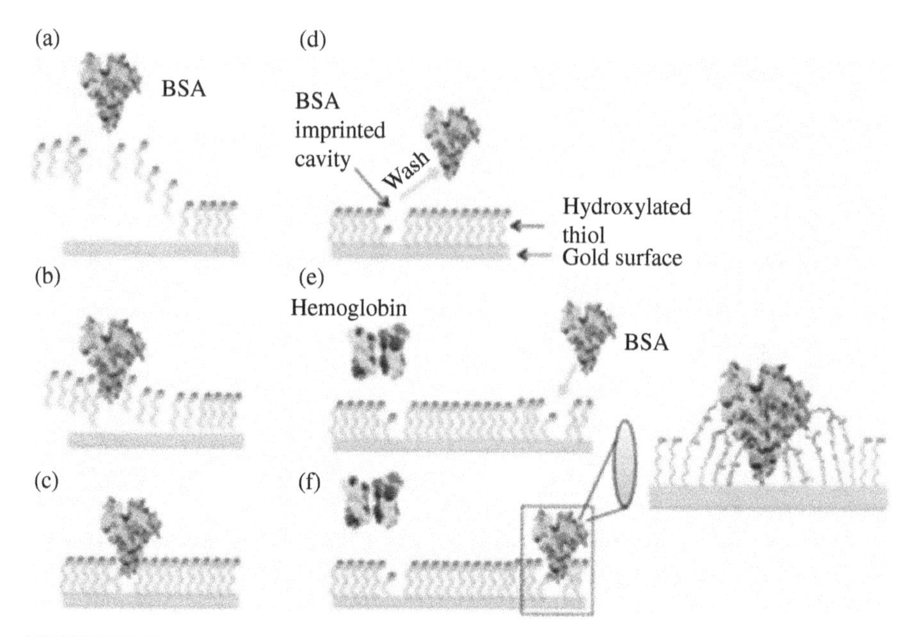

FIGURE 9.10 Surface imprinting process and detection of the analytes. (a–c) Formation of the layer, (d) wash, and (e–f) binding of the BSA analyte and nonbinding of the Hg control.

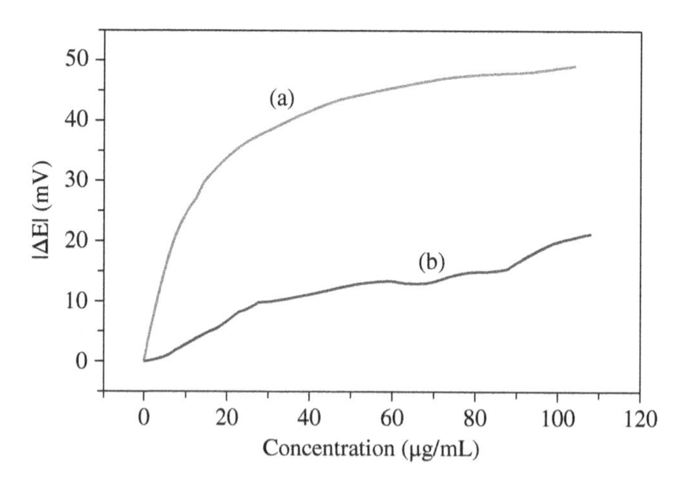

FIGURE 9.11 Potential response of myoglobin sensor to (a) the mixture of myoglobin, hemoglobin, and ovalbumin and (b) to hemoglobin and ovalbumin as a function of concentration of single component in the solution.

response confirming that neither chemical bonding nor redox reactions are involved in the binding (Fig. 9.11). Similarly, as the size difference between an analyte protein and the thickness of the partial SAM layer is too large for individual interactive forces, it can be assumed that the protein binding with the "hole in the SAM" completes the cooperative dipole effect on gold WF by closing, sealing the surface, and inducing cooperation, resembling a field effect on the surface. Tinberg et al. presented recently a novel computational design of high-affinity and selective ligand-binding proteins by first idealizing the binding sites with digoxigenin, a steroid used as the model ligand: hydrogen bonds between Tyr or His and the polar groups of digoxigenin and also hydrophobic interactions between the steroid ring and Tyr, Phe, or Trp. After this, they placed the ligand and the interacting residues in scaffold, which was a selected protein with unknown function. Eventually, they successfully probed the binding by yeast surface display and flow cytometry and were able to show that the binding activities were mediated by the designed interfaces [51]. Imprinting follows similar design aspects: first, the spatial arrangement of the local selective hydrogen bonding sites is "set in," then the scaffolding is built around the selective binding sites, and eventually, the macroscopic arrangement with surface confinement provides structural stability and allows the recognition via charge image formation. Thus, during the diagnostic readsorption, the cooperative intermolecular effect including the assembly of the proteins with the SAM enhances the surface dipole effect, which leads to the lowering of the WF (ionization of the gold) due to the increased thiol redox reaction with gold.

Several OCP results were shown earlier, emphasizing perhaps the easiness of the measurements but truly directing to the very beneficial measurement method using field effect as the measure. Field effect closely resembles open circuitry in a sense that in neither case the circuit is closed with the application of current/voltage control, but the voltage comes from the binding effect within the electrolytic solution. Often, a reference electrode is used to increase the potential effect, but extended floating gate probe offers the possibility to ignore the reference electrode and let the charged differences within the dielectric area affect the operational threshold voltage. Thus, in the future, one can use a FGFET chip for HTS immunoassays, for massive data analysis of brain slices, and for all *in vivo* applications where a stand-alone device is needed [52].

Yitzchaik et al. had already shown the cooperative effect of mixed dipole monolayer on WF [36] and have extended the investigation to molecular electronic tuning by the control of inductive effect (Hammett parameter) [52] and to direct detection of molecular recognition, by monitoring the effect of a conformational change of an enzyme during the selection binding with the substrate on the floating gate field effect transistor [53].

REFERENCES

1. Tuckett, A., Levon, K. Poster at BMES Annual Fall Meeting; 2005.
2. Roach, P., Farrar, D. and Perry, C. C. Interpretation of protein adsorption: surface-induced conformational changes, *J Am Chem Soc*. 127: 8168–8173; 2005.
3. Steele, J. G., Dalton, B. A., Johnson, G. and Underwood, P. A. Adsorption of fibronectin and vitronectin onto Primaria and tissue culture polystyrene and relationship to the mechanism of initial attachment of human vein endothelial cells and BHK-21 fibroblasts. *Biomaterials*. 16: 1057–1067; 1995.

4. Aota, S., Nomizu, M. and Yamada, K. M. The short amino acid sequence Pro-His-Ser-Arg-Asn in human fibronectin enhances cell-adhesive function. *J Biol Chem*. 269: 24756–24761; 1994.
5. Diener, A., Nebe, B., Luthen, F., Becker, P., Beck, U., Neumann, H. G. and Rychly, J. Control of focal adhesion dynamics by material surface characteristics. *Biomaterials*. 26: 383–392; 2005.
6. Vertegel, A. A., Siegel, R. W. and Dordick, J. S. Silica nanoparticle size influences the structure and enzymatic activity of adsorbed lysozyme. *Langmuir*. 20: 6800–6807; 2004.
7. Webster, T. J., Ergun, C., Doremus, R. H., Siegel, R. W. and Bizios, R. Enhanced functions of osteoblasts on nanophase ceramics. *Biomaterials*. 21: 1803–1810; 2000.
8. Hondroulis, E., Liu, C., and Li, C.-Z. Whole cell based electrical impedance sensing approach for a rapid nanotoxicity assay. *Nanotechnology*. 21: 315103 (9pp); 2010.
9. Xiao, C. D., Luong, J. H. T. On-line monitoring of cell growth and cytotoxicity using electric cell-substrate impedance sensing. *Biotechnol Prog*. 19: 1000–1005; 2003.
10. Luong, J. H. T., Habibi-Razaei, M., Meghrous, J., Xiao, C. D., Male, K. B., Kamen, A. Monitoring motility, spreading, and mortality of adherent insect cells using an impedance sensor. *Anal Chem*. 73: 1844–1848; 2001.
11. Xiao, C. D., Lachance, B., Sunahara, G., Luong, J. H. T. An in-depth analysis of electric cell-substrate impedance sensing to study the attachment and spreading of mammalian cells. *Anal Chem*. 74: 1333–1339; 2002.
12. Xiao, C. D., Lachance, B., Sunahara, G., Luong, J. H. T. Assessment of cytotoxicity using electric cell-substrate impedance sensing: concentration and time response function approach. *Anal Chem*. 74: 5748–5753; 2002.
13. Keese, C. R., Giaever, I. A biosensor that monitors cell morphology with electrical fields. *IEEE Eng Med Biol*, 13(3): 402–408; 1994.
14. Wegener, J., Keese, C. R., Giaever, I. Electric Cell-Substrate Impedance Sensing (ECIS) as a noninvasive means to monitor the kinetics of cell spreading to artificial surfaces. *Exp Cell Res*, 259: 158–166; 2000.
15. Opp, D., Wafula, B., Lim, J., Huang, E., Lo, J.-C., Lo C.-M. Use of electric cell–substrate impedance sensing to assess in vitro cytotoxicity. *Biosens Bioelectron*, 24: 2625–2629; 2009.
16. Wang, Y., Chen, Q., Zeng, X. Potentiometric biosensor for studying hydroquinone cytotoxicity in vitro. *Biosens Bioelectron*, 25: 1356–1362; 2010.
17. Konicki, R. Recent developments in potentiometric biosensors for biomedical analysis. *Anal Chim Acta*, 599(1): 7–15; 2007.
18. Potentiometric Cell-based Sensor Monitoring Cytotoxicity, Kolli, R., Kaivosoja, E., Levon, K., submitted to Electroanalysis.
19. Teichroeb, J. H., Forresta, J. A., and Jones L. W. Size-dependent denaturing kinetics of bovine serum albumin adsorbed onto gold nanospheres. *Eur Phys J E*, 26: 411–415; 2008.
20. Teichroeb, J. H., Forresta, J. A., Ngai, V., and Jones, L. W. Anomalous thermal denaturing of proteins adsorbed to nanoparticles. *Eur Phys J E*, 21: 19–24; 2006.
21. Linse, S., Cabaleiro-Lago, C., Xue, W. F., Lynch, I., Lindman, S., Thulin, E., Radford, S. E., and Dawson, K. A. Nucleation of protein fibrillation by nanoparticles. *Proc Natl Acad Sci U S A*, 104(21): 8691–8696; 2007.
22. Kaur, K., Forrest, J. A. Influence of particle size on the binding activity of proteins adsorbed onto gold nanoparticles. *Langmuir*, 28: 2736–2744; 2012.

23. Roach, P., Farrar D., and Perry, C. C. Surface tailoring for controlled protein adsorption: effect of topography at the nanometer scale and chemistry. *J Am Chem Soc*, 128: 3939–3945; 2006.
24. Ying, P., Viana, A.S., Abrantes, L. M., Jin, G. Adsorption of human serum albumin onto gold: a combined electrochemical and ellipsometric study, *J Colloid Interface Sci*, 279: 95–99; 2004.
25. Lori, J. A., and Hanawa, T. Adsorption characteristics of albumin on gold and titanium metals in Hanks' solution using EQCM, *Spectroscopy*, 18: 545–552; 2004.
26. Ferapontova, E., Grigorenko V. G., Egorov A. M., Börcers T., Ruzgas T., Gorton L. Mediatorless biosensor for H_2O_2 based on recombinant forms of horseradish peroxidase directly adsorbed on polycrystalline Gold. *Biosens Bioelectron*, 16: 147–157; 2001.
27. Broker, B., Blum, R.-P., Frisch, J. Vollmer, A., Hofmann, O. T., Rieger, R., Mullen, K. Rabe, J. P., Zojer, E., Koch, N. Gold work function reduction by 2.2 eV with an air-stable molecular donor layer. *Appl Phys Lett* 93: 243303; 2008.
28. Cohen-Atiya, M., Daniel Mandler, D. Studying thiol adsorption on Au, Ag and Hg surfaces by potentiometric measurements. *J Electroanal Chem*, 550/551: 267–276; 2003.
29. Paik, W., Eu, S., Lee, K., Chon, S., Kim, M. Electrochemical reactions in adsorption of organosulfur molecules on gold and silver: potential dependent adsorption. *Langmuir*, 16: 10198–10205; 2000.
30. Paik, W., Han S., Shin, W., and Kim, Y. Adsorption of carboxylic acids on gold by anodic reaction. *Langmuir*, 19(10): 4211–4216; 2003.
31. Tsai, D. H., DelRio, F. W., Keene, A. M., Tyner, K. M., MacCuspie, R. I., Cho, T. J., Zachariah, M. R., and Hackley, V. A. Adsorption and conformation of serum albumin protein on gold nanoparticles investigated using dimensional measurements and in situ spectroscopic methods. *Langmuir*, 27: 2464–2477; 2011.
32. Sen, T., Mandal, S., Haldar, S., Chattopadhyay, K. and Patra, P. Interaction of gold nanoparticle with Human Serum Albumin (HSA) protein using surface energy transfer. *J Phys Chem C*, 115: 24037–24044; 2011.
33. Bagus, P. S., Kafer, D., Witte, G., Woll, C. Work function changes induced by charged adsorbates: origin of polar asymmetry. *Phys Chem Lett*, 100(12): 126101; 2008.
34. Bagus, P. S., Staemmler, V., Woll, C. Exchange like effects for closed-shell adsorbates: interface dipole and work function. *Phys Chem Lett*, 89(9): 096104; 2002.
35. Bagus, P. S., Hermann, K., Woll, C. The interaction of C6H6 and C6H12 with noble metal surfaces: electronic level alignment and the origin of the interface dipole. *J Chem Phys*, 123: 184109; 2005.
36. Stef, R., Peor, N., Yitzchak, S. Experimental evidence for molecular cooperative effect in mixed parallel and antiparallel dipole monolayer. *J Phys Chem C*, 114: 20531–20538 2010.
37. Gershewitz, O., Grinstein, M., Sukenik, C. N., Regev, K., Ghabboun, J., Cahen, D. Effect of molecule-molecule interaction on the electronic properties of molecularly modified Si/SiO_2 surfaces. *J Phys Chem B*, 108: 664–672; 2004.
38. Lee, S.-H., Lin, W.-C., Chang, C.-J., Huang, C.-C., Liu, C.-P., Kuo, C.-H., Chang, H.-Y., You, Y.-W., Kao, W.-L., Yen, G.-J., Kuo, D.-Y., Kuo, Y.-T., Tsai, M.-H., Shyue, J.-J. Effect of the chemical composition on the work function of gold substrates modified by binary self-assembled monolayers. *Phys Chem Chem Phys*, 13: 4335–4339; 2011.
39. Alloway, D. M., Graham, A. L., Yang, X., Mudalige, A., Coloraduo, R., Wysocki, V. H., Pemberton, J. F., Lee, T. R., Wysocki, R. J., Armstrong, N. R. Tuning the effective work

function of gold and silver using omega-functionalized alkanethiols: varying the surface composition through dilution and choice of terminal groups. *J Phys Chem C*, 113: 20328–20334; 2009.

40. Wulff, G., Sarhan, A. Uber die Anwendung von enzymanalog gebauten Polymeren zur Racemattrennung. *Angew Chem*, 84: 364; 1972.
41. Arshady, R., Mosbach, M. Synthesis of substrate-selective polymers by host–guest polymerization. *Macromol Chem Phys Makromol Chem*. 182: 687–692; 1982.
42. Zhou, Y. X., Yu, B., Levon, K. Potentiometric sensor for dipicolinic acid. *Biosens Bioelectron*, 20: 1851–1855; 2005.
43. Wang, Y. T., Zhou, Y. X., Sokolov, J., Rigas, B., Levon, K., Rafailovich, M. H. A potentiometric protein sensor built with surface molecular imprinting method, *Biosens Bioelectron*, 24: 162–166; 2008.
44. Wang, Y., Zhang, Z., Jain, V., Yi, J., Mueller, S., Sokolov, J., Liu, Z., Levon, K., Rigas, B., Rafailovich, M. H. Potentiometric sensors based on surface molecular imprinting: detection of cancer biomarkers and viruses. *Sensors Actuators B*, 146: 381–387; 2010.
45. Marthur, A., Biais, B., Goparaju, G. M. V., Neubert, T. A., Pass, H., Levon, K. Development of a biosensor for detection of pleural mesothelioma cancer biomarker using surface imprinting. *PLoS One*, 8(3): e57681; 2013.
46. Jenik, M., Seifner, A., Krassnig, S., Seidler, K., Lieberzeit, P. A., Dickert, F., Jungbauer, C. Sensors for bioanalytes by imprinting—polymers mimicking both biological receptors and the corresponding bioparticles. *Biosens Bioelectron*, 25: 9–14; 2009.
47. Chun-Hua Lu, C. H., Zhang, Y., Tang, S. F., Fang, Z. B., Yang, H. H., Chen, X., Chen, G. N. Sensing HIV related protein using epitope imprinted hydrophilic polymer coated quartz crystal microbalance. *Biosens Bioelectron*, 31: 439–444; 2012.
48. Cohen, T., Starosvetsky, J., Cheruti, U., and Armon, R. Whole cell imprinting in sol-gel thin films for bacterial recognition in liquids: macromolecular fingerprinting. *Int J Mol Sci*, 11: 1236–1252; 2010.
49. Hoshino, Y., Koide, H., Urakami, T., Kanazawa, H., Kodama, T., Oku, N., and Shea, K. J. Recognition, neutralization, and clearance of target peptides in the bloodstream of living mice by molecularly imprinted polymer nanoparticles: a plastic antibody. *J Am Chem Soc*, 132: 6644–6645; 2010.
50. Zeng, Z., Hoshino, Y., Rodriguez, A., Yoo, H., and Shea, K. J. Synthetic polymer nanoparticles with antibody-like affinity for a hydrophilic peptide. *ACSNANO* 4(1): 199–204; 2010.
51. Tinberg, C. E., Khare, S. D., Dou, J., Doyle, L., Nelson, J. W., Schena, A., Jankowski, W., Kalodimos, C. G., Johnsson, K., Stoddard, B. L., Baker, D. Computational design of ligand-binding proteins with high affinity and selectivity. *Nature*, 501: 212–216; 2013.
52. Cohen, R., Zenou, N., Cahen, D., Yitzchaik, S. Molecular electronic tuning of Si surfaces. *Chem Phys Lett*. 279: 270–274; 1997.
53. Goykhman, I., Korbakov, N., Bartic, C., Borghs, G., Spira, M. E., Shappir, J., Yitzchaik, S. Direct detection of molecular biorecognition by dipole sensing mechanism. *J Am Chem Soc*, 131: 4788–4794; 2009.

10

ELECTROCHEMICAL SYSTEMS CONTROLLED BY ENZYME-BASED LOGIC NETWORKS: TOWARD BIOCHEMICALLY CONTROLLED BIOELECTRONICS

JAN HALÁMEK[2] AND EVGENY KATZ[1]

[1]*Department of Chemistry and Biomolecular Science, and NanoBio Laboratory (NABLAB), Clarkson University, Potsdam NY, USA*

[2]*Department of Chemistry, University at Albany, SUNY, Albany, NY, USA*

The recently developed concept of biocomputing with the use of enzyme-catalyzed reactions allows logic processing of multiple biochemical signals prior to their transduction in a biosensor device [1]. A new strategy applying digital multisignal biosensors became possible according to this approach. Novel biosensors based on concatenated logic gates [2] were already developed for the analysis of pathophysiological conditions related to different kinds of injuries [3, 4].

Signals generated by enzyme logic systems in the form of concentration changes of reactant species can be read out using various analytical techniques: optical [5–7], electrochemical [8], etc. Application of sensitive analytical methods and instruments is required because the concentration changes produced upon biocatalytic reactions usually take place at low levels (micromolar or even nanomolar concentration ranges). Application of highly sensitive techniques requires electronic transduction and amplification of the output signals produced by enzyme logic systems. This approach was actually applied to most of chemical computing systems based on nonbiological

Electrochemical Processes in Biological Systems, First Edition. Edited by Andrzej Lewenstam and Lo Gorton.

molecules [9–11]. While the majority of nonbiochemical computing systems is based on the application of noncatalytic reactions, the enzyme logic systems have the advantage of catalyzing biochemical transformations when continuous production of the output species could potentially produce substantial chemical changes in the systems even when the concentrations of the reacting catalytic species are low. The chemical changes in the enzyme biocatalytic systems can be coupled to signal-responsive materials resulting in their bulk properties changes, thus amplifying the chemical changes generated by enzymes. Signal-responsive materials coupled with enzyme logic systems could be represented by polymers responding to external chemical signals by restructuring between swollen and shrunken states [12–14]. The structural reorganization of the polymers will substantially amplify the chemical changes generated by the enzyme reactions, thus excluding the need of highly sensitive analytical techniques to observe the output signals from the biocomputing systems. Application of signal-responsive polymers in a biochemical environment has already a well-established background, thus allowing their integration with biocomputing systems [15].

This approach leads to the fabrication of "smart" multisignal-responsive materials equipped with built-in Boolean logic [16]. The systems will be capable of switching physical properties (such as optical, electrical, magnetic, wettability, permeability, etc.) upon application of some incoming chemical signals and according to the built-in logic program. Chemical reactions biocatalyzed by the enzyme-based systems will respond to the incoming chemical signals according to the Boolean logic and transfer the outgoing signal to the responsive polymeric support. The chemical coupling between the enzymatic systems and the polymeric supports can be based on electron or proton exchange between them. The electron exchange between the redox-active polymeric support and the enzymatic system will result in the reduction or oxidation of the redox polymer. The proton exchange between the polymeric support (polyelectrolyte in this case) and the enzyme system will yield different ionic states (protonated or unprotonated) of the polyelectrolyte support. The expected changes of the polymeric support in its oxidated or protonated states will result in the changes of the composition/structure of the polymeric matrix support. This will result in the corresponding changes of the matrix-specific properties to be considered as final outgoing signals (Fig. 10.1). It should be noted that two distinct parts of the integrated systems will be responsible for the process: (i) the biochemical (enzymatic) systems responsive to the external chemical incoming signals will provide the variation of the system composition/structure according to the built-in Boolean logic. (ii) The polymeric supports will transduce the composition/structure changes to the changes of the matrices properties, thus providing the transduction of the incoming chemical signals (addition of chemicals) into outgoing physical signals (change of the optical, electrical, and magnetic properties; change of wettability; permeability; etc. of the matrix). It should be emphasized that this transduction process will follow the logic operations provided by the biochemical systems. The macroscopic changes in the polymeric structure induced by the enzyme reactions can eventually be used to design chemical actuators controlled by the signals processed by the enzyme logic systems.

Recent studies have shown that chemical transformations occurring at various interfaces and in polymeric matrices (induced by photochemical [17], electrochemical

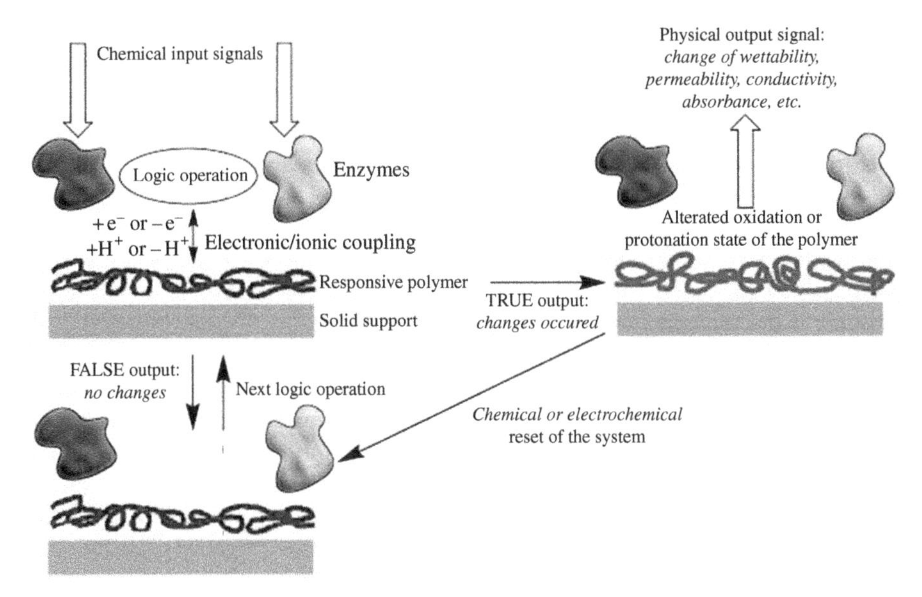

FIGURE 10.1 Scheme showing the general concept of interfacing enzyme-based logic systems with signal-responsive polymers operating as "smart" chemical actuators controlled by the gate output signals. Adopted from Ref. [16] with permission.© Springer.

[18], magnetic [19], or chemical/biochemical means [20]) can result in the substantial changes of the properties of the materials (optical density, reflectivity, electrical conductivity, porosity/permeability, density/volume, wettability, etc.). Mixed-polymer systems (specifically polymer "brushes") were shown as highly efficient responsive systems substantially changing their physical properties upon reconfiguration of the components included in the mixed system [21, 22]. Incorporation of enzymes, which operate as logic gates [2–7], in the polymeric matrices would allow Boolean treatment of the incoming (input) chemical signals and the respective changes of the material properties. For example, two different input signals coming to the system will change the material properties according to the Boolean treatment of the incoming signals and upon electronic or ionic coupling between the enzymes and the responsive polymer (exchange of electrons or protons resulting in the alteration of the oxidation or protonation state of the polymer) (Fig. 10.1). It should be noted that the changes schematically shown in Figure 10.1 will occur only when the output signal of the logic gate is "TRUE" (**1**), while the "FALSE" (**0**) will not result in any changes in the system. If the system is changed (upon a TRUE outcome signal), it might be reset to the original state by chemical or electrochemical means (addition of chemicals to the solution or application of a potential on the conductive solid support). Then, the system will be ready to response to the next incoming signals that may (or may not) change the polymer state according to the Boolean treatment of the new input signals.

Since many polymer systems are pH sensitive and switchable, we designed enzyme-based logic gates using enzymes as biocatalytic input signals, processing information according to the Boolean operations **AND** or **OR**, and generating pH changes as the output signals from the gates [23–26]. The **AND** gate performed a sequence of biocatalytic reactions (Fig. 10.2a): sucrose hydrolysis was biocatalyzed by invertase (Inv) yielding glucose, which was then oxidized by oxygen in the presence of glucose oxidase (GOx). The later reaction resulted in the formation of gluconic acid and therefore lowered the pH value of the solution. The absence of the enzymes was considered as the input signals "**0**," while the presence of them in the optimized concentrations was interpreted as the input signals "**1**." The biocatalytic reaction chain was activated only in the presence of both enzymes (Inv and GOx) (input signals **1**,**1**), resulting in the decrease of the solution pH value (Fig. 10.2b). The absence of any of two enzymes (input signals **0**,**1** or **1**,**0**) or both of them (input signals **0**,**0**) resulted in the inhibition of the gluconic acid formation, and thus, no pH changes were produced. Thus, the biocatalytic chain mimics the **AND** logic operation expressed by the standard truth table corresponding to **AND** Boolean operation (Fig. 10.2c). After completion of the biocatalytic reactions and reaching the final pH value, the system might be reset to the original pH by using another biochemical reaction in the presence of urease and urea, resulting in the production of ammonia and elevation of pH (Fig. 10.2a). The performance of the biochemical system can be described in terms of a logic circuitry with **AND/Reset** function (Fig. 10.2d). Another gate operating as Boolean **OR** logic function was composed of two parallel reactions (Fig. 10.2e): hydrolysis of ethyl butyrate and oxidation of glucose biocatalyzed by esterase (Est) and GOx, respectively, resulting in the formation of butyric acid and gluconic acid. Any of the produced acids and both of them together resulted in the formation of the acidic solution (Fig. 10.2f). Thus, in the absence of both enzymes (Est and GOx) (input signals **0**,**0**), both reactions were inhibited and the pH value was unchanged. When either enzyme (Est or GOx, input signals **0**,**1** or **1**,**0**) or both of them are together (input signals **1**,**1**), one or both of the reactions proceeded and resulted in the acidification of the solution. The features of the system correspond to the **OR** logic operation and can be expressed by the standard truth table (Fig. 10.2g). Similarly to the previous example, the system can be reset to the initial pH by the urease-catalyzed reaction allowing sequence of **OR/Reset** functions (Fig. 10.2h).

The enzyme logic systems producing pH changes were coupled with various pH-sensitive polymer-functionalized systems: membranes [23], nanoparticle (NP) suspensions [24, 25], water/oil emulsions [26], modified electrodes [27], and Si chips [28]. For example, the pH changes produced by the **AND/OR** enzyme logic gates shown in Figure 10.2 were coupled with a pH-responsive membrane, resulting in the opening/closing of the membrane pores, thus transducing the biochemical logic operation into the bulk material property change (Fig. 10.3) [23]. The membrane was prepared by salt-induced phase separation of sodium alginate and gelatin and cross-linked by $CaCl_2$ (Fig. 10.3A). The membrane operates by swelling its gel body in response to changes in pH; this leads to shrinkage of the pores and consequently to a change in its permeability. The membrane was deposited onto an ITO-glass electrode for electrochemical characterization or onto a porous substrate (track-etched polycarbonate membrane) for permeability measurements. The scanning probe microscopy (SPM) topography images obtained *in situ* in a liquid cell (Fig. 10.3B),

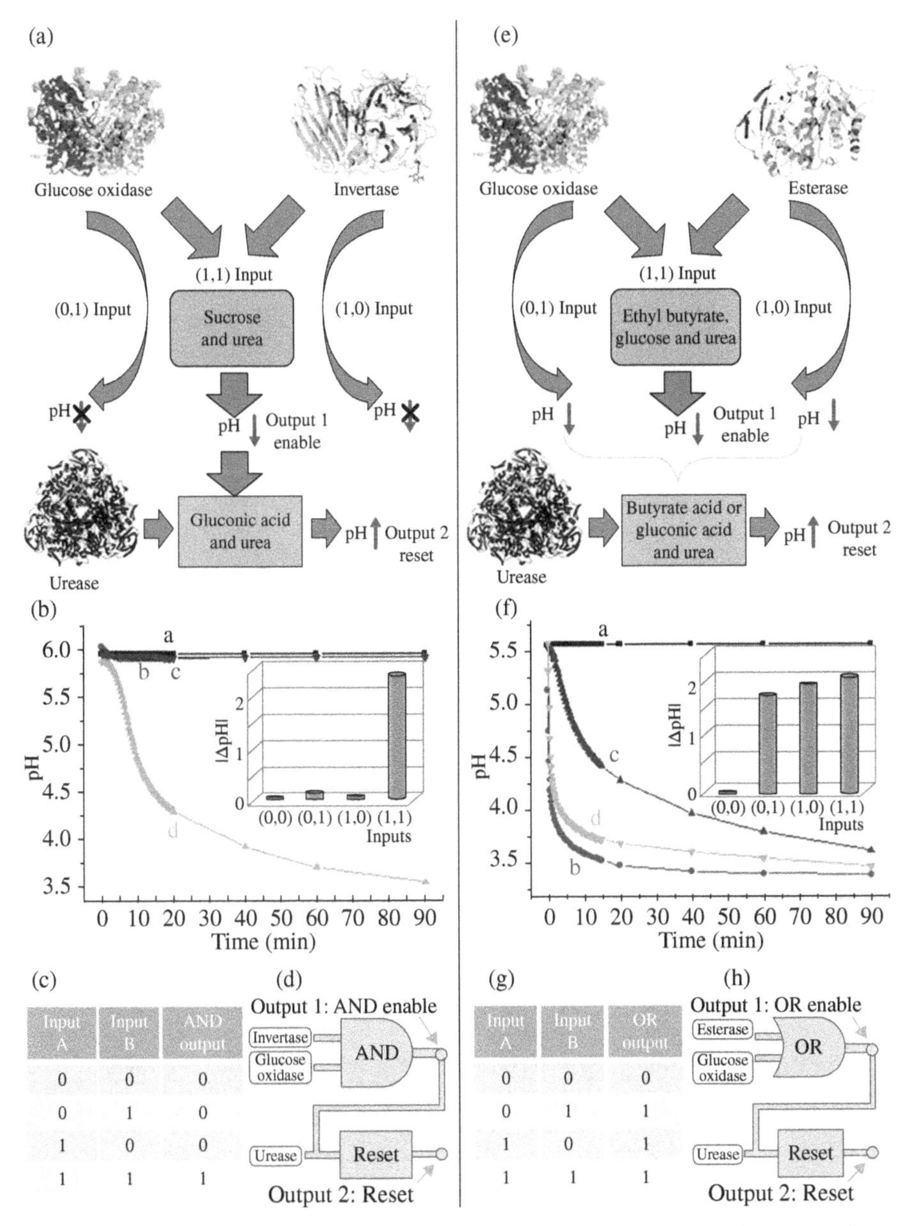

FIGURE 10.2 The biochemical logic gates with the enzymes used as input signals to activate the gate operation: the absence of the enzyme is considered as "**0**" and the presence as "**1**" input signals. The **Reset** function was catalyzed by urease. (a) The **AND** gate based on GOx- and Inv-catalyzed reactions. (b) pH changes generated *in situ* by the **AND** gate upon different combinations of the input signals: (a) "**0,0**"; (b) "**0,1**"; (b) "**1,0**"; and (c) "**1,1**." Inset: bar diagram showing the pH changes as the output signals of the **AND** gate. (c) The truth table of the **AND** gate showing the output signals in the form of pH changes generated upon different combinations of the input signals. (d) Equivalent electronic circuit for the biochemical **AND/Reset** logic operations. (e) The **OR** gate based on GOx- and Est-catalyzed reactions. (f–h) The same as (b–d) for the **OR** gate. Adopted from Ref. [23] with permission. © American Chemical Society, 2009.

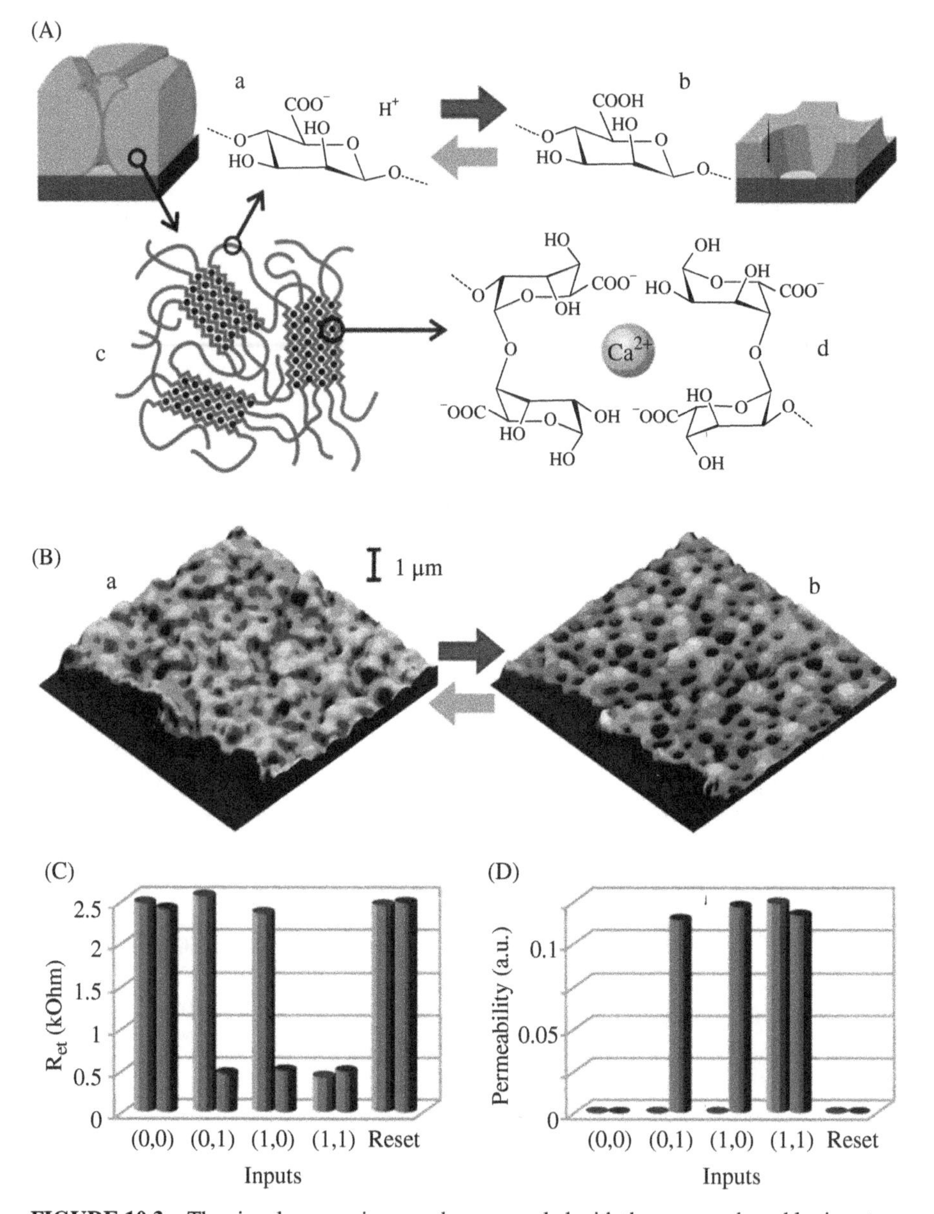

FIGURE 10.3 The signal-responsive membrane coupled with the enzyme-based logic gates. (A) The schematic representations of a single pore of the polyelectrolyte membrane switched between the closed (a) and open (b) states. (c) The structure of the alginate hydrogel constituted of D-mannuronic acid and L-guluronic acid residues cross-linked with divalent ions (Ca^{2+}) in (d) an egg box-like conformation. The swelling and shrinking of the hydrogel are attributed to the ionization (a) and protonation (b) of the unbound carboxyl groups at $pH > 5$ and $pH < 4$, respectively. (B) SPM topography images ($10 \times 10\ \mu m^2$) of the swollen (a) and shrunken (b) membrane. (C) The electron transfer resistance, R_{et}, of the membrane deposited on the electrode surface derived from the impedance spectroscopy measurements obtained upon different combinations of input signals. (D) The permeability (ratio of the membrane permeability deposited on the supporting filter to the permeability of the filter with no membrane) for rhodamine B obtained upon different combinations of the input signals. The left and right bars in each pair correspond to the **AND** and **OR** gates, respectively. Adopted from Ref. [23] with permission. © American Chemical Society, 2009.

electrochemical impedance spectroscopy (Fig. 10.3C), and probe molecules (fluorescent dye rhodamine B) diffusivity through the polyelectrolyte membrane (Fig. 10.3D) were explored to monitor behavior of the membrane coupled with the enzyme-based logic gates. The membrane showed a strong dependence of the swelling of the polyelectrolyte membrane on pH: the pores are open at $pH < 4$ and completely closed at $pH > 5$. The pH changes were induced *in situ* using the enzyme logic gates shown in Figure 10.2. To activate the **AND** biochemical logic gate (Fig. 10.2a), one or both enzymes (GOx and Inv) were added as input signals to the solution with dissolved sucrose, O_2, and urea, which wet the membrane at pH 6. In the absence of each enzyme in the system (input signals **0,0**), the membrane pores were closed (Figs. 10.3A-a and 10.3B-a); the impedance measurements on the membrane deposited on the electrode showed an electron transfer resistance, R_{et}, of ca. 2.5 kΩ for a diffusional redox probe, $[Fe(CN)_6]^{3-/4-}$ (Fig. 10.3C); and no diffusion of the dye was detected through a membrane deposited on the porous substrate (Fig. 10.3D). Obviously, the same behavior of the membrane was documented if only one of the enzymes (input signals **0,1**; **1,0**) was added to the system. However, if both enzymes were added (input signals **1,1**), the enzymatic reactions resulted in a pH decrease from pH 6 to pH 4 and an opening of the pores of the membrane (Fig. 10.3A-b, B-b), R_{et} dropped down to ca. 0.5 kΩ (Fig. 10.3C), and the dye fluorescent spectra were detected in the filtration chamber from the inverse side of the membrane, indicating diffusion of the dye through the open pores (Fig. 10.3D). The state of the system was reset by adding urease. Hydrolysis of urea resulted in elevation of pH to the original value of pH 6. In the experiment, we observed closed pores, a reversible increase of R_{et} to ca. 2.5 kΩ (Fig. 10.3C), and a dramatic decrease of transport of the dye through the membrane (Fig. 10.3D). A similar experiment was conducted with the **OR** enzyme gate shown in Figure 10.2e. In this case, all input signals—**0,1**; **1,0**; and **1,1**—resulted in open pores of the membrane, while reset of the system returned the membrane to the state of closed pores, resulting in the respective changes in R_{et} in the impedance measurements (Fig. 10.3C) and the membrane permeability for the dye (Fig. 10.3D).

Similarly to the logically controlled switchable membrane, other signal-responsive nanostructured materials were functionally coupled with enzyme logic systems. Reversible aggregation–dissociation of polymer-functionalized NPs [24] and inversion of water–oil emulsion stabilized with NPs [26] were controlled by the enzyme signals processed according to the **AND/OR** Boolean logic outlined in Figure 10.2. All studied signal-responsive systems revealed Boolean logic operations encoded in the biochemical systems (it should be noted that in the examples outlined earlier, the enzymes were used as input signals activating biocatalytic transformations).

A variety of applications could be envisaged for the developed hybrid systems, including signal-controlled drug delivery. A diverse range of "smart" (stimuli-responsive) materials, with switchable physical properties, has been developed for *in vivo* drug delivery [29]. The new hybrid materials with built-in Boolean logic will be capable of switching physical properties in response to the output of the enzyme logic system toward autonomous on-demand drug delivery. The gate output signals will activate "smart" chemical actuators, resulting, for example, in the opening of a membrane releasing a drug (such as mannitol, if the logic states dictated need for a reduced intracranial pressure [30]). The demonstrated approach for the interfacing of biomolecular computing systems with signal-responsive materials enables the use of various biocatalytic reactions to control transition of the properties of responsive materials and systems with built-in Boolean logic. This approach would be an efficient way to fabricate "smart" multisignal-responsive drug delivery systems, sensors, miniaturized switchers, microfluidic devices, etc., which can serve without communication to an external computer.

pH-switchable materials immobilized on interfaces of electronic/electrochemical transducers (e.g., Si-based chips [28] or conducting electrodes [27, 31]) were coupled with enzyme logic systems producing pH changes in solutions as logic responses to input signals. This allowed electronic transduction of the generated output signals, converting the systems into multisignal biosensors chemically processing various patterns of the input signals using logic "programs" built-in in the enzyme systems. For example, enzyme logic systems mimicking Boolean **AND/OR** logic operations and producing the output signal in the form of solution pH changes were coupled with charging–discharging organic shells around Au NPs associated with a Si-chip surface (Fig. 10.4a) [28]. This resulted in the capacitance changes at the modified interface, allowing electronic transduction of the biochemical signals processed by the enzyme logic systems (Fig. 10.4b, c). The **AND** logic operation was performed by the concerted operation of Inv and GOx (operating in this case as a biocomputing "machinery") being activated by sucrose and oxygen as input signals. Sucrose was converted to glucose by Inv and then oxidized by GOx in the presence of oxygen. The final product, gluconic acid, was generated only in the presence of both input signals (combination **1,1**), resulting in the acidic pH value and yielding the protonated (neutral) carboxylic groups in the NPs shells. Otherwise, when any or both inputs were missing (signal combinations **0,0**; **0,1**; **1,0**), the initial neutral pH was preserved keeping dissociated (negatively charged) carboxylic groups, thus mimicking the **AND** operation. After a set of the input signals was applied to the system, the reset function was activated to bring the pH value to the initial value if it was altered. The reset was achieved by urea hydrolysis biocatalyzed in the presence of urease (Fig. 10.4b). Similarly, **OR** logic was realized by two parallel reactions biocatalyzed by Est and GOx being activated by ethyl butyrate and glucose and yielding an acid (butyric acid or gluconic acid) in the presence of either or both inputs (input combinations **0,1**; **1,0**; **1,1**) (Fig. 10.4c). The reset function was applied at the end to return the initial pH value as it is described previously.

Another approach to the electrochemical transduction of the output signals generated by enzyme logic systems in the form of pH changes was based on the application of polyelectrolyte-modified electrode surfaces [27, 31]. Polyelectrolytes covalently bound to electrode surfaces as polymer brushes reveal pH sensitivity, allowing control of the electrode interfacial properties by varying pH values. Charged states of the polymer brushes produce hydrophilic swollen thin films on the electrode surfaces resulting in their high permeability for soluble redox probes to the conducting supports, thus yielding the electrochemically active states of the modified electrodes. Upon discharging the polymer chains, the produced hydrophobic shrunken states isolated the conducting supports yielding the inactive states of the modified electrodes. Switching

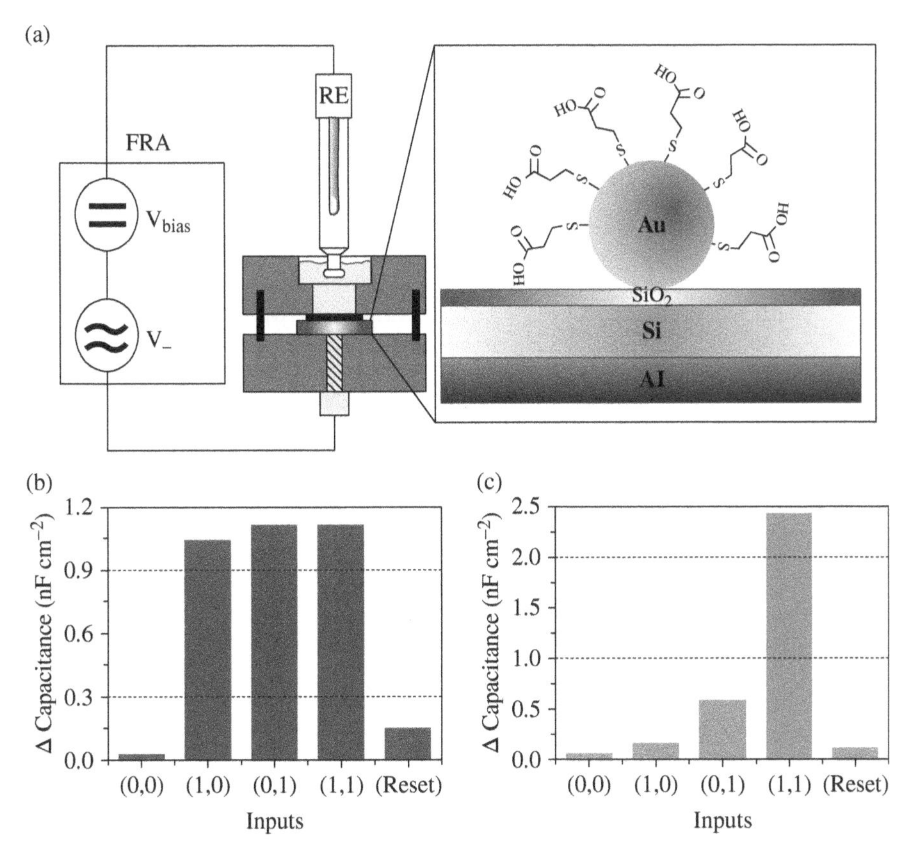

FIGURE 10.4 The electronic scheme (a) of the signal-transducing device based on the Si chip modified with Au nanoparticles coated with a pH-sensitive organic shell (FRA, frequency response analyzer; RE, reference electrode). The bar diagrams showing the output signals generated by **OR** (b)/**AND** (c) enzyme logic gates and transduced by the Si chip in the form of capacitance changes. The dash lines correspond to the threshold values: the output signals located below the first threshold were considered as "**0**," while the signals higher than the second threshold were treated as "**1**." Adopted from Ref. [28] with permission. © American Chemical Society, 2009.

between the ON and OFF states of the electrode modified with the polymer brush was achieved by varying the pH value of the solution. This property of the polymer brush-functionalized electrodes was used to couple them with an enzyme logic system composed of several networked gates [31]. The logic network composed of three enzymes (alcohol dehydrogenase, glucose dehydrogenase, and GOx) operating in concert as four concatenated logic gates (**AND/OR**) was designed to process four different chemical input signals (NADH, acetaldehyde, glucose, and oxygen) (Fig. 10.5). The cascade of biochemical reactions resulted in pH changes controlled by the pattern of the applied biochemical input signals. The "successful" set of the inputs produced gluconic acid as the final product and yielded an acidic medium, lowering the pH of a

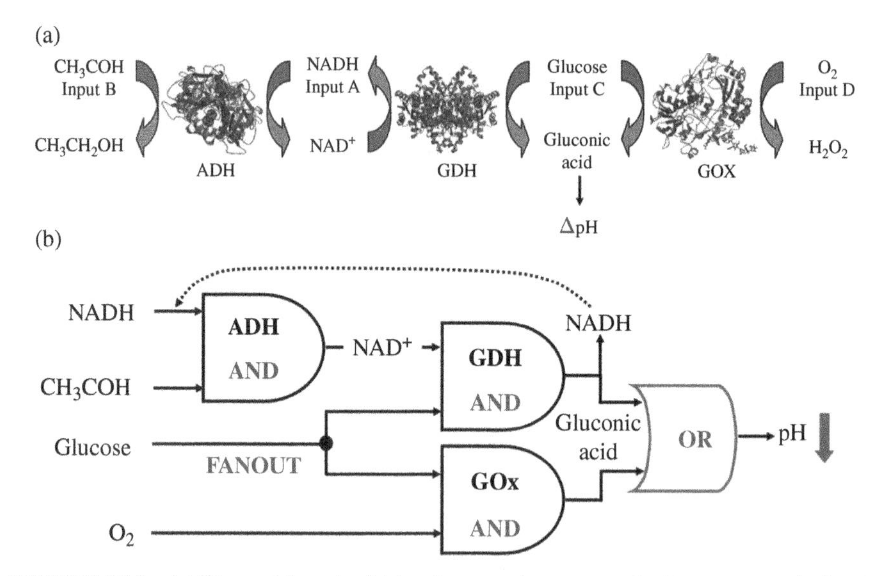

FIGURE 10.5 (a) The multigate/multisignal processing enzyme logic system producing *in situ* pH changes as the output signal. (b) The equivalent logic circuitry for the biocatalytic cascade. Adopted from Ref. [31] with permission. © American Chemical Society, 2009.

solution from its initial value of pH 6–7 to the final value of ca. 4, thus switching ON the interface for the redox process of a diffusional redox probe, $[Fe(CN)_6]^{3-/4-}$. The chemical signals processed by the enzyme logic system and transduced by the sensing interface were read out by electrochemical means using cyclic voltammetry (Fig. 10.6a). Reversible activation–inactivation of the electrochemical interface was achieved upon logic processing of the biochemical input signals and then by the reset function activated in the presence of urease and urea (Fig. 10.6a, inset). The whole set of the input signal combinations included 16 variants, while only **0,0,1,1**; **0,1,1,1**; **1,0,1,1**; **1,1,1,0**; and **1,1,1,1** combinations resulted in the ON state of the electrochemical interface (Fig. 10.6b). The present system exemplified a multigate/multisignal processing enzyme logic system associated with an electrochemical transduction readout of the output signal.

More sophisticated electrochemical properties were found for the polymer brush bound to an electrode surface and functionalized with redox species [32]. Os(dmo-bpy)$_2$ redox groups (dmo-bpy = 4,4′-dimethoxy-2,2′-bipyridine) were covalently bound to poly(4-vinyl pyridine) (P4VP) polymer chains grafted on an ITO electrode. The polymer-bound redox species were found to be electrochemically active at pH < 5 when the polymer is protonated and swollen and the chains are flexible. Upon changing pH to values greater than pH 6, the polymer chains are losing their charges and the produced shrunken state of the polymer does not show substantial electrochemical activity. This was explained by the poor mobility of the polymer chains in the shrunken state restricting the direct contact between the Os-complex units and the conducting support. A low density of Os complex in the polymer film does not allow

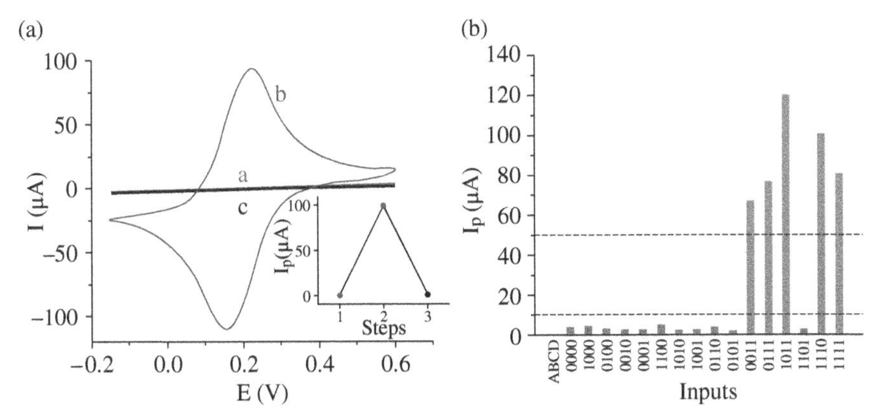

FIGURE 10.6 (a) Cyclic voltammograms obtained for the ITO electrode modified with the P4VP polymer brush in (a) the initial OFF state, pH ca. 6.7; (b) the ON state enabled by the input combinations resulting in acidifying of the solution to pH ca. 4.3; and (c) *in situ* reset to the OFF state, pH ca. 8.8. Inset: reversible current changes upon switching the electrode ON–OFF. Deoxygenated unbuffered solution of 0.1 M Na_2SO_4, 10 mM $K_3Fe(CN)_6$, and 10 mM $K_4Fe(CN)_6$ also contained ADH, GDH, and GOx, 10 units·mL^{-1} each. Input A was 0.5 mM NADH, input B was 5 mM acetaldehyde, and input C was 12.5 mM glucose. Potential scan rate of 100 mV s^{-1}. (b) Anodic peak currents, I_p, for the 16 possible input combinations. The dotted lines show threshold values separating logic **1**, undefined, and logic **0** output signals. Adopted from Ref. [31] with permission. © American Chemical Society, 2009.

the electron hopping between the redox species, and their electrochemical activity can be achieved only upon quasidiffusional translocation of the polymer chains in the swollen state bringing the redox species to a short distance from the conducting support. The pH-controlled reversible activation–inactivation of the interface-confined redox species was used to read out the pH signals produced *in situ* by enzyme systems performing **AND/OR** logic operations (Fig. 10.7). The electrochemical signal produced by the modified interface in response to the pH changes generated by the enzyme reactions was further amplified by electrocatalytic oxidation of NADH being switched ON upon activation of the redox species on the electrode surface [27].

The logically controlled Os-complex P4VP modified electrode was applied to mediate oxygen reduction in the presence of laccase—an enzyme reducing O_2 to water. This electrode was integrated into an enzyme-based biofuel cell, serving there as a switchable biocatalytic cathode [33–35] (Fig. 10.8). Glucose oxidation in the presence of soluble GOx and methylene blue (MB) mediating electron transport to a bare ITO electrode was used as an anodic process. The biofuel cell was switched ON on demand upon processing biochemical signals by enzyme logic gates and reset back to the OFF state by another biocatalytic reaction in the presence of urea and urease. The coupling of the *in situ* enzyme reactions with the switchable electrochemical interface was achieved by pH changes generated in course of the enzymatic reactions, resulting in swelling/shrinking of the redox polymer at the electrode surface as described earlier. At the beginning, single logic gates **AND/OR** were used to control

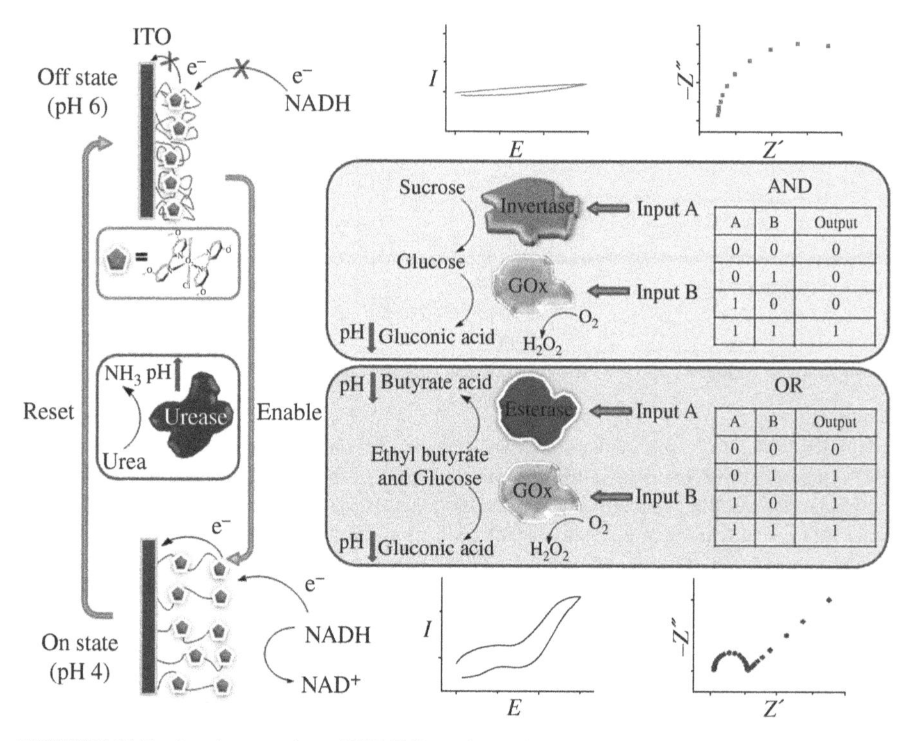

FIGURE 10.7 Logic operations **AND/OR** performed by the enzyme-based systems resulting in the **ON** and **OFF** states of the bioelectrocatalytic interface followed by the **Reset** function to complete the reversible cycle. Schematically shown cyclic voltammograms and impedance spectra correspond to the **ON** and **OFF** states of the bioelectrocatalytic electrode applied for the NADH oxidation. Adopted from Ref. [27] with permission. © American Chemical Society, 2009.

the biofuel cell power production [34]. Higher-complexity biocatalytic system based on the concerted operation of four enzymes activated by four chemical input signals was designed (Fig. 10.9a) to mimic a logic network composed of three logic gates (**AND/OR** connected in parallel generating two intermediate signals for the final **AND** gate) [35] (Fig. 10.9b). The switchable biofuel cell was characterized by measuring polarization curves at its "mute" and active states. A low voltage-current production was characteristic of the initial inactive state of the biofuel cell at pH ca. 6 (Fig. 10.10a, curve a). Upon receiving an output signal in the form of a pH decrease from the enzyme logic network, the voltage-current production by the biofuel cell was dramatically enhanced when pH reached ca. 4.3 (Fig. 10.10a, curve b). When the activation of the biofuel cell was achieved, another biochemical signal (urea in the presence of urease) resulted in the increasing pH, thus resetting the cell to its inactive state with the low voltage-current production (Fig. 10.10a, curve c). The cyclic operation of the biofuel cell upon receiving biochemical signals can be followed by the reversible changes of the current production (Fig. 10.10a, inset). The biofuel cell switching from

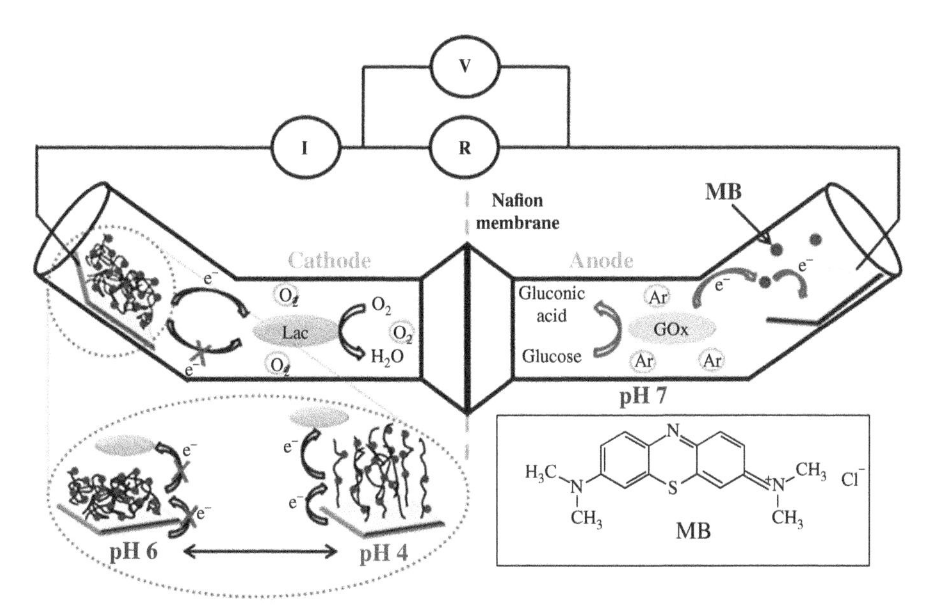

FIGURE 10.8 The biofuel cell composed of the pH-switchable logically controlled biocatalytic cathode and glucose-oxidizing anode. Adopted from Ref. [35] with permission.

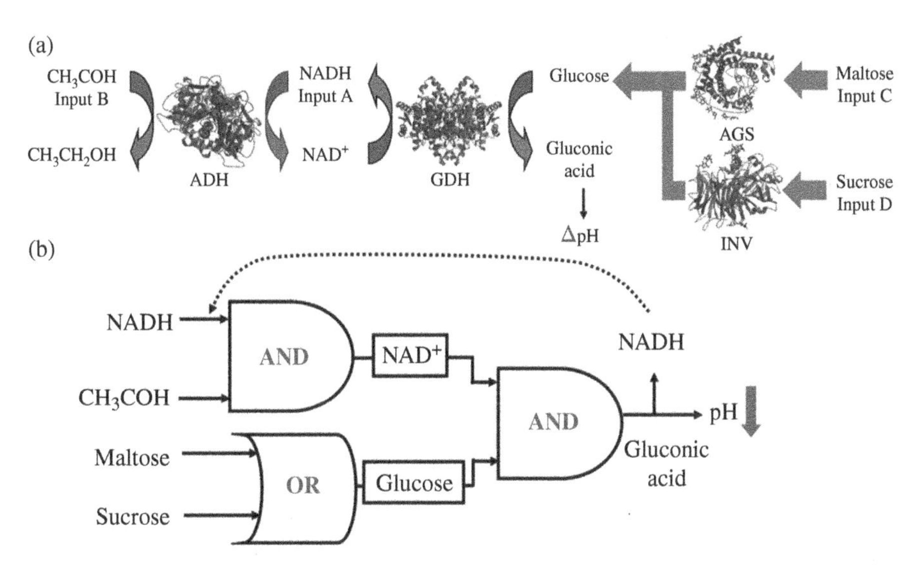

FIGURE 10.9 (a) The cascade of reactions biocatalyzed by alcohol dehydrogenase (ADH), amyloglucosidase (AGS), invertase (Inv) and glucose dehydrogenase (GDH) and triggered by the chemical input signals NADH, acetaldehyde, maltose, and sucrose added in different combinations. (b) The logic network composed of three concatenated gates and equivalent to the cascade of enzymatic reactions outlined in (a). Adopted from Ref. [35] with permission.

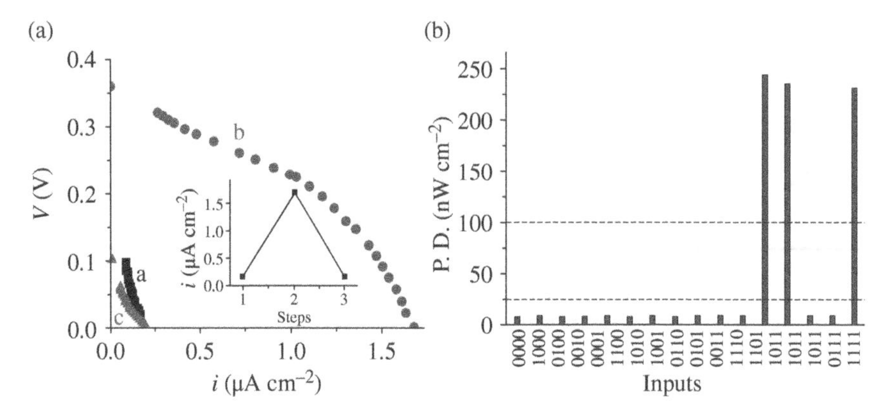

FIGURE 10.10 (a) V–i polarization curves obtained for the biofuel cell with different load resistances: (a) in the inactive state prior to the addition of the biochemical input signals (pH value in the cathodic compartment ca. 6), (b) in the active state after the cathode was activated by changing pH to ca. 4.3 by the biochemical signals, (c) after the **Reset** function activated by the addition of 5 mM urea. Inset: switchable i_{sc} upon transition of the biofuel cell from the mute state to the active state and back performed upon biochemical signals processed by the enzyme logic network. (b) The bar diagram showing the power density produced by the biofuel cell in response to different patterns of the chemical input signals. Dashed lines show thresholds separating digital **0**, undefined, and **1** output signals produced by the system. Adopted from Ref. [35] with permission.

the "mute" state with a low activity to the active state was achieved upon appropriate combination of the input signals processed by the enzyme logic network. Only three combinations of the input signals—**1,1,1,0**; **1,1,0,1**; and **1,1,1,1**—from all 16 possible variants resulted in the solution pH change, thus switching the biofuel cell to its active state (Fig. 10.10b). The studied biofuel cells exemplify a new kind of bioelectronic devices where the bioelectronic function is controlled by a biocomputing system. Such devices will provide a new dimension in bioelectronics and biocomputing benefiting from the integration of both concepts.

The enzyme-based biochemical networks demonstrate robust error-free processing of biochemical signals upon appropriate optimization of their components and interconnections [36–38]. However, the limit of the biocomputing network complexity is set by the cross-reactivity of the enzyme-catalyzed reactions. Only enzymes belonging to different biocatalytic classes (oxidases, dehydrogenases, peroxidises, hydrolases, etc.) could operate in a homogeneous system without significant cross-reactivity. If chemical reasons require the use of cross-reacting enzymes in the system, they must be compartmentalized using patterning on surfaces or applied in microfluidic devices. Application of more selective biomolecular interactions would be an advantage to make biocomputing systems more specific to various input signals and less cross-reactive in the chemical signal processing. This aim can be achieved by the application of highly selective biorecognition (e.g., immune) interactions for biocomputing [39]. One of the novel immune-based biocomputing systems was already applied for switching the biofuel cell activity by the logically processed antibody signals [40].

A surface functionalized with a mixed monolayer of two different antigens, 2,4-dinitrophenyl (DNP) and 3-nitro-l-tyrosine (NT), loaded on human serum albumin (HSA) and bovine serum albumin (BSA), respectively, was used to analyze the input signals of the corresponding antibodies: anti-dinitrophenyl (anti-DNP IgG polyclonal from goat) and anti-nitrotyrosine (anti-NT IgG from rabbit) [40]. After binding to the surface, the primary antibodies were reacted with the secondary antibodies: anti-goat IgG HRP and anti-rabbit IgG HRP (mouse origin IgG against goat immunoglobulin and mouse origin IgG against rabbit IgG, both labeled with horseradish peroxidise, HRP) to attach the biocatalytic HRP tag to the immune complexes generated on the surfaces (Fig. 10.11a). The primary anti-DNP and anti-NT antibodies (signals *A* and *B*, respectively) were applied in four different combinations: **0**,**0**; **0**,**1**; **1**,**0**; and **1**,**1**, where the digital value **0** corresponded to the absence of the antibody and value **1** corresponded to their presence in the optimized concentrations. The secondary antibody labeled with the HRP biocatalytic tag was bound to the surface only if the respective primary antibody was already there. Since both secondary antibodies were labeled with HRP, the biocatalytic entity appeared on the surface upon application of **0**,**1**; **1**,**0**; and **1**,**1** signal combinations. Only in the absence of both primary antibodies (signals **0**,**0**) the secondary antibodies were not bound to the surface and the HRP biocatalyst did not appear there, thus resembling the **OR** logic operation. The assembled functional interface was reacted with 2,2′-azino-bis(3-ethylbenzothiazoline-6-sulfonic acid) (ABTS) and H_2O_2. The biocatalytic oxidation of ABTS and concomitant reduction of H_2O_2 resulted in the increase of the solution pH only when the biocatalytic HRP tag was present on the surface (Fig. 10.11b). This happened when the primary antibody signals were applied in the combinations **0**,**1**; **1**,**0**; and **1**,**1**. The pH increase generated *in situ* by the enzyme reaction coupled with the immune-recognition system yielded the inactive shrunken state of the polymer brush-modified electrode, thus deactivating the entire biofuel cell. It should be noted that for simplicity, the cathode was represented by a model redox system with a ferricyanide solution instead of the oxygen system (Fig. 10.11c). The biofuel cell being active at pH 4.5 (Fig. 10.12a, b, curves a) was partially inactivated (curves b) by the pH increase up to 5.8 generated by the immune-based logic system. Since the output signal **1** from the logic system resulted in the inactivation of the biofuel cell (operating as the inverter producing **0** output for input **1**, and vice versa), the system modeled an **NOR** logic gate (Fig. 10.12b, inset). After the biofuel cell inactivation, the next cycle was started by the reset to the initial pH value activating the switchable electrode again. To activate the biofuel cell, GOx and glucose were added to the cathodic compartment, resulting in the pH decrease to ca. 4.2 due to the biocatalytic oxidation of glucose and formation of gluconic acid.

It should be noted however that all biomolecular logic systems described earlier produce bulk pH changes in order to switch the electrode interface between active and inactive states. This cannot be used in many biomedical applications when bulk pH cannot be altered in physiological liquids, particularly in implantable switchable

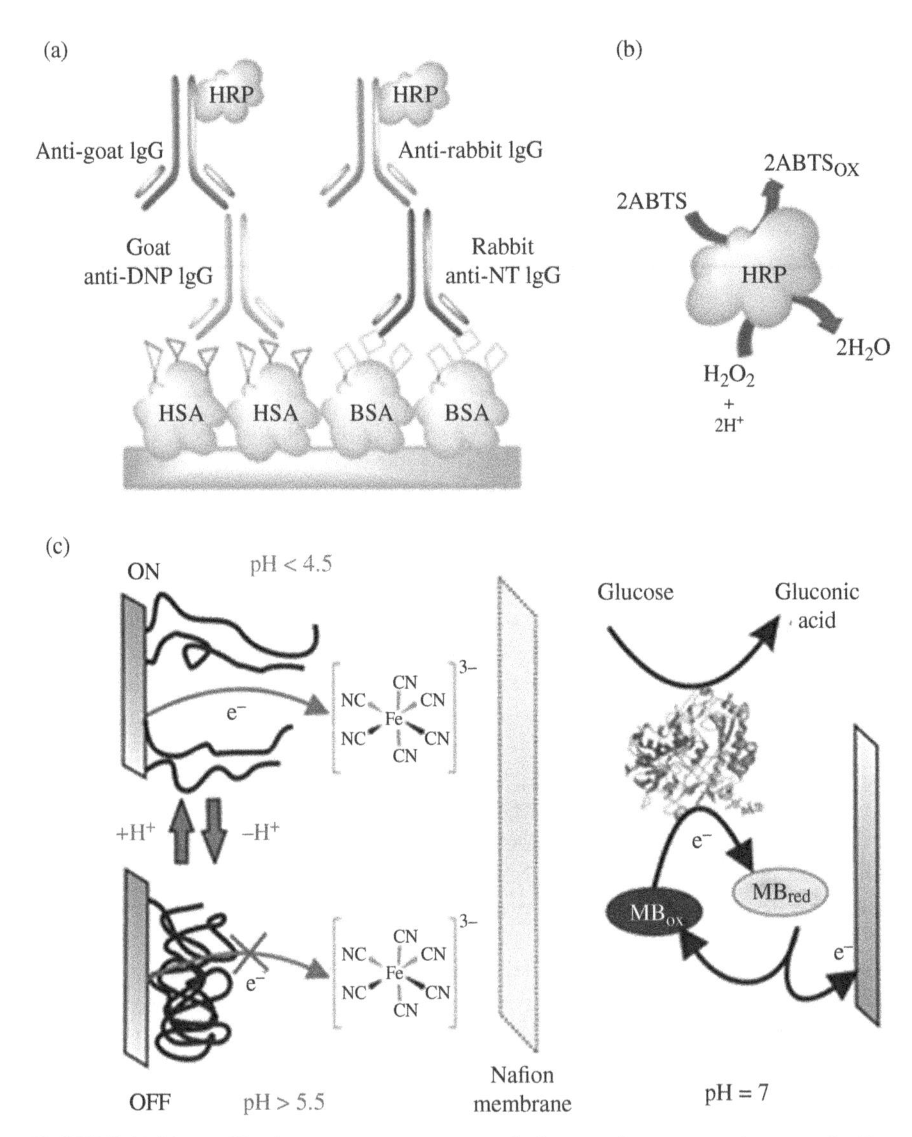

FIGURE 10.11 (a) The immune system composed of two antigens, two primary antibodies, and two secondary antibodies labeled with horseradish peroxidise (HRP) biocatalytic tag used for the **OR** logic gate. (b) The biocatalytic reaction producing pH changes to control the biofuel cell performance. (c) The biofuel cell controlled by the immune **OR** logic gate due to the pH-switchable $[Fe(CN)^6]^{3-}$-reducing cathode. MB_{ox} and MB_{red} are oxidized and reduced states of the mediator methylene blue. Adapted from Ref. [40] with permission. © American Chemical Society, 2009.

biosensors and bioactuators operating *in vivo*. Thus, development of switchable electrode interfaces controlled by local pH changes is an important goal. One of the first systems allowing switchable functioning of electrode interfaces upon pH changes localized at the electrode surface has been developed recently [41]. Magnetic NPs

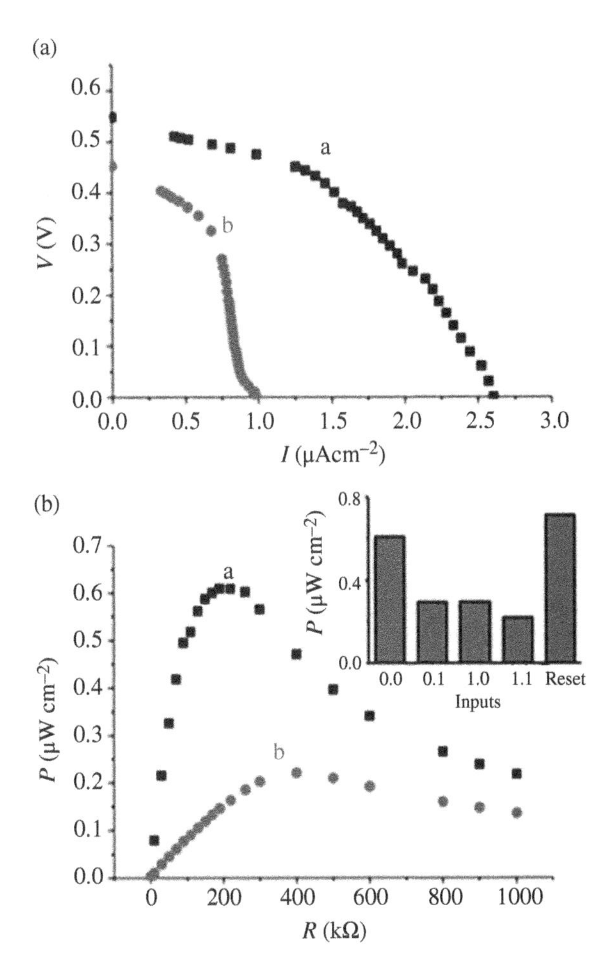

FIGURE 10.12 (a) The polarization curves of the biofuel cell with the pH-switchable cathode obtained at different pH values generated *in situ* by the immune **OR** logic gate: (a) pH 4.5, (b) pH 5.8. (b) Electrical power density generated by the biofuel cell on different load resistances at different pH values generated *in situ* by the immune **OR** logic gate: (a) pH 4.5, (b) pH 5.8. Inset: the maximum electrical power density produced by the biofuel cell upon different combinations of the immune input signals. Adapted from Ref. [40] with permission. © American Chemical Society, 2009.

covalently modified with GOx (Fig. 10.13a) were used to produce acidic pH value upon biocatalytic oxidation of glucose, resulting in the formation of gluconic acid. The biocatalytic NPs suspended in a bulk solution were collected on an electrode surface functionalized with a pH-switchable polymer layer (P4VP brush) and produced local pH decrease when glucose was added to the solution (Fig. 10.13b). It is very important to note that the local pH decrease was not accompanied by any pH changes in the bulk solution and the electrode interface was open for electrochemical reactions

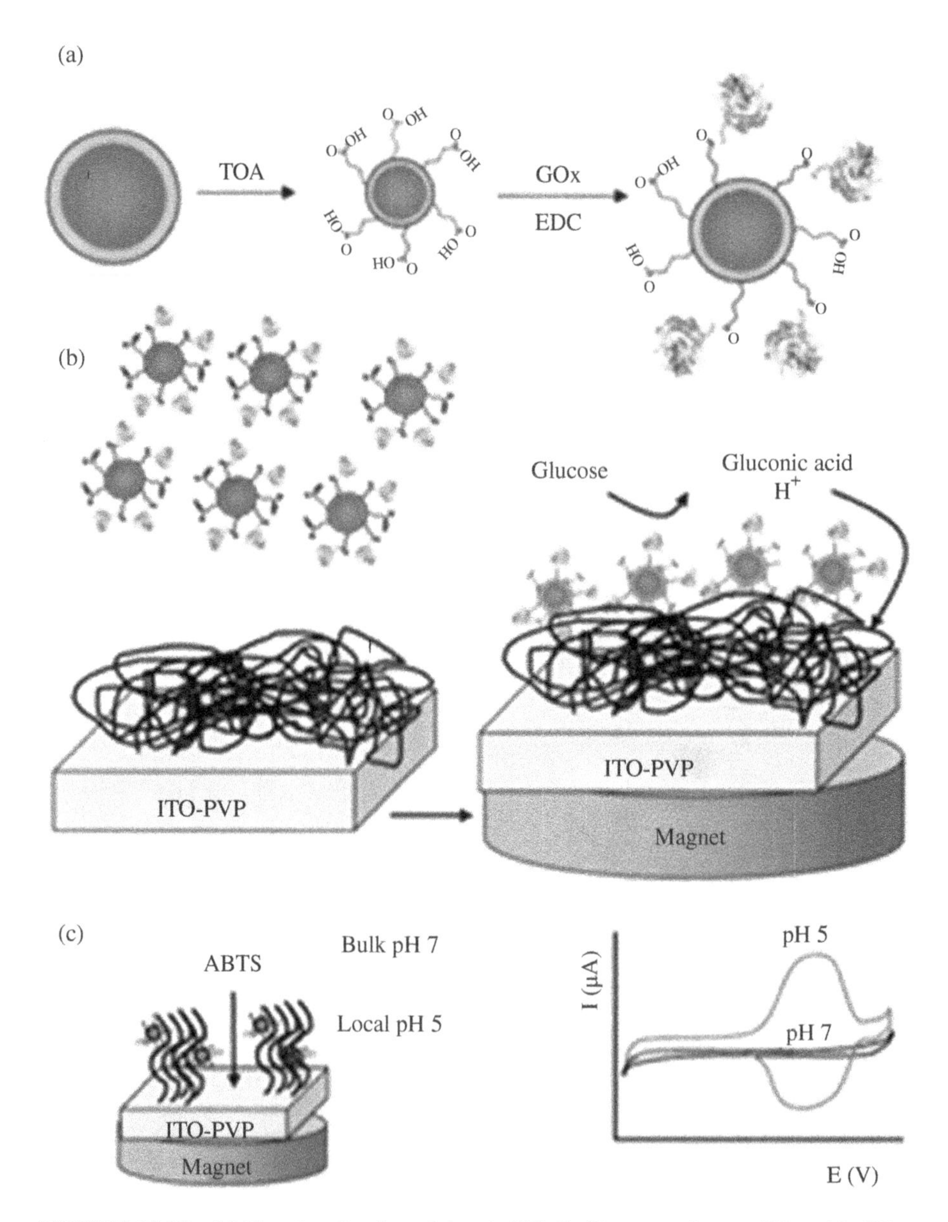

FIGURE 10.13 (a) Functionalization of Au-shell/$CoFe_2O_4$-magnetic core NPs with GOx. (b) Magneto-assisted concentration of the GOx NPs on the electrode surface modified with the P4VP brush to perform glucose oxidation at the interface. (c) Opening of the P4VP brush for the electrochemical reaction at acidic pH generated at the interface upon the biocatalytic reaction. Adopted from Ref. [41] with permission. © American Chemical Society, 2009.

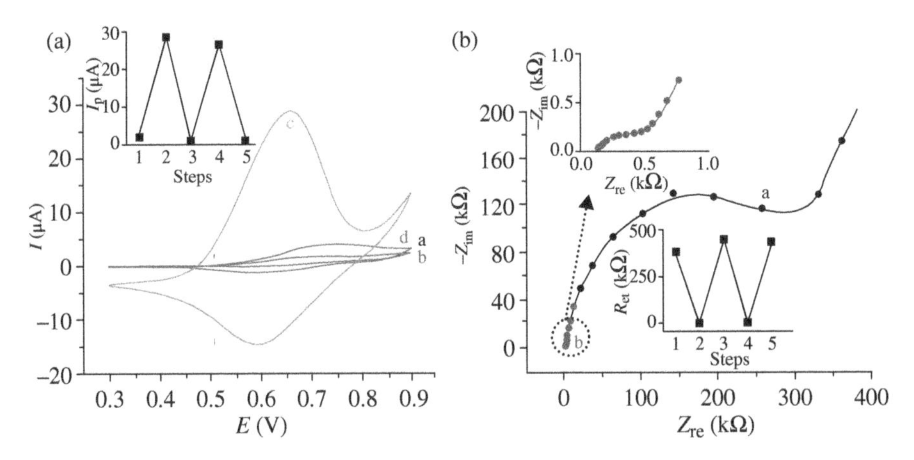

FIGURE 10.14 (a) Cyclic voltammograms obtained on the P4VP-modified electrode: (a) GOx NPs in the solution in the absence of glucose, (b) GOx NPs confined at the electrode in the absence of glucose, (c) GOx NPs confined at the electrode in the presence of glucose, (d) GOx NPs redispersed in the solution in presence of glucose. Inset: reversible ON–OFF electrode switching by adding and removing glucose, while the GOx NPs are confined at the electrode surface. (b) Impedance spectra (Nyquist plots) obtained on the P4VP-modified electrode with the GOx NPs confined at the electrode: (a) in the absence of glucose, (b) in the presence of glucose (also shown at a smaller scale). Inset: reversible switching of the R_{et} by adding and removing glucose. Solution: 0.1 mM ABTS in 0.1 M Na_2SO_4, pH 7; GOx NPs, 0.3 mg mL^{-1}; glucose addition, 10 mM; cyclic voltammograms, 100 mV s^{-1}; impedance bias potential, 0.62 V. Adopted from Ref. [41] with permission. © American Chemical Society, 2009.

when only local pH changes were achieved (Fig. 10.13c). Reversible confinement–removal of the biocatalytic NPs to and from the electrode surface as well as addition and washing out of glucose resulted in the reversible activation and inactivation of the switchable electrode interface controlled exclusively by local interfacial pH changes. The switchable behavior of the electrode interface was followed by cyclic voltammetry and Faradaic impedance spectroscopy (Fig. 10.14). The present results demonstrate the possibility to control the activity of switchable surfaces by producing local pH changes upon biocatalytic reactions running at the interfaces. The external signals activating the interfaces might be chemical (e.g., glucose) or physical (magnetic field) or both of them together (**AND** logic gate).

The developed approach exemplified by several systems described in the present paper paves the way to the novel digital biosensors and bioelectronic devices processing multiple biochemical input signals and producing a combination of output signals dependent on the whole pattern of various input signals. The biochemical signals are processed by chemical means based on the enzyme logic system prior to their electronic transduction, hence obviating the need for computer analysis of the biosensing information. In addition to the analysis of the data, the output signals might be directed to chemical actuators (e.g., signal-responsive membranes) leading to an on-demand drug release. We anticipate that such biochemical logic gates connected with bioelectronic sensing and actuating devices will find numerous biomedical applications. They will facilitate decision-making in connection to an autonomous feedback-loop drug delivery system and will revolutionize the monitoring and treatment of patients.

ACKNOWLEDGMENT

This research was supported by the National Science Foundation (grants DMR-0706209 and CCF-1015983), by ONR (grant N00014-08-1-1202), and by the Semiconductor Research Corporation (award 2008-RJ-1839G).

REFERENCES

1. E. Katz, V. Privman, *Chem. Soc. Rev.* 2010, 39(5), 1835–1857.
2. G. Strack, M. Pita, M. Ornatska, E. Katz, *ChemBioChem* 2008, 9, 1260–1266.
3. K.M. Manesh, J. Halámek, M. Pita, J. Zhou, T.K. Tam, P. Santhosh, M.-C. Chuang, J.R. Windmiller, D. Abidin, E. Katz, J. Wang, *Biosens. Bioelectron.* 2009, 24, 3569–3574.
4. M. Pita, J. Zhou, K.M. Manesh, J. Halámek, E. Katz, J. Wang, *Sens. Actuat. B* 2009, 139, 631–636.
5. R. Baron, O. Lioubashevski, E. Katz, T. Niazov, I. Willner, *Org. Biomol. Chem.* 2006, 4, 989–991.
6. R. Baron, O. Lioubashevski, E. Katz, T. Niazov, I. Willner, *J. Phys. Chem. A* 2006, 110, 8548–8553.
7. R. Baron, O. Lioubashevski, E. Katz, T. Niazov, I. Willner, *Angew. Chem. Int. Ed.* 2006, 45, 1572–1576.
8. M. Pita, E. Katz, *J. Am. Chem. Soc.* 2008, 130, 36–37.
9. A.P. de Silva, S. Uchiyama, T.P. Vance, B. Wannalerse, *Coord. Chem. Rev.* 2007, 251, 1623–1632.
10. A.P. de Silva, S. Uchiyama, *Nat. Nanotechnol.* 2007, 2, 399–410.
11. K. Szacilowski, *Chem. Rev.* 2008, 108, 3481–3548.
12. I. Tokarev, S. Minko, *Soft Matter* 2009, 5, 511–524.
13. S.K. Ahn, R.M. Kasi, S.C. Kim, N. Sharma, Y.X. Zhou, *Soft Matter* 2008, 4, 1151–1157.
14. K. Glinel, C. Dejugnat, M. Prevot, B. Scholer, M. Schonhoff, R.V. Klitzing, *Colloids Surf. A* 2007, 303, 3–13.
15. P.M. Mendes, *Chem. Soc. Rev.* 2008, 37, 2512–2529.
16. M. Pita, S. Minko, E. Katz, *J. Mater. Sci. Mater. Med.* 2009, 20, 457–462.
17. I. Willner, A. Doron, E. Katz, *J. Phys. Org. Chem.* 1998, 11, 546–560.
18. V.I. Chegel, O.A. Raitman, O. Lioubashevski, Y. Shirshov, E. Katz, I. Willner, *Adv. Mater.* 2002, 14, 1549–1553.
19. E. Katz, L. Sheeney-Haj-Ichia, B. Basnar, I. Felner, I. Willner, *Langmuir* 2004, 20, 9714–9719.

20. X. Wang, Z. Gershman, A.B. Kharitonov, E. Katz, I. Willner, *Langmuir* 2003, 19, 5413–5420.
21. I. Luzinov, S. Minko, V.V. Tsukruk, *Prog. Polym. Sci.* 2004, 29, 635–698.
22. S. Minko, *Polymer Rev.* 2006, 46, 397–420.
23. I. Tokarev, V. Gopishetty, J. Zhou, M. Pita, M. Motornov, E. Katz, S. Minko, *ACS Appl. Mater. Interfaces* 2009, 1, 532–536.
24. M. Motornov, J. Zhou, M. Pita, V. Gopishetty, I. Tokarev, E. Katz, S. Minko, *Nano Lett.* 2008, 8, 2993–2997.
25. M. Pita, M. Krämer, J. Zhou, A. Poghossian, M.J. Schöning, V.M. Fernández, E. Katz, *ACS Nano* 2008, 2, 2160–2166.
26. M. Motornov, J. Zhou, M. Pita, I. Tokarev, V. Gopishetty, E. Katz, S. Minko, *Small* 2009, 5, 817–820.
27. J. Zhou, T.K. Tam, M. Pita, M. Ornatska, S. Minko, E. Katz, *ACS Appl. Mater. Interfaces* 2009, 1, 144–149.
28. M. Krämer, M. Pita, J. Zhou, M. Ornatska, A. Poghossian, M.J. Schöning, E. Katz, *J. Phys. Chem. B* 2009, 113, 2573–2579.
29. D.A. LaVan, T. McGuire, R. Langer, *Nature Biotechnol.* 2003, 21, 1184–1191.
30. J. Ghajar, *Lancet* 2000, 356, 923–929.
31. M. Privman, T.K. Tam, M. Pita, E. Katz, *J. Am. Chem. Soc.* 2009, 131, 1314–1321.
32. T.K. Tam, M. Ornatska, M. Pita, S. Minko, E. Katz, *J. Phys. Chem. C* 2008, 112, 8438–8445.
33. E. Katz, M. Pita, *Chem. Eur. J.* 2009, 15, 12554–12564.
34. L. Amir, T.K. Tam, M. Pita, M.M. Meijler, L. Alfonta, E. Katz, *J. Am. Chem. Soc.* 2009, 131, 826–832.
35. T.K. Tam, M. Pita, M. Ornatska, E. Katz, *Bioelectrochemistry* 2009, 76, 4–9.
36. D. Melnikov, G. Strack, M. Pita, V. Privman, E. Katz, *J. Phys. Chem. B* 2009, 113, 10472–10479.
37. V. Privman, M.A. Arugula, J. Halámek, M. Pita, E. Katz, *J. Phys. Chem. B* 2009, 113, 5301–5310.
38. V. Privman, G. Strack, D. Solenov, M. Pita, E. Katz, *J. Phys. Chem. B* 2008, 112, 11777–11784.
39. G. Strack, S. Chinnapareddy, D. Volkov, J. Halámek, M. Pita, I. Sokolov, E. Katz, *J. Phys. Chem. B* 2009, 113, 12154–12159.
40. T.K. Tam, G. Strack, M. Pita, E. Katz, *J. Am. Chem. Soc.* 2009, 131, 11670–11671.
41. M. Pita, T.K. Tam, S. Minko, E. Katz, *ACS Appl. Mater. Interfaces* 2009, 1, 1166–1168.

INDEX

Note: Page numbers in *italics* refer to Figures; those in **bold** to Tables.

Electrochemical Processes in Biological Systems, First Edition. Edited by Andrzej Lewenstam and Lo Gorton.